CONTEMPORARY RESEARCH ON BUSINESS AND MANAGEMENT

PROCEEDINGS OF THE INTERNATIONAL SEMINAR OF CONTEMPORARY RESEARCH ON BUSINESS AND MANAGEMENT (ISCRBM 2020), 25–27 NOVEMBER 2020, SURABAYA, INDONESIA

Contemporary Research on Business and Management

Edited by

Siska Noviaristanti
Telkom University, Indonesia

CRC Press/Balkema is an imprint of the Taylor & Francis Group, an informa business

Typeset in Times New Roman by MPS Limited, Chennai, India

Library of Congress Cataloging-in-Publication Data
A catalog record has been requested for this book

Published by: CRC Press/Balkema
Schipholweg 107C, 2316 XC Leiden, The Netherlands
e-mail: enquiries@taylorandfrancis.com
www.routledge.com – www.taylorandfrancis.com

ISBN: 978-1-032-05097-3 (hbk)
ISBN: 978-1-032-05103-1 (pbk)
ISBN: 978-1-003-19601-3 (ebk)
DOI: 10.1201/9781003196013

Contemporary Research on Business and Management – Noviaristanti (Ed.)
© 2022 copyright the Author(s), ISBN 978-1-032-05097-3.
Open Access: www.taylorfrancis.com, CC BY-NC-ND 4.0 license

Table of contents

Contemporary Research on Business and Management – Noviaristanti (Ed.)
© 2022 copyright the Author(s), ISBN 978-1-032-05097-3.
Open Access: www.taylorfrancis.com, CC BY-NC-ND 4.0 license

Background

Since 2018, the Indonesian Master of Management Program Alliance (APMMI) annually held a seminar called International Seminar of Contemporary Research on Business and Management. The first one was in Surabaya, hosted by the Technology Institute of Surabaya. The second, in Marwadewa University, Denpasar Bali. The third seminar was in Indonesia University, Jakarta, and recently, due to the pandemic situation, the conference was held online, hosted by Airlangga University in collaboration with Petra University, Surabaya University, STIE Perbanas Surabaya and Trunojoyo University, Madura.

The Association of Indonesian Master of Management Programs (APMMI) is positioning itself as an institution that provides guidance as the best practices of MM education in order to improve the quality of MM education in Indonesia.

APMMI is expected to be an institution that can provide motivation and set operational guidelines to its members to achieve the national quality standard of MM education, capable of producing graduates that are in accordance with the times and meet the needs of business practices and national management.

In the future, APMMI is expected to be able to cooperate with the international MBA program accreditation institution (such as AACSB International) to accredit the MM program in Indonesia, so as to encourage national MM programs to meet international quality standards in accordance with the purpose and objectives of the establishment of APMMI.

Contemporary Research on Business and Management – Noviaristanti (Ed.)

Preface from the chair committee of ISCRBM 2020

It is our pleasure to welcome you to the Fourth International Seminar of Contemporary Research on Business and Management (ISCRBM). The major goals of this event are to bring APMMI's members and industry researchers together to exchange and share their experiences and research results, discuss practical challengenges encountered and the solution adopted and also developing their professional network. On the other side, this conference aims to increase the quality and quantity of research publications in the management and business area. This is a good occasion to meet old friends and make new ones.

This is an annual event of the Indonesian Master of Management Program Alliance (APMMI). This year, the conference was held online and hosted by the Master of Management Program, Airlangga University, Surabaya on November 25–27, 2020. Managing Business and Innovation in the New Normal and Beyond was the conference theme. The distinguished speakers were invited from professional and academic backgrounds to share their knowledge and experience.

There are 65 selected papers published in this proceeding. Most papers are authored by APMMI member students and faculty staff with various topic in operation management, marketing management, human resource management, finance, strategic management, and entrepreneurship. Through digiseminar platform, all conference participants had a new experience in attending the online conference and accessing the presentations material.

Finally, we would like to record our thanks to the conference committee for their work, APMMI members who actively contributed to this event, and the Master of Management Program Airlangga University in collaboration with Petra University, Surabaya University, STIE Perbanas Surabaya and Trunojoyo University, Madura who nicely hosted this year's conference. Without their support, the conference could not have been the success that it was. We also acknowledge the authors for their contribution. Hopefully, the success of the conference will continue next year.

Dr. Gancar C Premananto

Contemporary Research on Business and Management – Noviaristanti (Ed.)

Advisory board

Contemporary Research on Business and Management – Noviaristanti (Ed.)

Scientific committee

Contemporary Research on Business and Management – Noviaristanti (Ed.)

Conference committee

Chair:
Dr. Gancar Candra Premananto, SE., M.Si
Airlangga University

Co-Chair:
Dr. Werner Ria Murhadi S.E., M.M., CSA.
Surabaya University

Members:
Serli Wijaya, S.E., M.Bis., Ph.D
PETRA University, Surabaya

Dr. Ronny, S.Kom., M.Kom., M.H.
STIE Perbanas

Dr. Ir. Nurita Andriyani, M.M
Trunojoyo University Madura

Contemporary Research on Business and Management – Noviaristanti (Ed.)

Support vector machine for predicting the Indonesia Stock Exchange Composite Index (IDX Composite) using domestic and international economic factors

M.D. Wahono & Z.A. Husodo
Master of Management, Universitas Indonesia, Jakarta, Indonesia

ABSTRACT: Economic crises in the past have raised questions about the validity of the efficient market hypothesis and led to the development of models that can predict the stock price. The efficiency of the stock markets is known to decline, which means that the stock market's performance is unpredictable. One of the developing models is a prediction based on economic components known to affect the IDX composite index and processing it by machine learning techniques. Support vector machines are known to have the ability to handle high-dimensional data and have advantages over other algorithms. To determine what economic components are used, this study begins with identifying the influence of domestic and international economic components on the future IDX composite index using Pearson's correlation coefficient. The identification result explains that eight domestic and six international components have influences on the IDX composite index. Using these components as IDX composite index-influencing factors, this study builds and evaluates a machine learning-based stock index prediction model. This research used two error parameters: mean absolute error (MAE) and root mean square error (RMSE). The SVM model's performance was compared to the most used machine learning algorithm: artificial neural networks (ANN) and the classic algorithm, multiple linear regression (MLR). The prediction results showed that SVM showed the best performance in predicting next day's stock index prices (t + 1).

1 INTRODUCTION

After the global economic crisis, the efficient market hypothesis (EMH) theory began to be questioned. This phenomenon is supported by findings, which state that the efficiency of the American stock market from 1929–2002 varied over time due to events affecting the financial, economic, and social systems, and it was also suggested that in the last ten years, market efficiency tended to decline (Alvarez-Ramirez *et al.* 2012). This question led to the emergence of various empirical studies relating to predicting stock price movements using a non-classical approach.

Many financial and economic theories have been developed and research has been conducted, which argued that the stock market is influenced by economic factors, including domestic components represented by macroeconomics and international components (Al-Majali and Al-Assaf 2014). The global economic crisis showed that both domestic and international economic conditions affect the stock market (Adesanmi and Jatmiko 2017). Macroeconomics is a component of the economy that affects the stock market in various ways. These things are shown in the conclusions of several studies on the influence of macroeconomic components on the composite stock index conducted in the past (Celebi and Hönig 2019, Demir 2019, Alshammari *et al.* 2020, Beh and Yew 2020). The use of daily data for the prediction can increase the prediction model's performance because the amount of available data is large, and this large amount of data can be used to maximize the machine learning process.

The machine learning method is considered to have potential in predicting stock prices and can overcome some of the shortcomings of other methods. The machine learning algorithms widely

DOI 10.1201/9781003196013-1

used today include artificial neural networks and the classical method, multiple linear regression (Henrique *et al.* 2019). The support vector machine is another machine learning algorithm that has a great potential in predicting stock prices. SVM has the advantage of the ability to process high-dimensional data. SVM has advantages over ANN, unlike the ANN training process, which requires nonlinear optimization with the potential danger of being stuck at the local minima. Several studies in economics conducted in the past that utilized SVM support this opinion and report ANN and SVM's success in predicting the stock prices (Weng *et al.* 2017, Zhong and Enke 2017).

This study aims to build a prediction model involving the supporting components based on the potential of machine learning, especially SVM, in predicting and the strong influence of economic components on stock prices. A machine learning method with the SVM algorithm was used, whose performance was compared with those of ANN and MLR. It showed the accuracy of the model assessed in terms of error. SVM predicted the IDX Composite by using historical economic data as the input.

2 LITERATURE REVIEW

2.1 *Economic component*

In analyzing the capital market, this study was started by analyzing the economic environment in which the capital market is located, for example, the domestic and international economic environment. The economy of a country depends on the daily business activities carried out by various companies (Alamsyah and Zahir 2018). Changes in economic components tend to influence the condition of the capital market. So investors, in general, will consider and predict future economic conditions before making investment decisions. This consideration is taken because economic growth is primarily determined by macroeconomic components' volatility (Wahyudi *et al.* 2017). Also, there is a relationship between the value of share prices and India's economic component (Ahmed 2008). This result explains that there is an interplay between stock prices and macroeconomic components. Thus, macroeconomics has a significant role in determining the future of the capital market.

There are various types of macroeconomic data: domestic and international; annual, monthly, daily, and hourly. The majority of the data used are monthly data, such as money supply, inflation rate, interest rates, and others from previous studies. In Indonesia, a study was conducted using an artificial neural network to predict the IDX Composite using macroeconomic variables (Alamsyah and Zahir 2018). In addition to the domestic economic component represented by macroeconomics, this study uses several international components. Four types of economic components are used as an input in this study: exchange rate, government bond, commodity, and foreign stock price index. The exchange rate and government bond are categorized as domestic economic components because they directly impact the economy and reflect economic conditions. Commodity and foreign stock prices are categorized as an international economic component.

3 METHOD

3.1 *Data collection*

For this research, the historical data were obtained from online datastreams: Refinitiv Eikon, AsianBondsOnline, and Investing.com. The data obtained are daily data, the period of which is from 2006 to 2019. It is hoped that large amounts of data will increase the number of training sets, which are nearly related to increasing the model's accuracy. The data obtained were still stored separately and contained outliers. The data collected consist of 26 economic components that were filtered. These variables were categorized into four groups: exchange rates, government bonds, commodities, and the foreign stock composite index.

3.2 *Preprocessing*

The data collected were compiled into one CSV file. At this stage, the data were conditioned or prepared to proceed to the modeling stage. The preparation included winsorizing, which eliminated the negative effect of outlier and feature selection that helped reduce the computational cost of modeling and improved the model's performance. Winsorizing was done by detecting an outlier and replacing the outlier value with another value that had been determined. For feature selection, Pearson's correlation coefficient was used to calculate the correlation between each economic component and IDX composite and to determine which economic components could be predictor variables. Only the economic components that had a strong correlation with the IDX Composite passed the feature selection.

3.3 *Modeling*

The data that passed the preprocessing stage were ready to be processed. Weka is an open-source machine learning software. This software helped in building models for this research. Three models were built, including three algorithms: SVM, ANN, and MLR. Some parameter settings were judgmental, and other parameters were set based on recommendations from references.

3.4 *Evaluation*

An evaluation stage was needed to understand how well the model performs. For this research evaluation, there are two essential components, namely, 10-fold cross-validation and error metric. 10-fold cross-validation divided the data into two parts, training dataset and testing dataset, and evaluated it. To assess the performance of a predictive model, the metric was needed. The error matrix is considered most appropriate in assessing the regression model's performance, especially for this study, particularly in terms of mean absolute error and root mean square error.

4 ANALYSIS

As predictor variables, 15 variables passed the feature selection, including the economic component and the IDX composite. Machine learning models for three algorithms were built using these 15 variables. Table 1 shows the result of the prediction process by three models formed by three different algorithms.

Table 1. Prediction one day ahead result ($t+1$).

Parameter		**Domestic + International**
ANN	MAE	36.3662
	RMSE	49.0389
SVM	MAE	30.8277
	RMSE	43.5466
MLR	MAE	30.9643
	RMSE	43.5186

From Table 1, we can see that ANN's performance is the worst among the three models. This model has the largest error compared to the other models, namely, MAE of 36.3662 and RMSE of 49.0389. For error reading, the MAE error illustrates that the model can predict the IDX Composite with a sizeable average deviation of 36.3662 either negatively or positively. Meanwhile, based on RMSE, the ANN model can predict with the residual magnitude of 49.0389. SVM's performance is better than ANN and almost the same as MLR, with the MAE value of 30.8277, which is 0.1366 higher, and RMSE of 43.5186, which is 0.0280 lower than those of MLR. This test's results provide an initial overview of each model's performance with the tentative conclusion that SVM and MLR show a better performance than ANN in predicting the next day IDX Composite Index. The input or predictor variables consist of 14 economic components and 14-year historical IDX Composite Index data. These results also tell us that the combination of the predictor variable and the target variable gives good results. The best model (SVM) predicts with error or deviation of 30.8277 for MAE and 43.5466 for RMSE from the average IDX Composite: 4011.28 (for the 14-year period) of 0.75% for MAE and of 1.09% for RMSE.

5 CONCLUSIONS

Prediction of the composite stock price index is essential for various economic agents in carrying out their roles. Successful prediction helps the investors build an optimal portfolio and help the government make appropriate policies. This study aims to build an optimal model to predict the IDX composite index using a potential machine learning algorithm, namely, SVM.

In the process, this study attempted to find the best predictor components and focused on the economy's domestic and international components. Based on their correlation with the future IDX composite index, 15 components (economic components and historical IDX composite index) were selected as predictor variables for prediction. From this research, for $t + 1$ prediction, we found that SVM and MLR show quite the same performance and are better than ANN. The error magnitude shows that the models can predict the next day IDX Composite well, and SVM is the best model.

REFERENCES

Adesanmi, A.A. and Jatmiko, D.P., 2017. The impact of macroeconomic variables on an emerging economy stock market: Evidence from Jakarta composite index, Indonesia. *International Journal of Economic Perspectives*, 11 (2), 665–684.

Ahmed, S., 2008. Aggregate Economic Variables and Stock Markets in India. *International Research Journal of Finance and Economics*, 14 (14), 1450–2887.

Al-Majali, A.A. and Al-Assaf, G.I., 2014. Long-Run and Short-Run Relationship Between Stock Market Index and Main Macroeconomic Variables Performance in Jordan. *European Scientific Journal*, 1010 (1010), 1857–7881.

Alamsyah, A. and Zahir, A.N., 2018. Artificial Neural Network for predicting Indonesia stock exchange composite using macroeconomic variables. *2018 6th International Conference on Information and Communication Technology, ICoICT 2018*, 0 (c), 44–48.

Alshammari, A.A., Altarturi, B., Saiti, B., and Munassar, L., 2020. The impact of exchange rate, oil price and gold price on the Kuwaiti stock market: A wavelet analysis. *European Journal of Comparative Economics*, 17 (1), 31–54.

Alvarez-Ramirez, J., Rodriguez, E., and Espinosa-Paredes, G., 2012. Is the US stock market becoming weakly efficient over time? Evidence from 80-year-long data. *Physica A: Statistical Mechanics and its Applications*, 391 (22), 5643–5647.

Beh, W.L. and Yew, W.K., 2020. Macroeconomic factors and stock markets interdependencies: Evidence from United States and China. *Journal of Critical Reviews*, 7 (5), 68–74.

Celebi, K. and Hönig, M., 2019. The impact of macroeconomic factors on the german stock market: Evidence for the crisis, pre-and post-crisis periods. *International Journal of Financial Studies*, 7 (2).

Demir, C., 2019. Macroeconomic determinants of stock market fluctuations: The case of BIST-100. *Economies*, 7 (1).

Henrique, B.M., Sobreiro, V.A., and Kimura, H., 2019. Literature review: Machine learning techniques applied to financial market prediction. *Expert Systems with Applications*, 124, 226–251.

Wahyudi, S., Hersugondo, H., Laksana, R.D., and Rudy, R., 2017. Macroeconomic Fundamental and Stock Price Index in Southeast Asia Countries: A Comparative Study. *International Journal of Economics and Financial Issues*, 7 (2), 182–187.

Weng, B., Ahmed, M.A., and Megahed, F.M., 2017. Stock market one-day ahead movement prediction using disparate data sources. *Expert Systems with Applications*, 79, 153–163.

Zhong, X. and Enke, D., 2017. Forecasting daily stock market return using dimensionality reduction. *Expert Systems with Applications*, 67, 126–139.

Contemporary Research on Business and Management – Noviaristanti (Ed.)

Analysis of the impact of foreign investor trading activity on return, liquidity, and volatility of the Indonesian Stock Market before and during the COVID-19 crisis period

F.R. Nandaru & B. Wibowo
Universitas Indonesia, Indonesia

ABSTRACT: The COVID-19 pandemic caused a decline in the Indonesian stock market performance. This paper analyzes how much of the decline was due to foreign investor trading activity and how much its impact differed prior to the crisis period. Stock market performance is proxied by return, liquidity, and volatility. Meanwhile, foreign investor trading activity is proxied by buying and selling volume, value, and frequency. Trading data for each period were processed through OLS regression with Newey–West's HAC procedure. The results reveal that foreign investor trading activity caused a bi-directional impact on return and liquidity, where the impact on both was greater during the crisis period. Additionally, foreign investor trading activity impacted volatility significantly during the non-crisis period, while an increase in the trading activity increased volatility during the crisis period.

1 INTRODUCTION

The stock market is an integral part of Indonesia's economic development. Its growth and return potential have attracted not only domestic investors, but also the foreign ones. Foreign investors investing in the Indonesian Stock Exchange (IDX) have grown to a sizable population, carrying out 46% of the total trades in the second quartile of 2015 (IDX 2019). However, during the COVID-19 crisis, foreign investors became passive traders. Considering the large number of foreign investors, decline in their trading activities may have a significant impact on the IDX performance. This paper analyzes the impact of foreign investor trading activity during the crisis period and its changes prior to the COVID-19 crisis.

Due to the pandemic and regulations such as social distancing and lockdowns from governments worldwide, the financial prospects of businesses all over the world and investor confidence declined. Liu et al. (2020) observed that stock markets worldwide provided negative abnormal returns, particularly those in Asia due to the strong spillover effect. Many studies have analyzed the impact of foreign investor trading activity on the stock market performance during non-crisis and crisis periods. However, the results vary due to the differences in location, period, or types of trading data used. This study is useful, as the results and a more robust understanding of the role of foreign investors in Indonesia during crises can be beneficial to prevent and counteract potential crises that may arise in the future.

2 LITERATURE REVIEW

Multiple researchers have conducted studies that are relevant to this study. Vo (2017), Nguyen (2017), and Samarakoon (2009) found that foreign investor trading activity caused a bi-directional impact on the stock market return, while Junior and Eid (2017) showed that the impact doubled during a crisis. Peranginangin et al. (2016); Agudelo (2010); and Liew, Lim, and Goh (2018)

DOI 10.1201/9781003196013-2

observed that trading inactivity reduced liquidity, and its impact was greater during a crisis. Choe, Kho, and Stulz (1999) stated that foreign investors generally were unwilling to trade during a crisis, thereby reducing market liquidity substantially. Meanwhile, Umutlu and Shackleton (2015); Che (2018); and Wang (2007) revealed that trading activity increased volatility.

Regarding the crisis period, Rhee and Wang (2009) and Peranginangin (2016) found that foreign investors in Indonesia tend to utilize the buy and hold strategy, sell small stocks, and buy stocks with a large market capitalization to secure a less risky portfolio. Not only that, foreign investors, who were net buyers, became net sellers. Because of this, the trading volume of foreign investors investing in Indonesia fell from 45% to 18% of total trades before it rose again after the crisis was over. Richards (2005) and Agarwal et al. (2009) implied that foreign investors in Indonesia were more risk-averse and had a significant impact on the stock market performance during crises, especially on liquidity.

3 DATA AND METHODOLOGY

3.1 *Data*

The daily historical price data of IDX from January 2019 to September 2020 were obtained through *Thompson Reuters*, while the data of foreign investor trading activity were obtained through *TICMI*. The data were then split into two period categories: normal (before 30 January 2020) and crisis (after 30 January 2020). The normal period contains data for 248 trading days, while the crisis period contains data for 162 trading days.

3.2 *Variables and methodology*

This research assessed the stock market performance using dependent variables, including return, liquidity, and volatility. The stock return was calculated by subtracting the current stock price from the previous day's stock price and dividing it by the previous day's stock price. Furthermore, the liquidity was measured by a high–low spread estimator procedure using the daily highest and lowest price of a stock (Corwin and Schultz 2012). The procedure is illustrated by the following equations:

$$S = \frac{2(e^{\alpha} - 1)}{1 + e^{\alpha}} \tag{1}$$

$$\alpha = \frac{\sqrt{2\beta} - \sqrt{\beta}}{3 - 2\sqrt{2}} - \sqrt{\frac{\gamma}{3 - 2\sqrt{2}}} \tag{2}$$

$$\beta = \left[\ln\left(\frac{H_t}{L_t}\right)\right]^2 + \left[\ln\left(\frac{H_{t+1}}{L_{t+1}}\right)\right]^2 \tag{3}$$

$$\gamma = \left[\ln\left(\frac{\max\{H_t, H_{t+1}\}}{\min\{L_t, L_{t+1}\}}\right)\right] \tag{4}$$

where is the liquidity measure, H_t is the highest price at day t, and L_t is the lowest price at day t. In terms of volatility, this study used the average of the 14-day average true range (ATR) and standard deviation. The true range was calculated by taking the greatest difference between the highest and lowest price of the day, the absolute value of the difference between the highest price of the day and the closing price of the previous day, and the absolute value of the difference between the lowest price of the day and the closing price of the previous day. The ATR was then calculated by taking the 14-day average of true range.

The independent variables of this research are buying and selling volume, value, and frequency of foreign investors. Volume is defined by the number of stocks traded. Value is the nominal amount traded, while frequency is the number of transactions that occurred. The variables were then processed through OLS regression with Newey–West's HAC procedure.

4 RESULTS AND DISCUSSION

Table 1. Regression results

	Constant	S_Volume	S_Value	S_Frequency	B_Volume	B_Value	B-Frequency	R-Squared	F-statistic
Panel A: Normal Period									
Return	0.000649	1.52E-13	2.29E-16	−1.47E-07	−3.37E-14	1.42E-16	1.42E-07	0.198	0.000000
	(−0.38)	(−0.33)	(−0.79)	(−6.63)***	(−0.20)	(2.85)***	(6.67)***		
Liquidity	−0.205716	6.77E-12	−5.22E-15	−8.63E-07	8.73E-13	−5.19E-16	6.76E-07	0.115	0.000214
	(−15.22)***	(2.33)**	(−2.47)**	(−3.51)***	(−0.59)	(−1.17)	(2.90)***		
Volatility	0.0006606	−1.69E-13	1.19E-16	3.29E-08	−1.69E-13	4.65E-17	−2.23E-08	0.135	0.000035
	(8.40)***	(−1.16)	(−0.76)	(3.67)***	(−2.63)***	(2.36)***	(−2.47)**		
Panel B: Crisis period									
Return	−0.014747	9.76E-12	−4.38E-15	−1.66E-07	−6.14E-12	2.91E-15	3.53E-07	0.251	0.000032
	(−1.56)	(3.48)***	(−1.43)	(−3.32)***	(−1.63)	(−0.89)	(3.34)***		
Liquidity	−0.157945	4.10E-11	−4.66E-14	−5.21E-07	−2.74E-11	2.74E-14	−7.05E-07	0.286	0.000000
	(−5.15)***	(2.35)**	(−2.21)**	(−1.20)	(−1.41)	(−1.54)	(−1.92)*		
Volatility	0.000898	−1.01E-12	−1.38E-15	1.25E-07	1.95E-12	−3.34E-16	4.96E-08	0.159	0.000283
	(−0.26)	(−0.42)	(−.0.59)	(3.24)***	(−1.13)	(−0.21)	(−1.13)		

***, **, and * indicate statistical significance at the 1%, 5%, and 10% level, respectively

As expected, this study's findings differ from the previous studies' results since the variables studied were different. The findings imply that whether it was buying or selling activity, each characteristic of the said activities, namely, frequency, volume, and value, had different impacts on the stock market performance in different periods. Furthermore, it is also implied that the impact was greater during crises, especially on liquidity, where the impact was six to nine times greater than the impact during the normal period. This highlights the importance of foreign investors in Indonesia during crises.

As stated by Rhee and Wang (2009) and Peranginangin (2016), foreign investors became net sellers and less active in trading during crises. Combined with the implication that foreign investors have a stronger impact during crises, it can be said that foreign investors played an important role in the decline of IDX's performance. This is probably due to the decline in foreign investors' confidence regarding the financial prospect of the Indonesian stock market during the crisis. This is because of the Indonesian government's inability to handle the pandemic or Indonesian businesses' incapability to maintain their operations at a profitable level.

5 CONCLUSIONS

Through this research, the impact of foreign investor trading activity on IDX's performance could be understood. In line with previous studies, an increase in trading activity had a bi-directional impact on returns, liquidity, and volatility, while during crises, an increase in trading activity increased volatility. The findings show that each characteristic of trading activities had different impacts on the stock market performance, and the impact was greater during crises.

The decline in foreign investors' confidence played an important part in the IDX's performance during the COVID-19 crisis. To uplift their confidence, the Indonesian government should be able to handle the spread of the virus more efficiently and give financial stimulus to citizens and struggling businesses. Therefore, the level of consumption could be maintained and businesses could still operate. If the government can maintain a healthy business environment, foreign investors' confidence can also be maintained, leading to a better performance of the stock market. Alternatively, they could try to reduce dependence on foreign investors with the increase of the population size and influence of domestic investors. This could be achieved by improving the level of public financial literacy. Therefore, the Indonesian government may be able to prevent or at least minimize the impact of crises that may arise in the future.

This study has several limitations that can make the findings to be less than ideal. First of all, the number of observations for the crisis period is limited since the crisis was still ongoing when this study was conducted. For liquidity, Liew, Lim, and Goh (2018) stated that the best proxy is CPQS, which requires bid-ask data. Unfortunately, IDX did not provide such data, so this study had to switch to another proxy. For further research, it will be better if the number of observations during the crisis period is increased. It is recommended to wait until the COVID-19 crisis is declared over, so the impact of foreign investors after the crisis could also be analyzed. The research could also be conducted with domestic investor trading activity to compare and contrast the impact caused by domestic and foreign investors.

REFERENCES

Agarwal, S., Faircloth, S., Liu, C., and Ghon Rhee, S., 2009. Why do foreign investors underperform domestic investors in trading activities? Evidence from Indonesia. *Journal of Financial Markets*, 12 (1), 32–53.

Agudelo, D.A., 2010. Friend or Foe? Foreign Investors and the Liquidity of Six Asian Markets. *Asia-Pacific Journal of Financial Studies*, 39 (3), 261–300.

Che, L., 2018. Investor types and stock return volatility. *Journal of Empirical Finance*, 47, 139–161.

Choe, H., Kho, B.-C., and Stulz, R.M., 1999. Do foreign investors destabilize stock markets? The Korean experience in 1997. *Journal of Financial Economics*, 54 (2), 227–264.

Corwin, S.A. and Schultz, P., 2012. A Simple Way to Estimate Bid-Ask Spreads from Daily High and Low Prices. *The Journal of Finance*, 67 (2), 719–760.

IDX, 2019. IDX Annually Statistic [online]. Available from: https://www.idx.co.id/media/8473/idx_annually-statistic_2019.pdf.

Junior, W. and Eid, W., 2017. Sophistication and price impact of foreign investors in the Brazilian stock market. *Emerging Markets Review*.

Liew, P.-X., Lim, K.-P., and Goh, K.-L., 2018. Foreign equity flows: Boon or bane to the liquidity of Malaysian stock market? *The North American Journal of Economics and Finance*, 45, 161–181.

Liu, H., Manzoor, A., Wang, C., Zhang, L., and Manzoor, Z., 2020. The COVID-19 Outbreak and Affected Countries Stock Markets Response. *International Journal of Environmental Research and Public Health*, 17, 2800.

Nguyen, T., 2017. The Impact of Foreign Investor Trading Activity on Vietnamese Stock Market. *International Journal of Marketing Studies*, 9, 109.

Peranginangin, Y., Ali, A., Brockman, P., and Zurbruegg, R., 2016. The impact of foreign trades on emerging market liquidity. *Pacific-Basin Finance Journal*, 40.

Rhee, S. and Wang, J., 2009. Foreign Institutional Ownership and Stock Market Liquidity: Evidence from Indonesia. *Journal of Banking & Finance*, 33, 1312–1324.

Richards, A., 2005. Big Fish in Small Ponds: The Trading Behavior and Price Impact of Foreign Investors in Asian Emerging Equity Markets. *The Journal of Financial and Quantitative Analysis*, 40 (1), 1–27.

Samarakoon, L.P., 2009. The relation between trades of domestic and foreign investors and stock returns in Sri Lanka. *Journal of International Financial Markets, Institutions and Money*, 19 (5), 850–861.

Umutlu, M. and Shackleton, M.B., 2015. Stock-return volatility and daily equity trading by investor groups in Korea. *Pacific-Basin Finance Journal*, 34, 43–70.

Vo, X.V., 2017. Trading of foreign investors and stock returns in an emerging market – Evidence from Vietnam. *International Review of Financial Analysis*, 52, 88–93.

Wang, J., 2007. Foreign equity trading and emerging market volatility: Evidence from Indonesia and Thailand. *Journal of Development Economics*, 84 (2), 798–811.

Contemporary Research on Business and Management – Noviaristanti (Ed.)

Continuous intention determinants in the use of food delivery application in Surabaya, Indonesia

S. Hariadi & S. Rahayu
Universitas Indonesia, Indonesia

ABSTRACT: This research aims to recognize and investigate the influence of information quality and UTAUTS on continuous intention in the use of FDA in Surabaya, Indonesia. 150 respondents, who have been users of FDAs for the last 6 months in Surabaya, filled out an online questionnaire as the research sample. This study customizes the SEM scheme with AMOS 22 and concludes that continuous intention is positively influenced only by hedonic motivation and not by other factors such as information quality, performance and effort expectancy, price value, facilitating conditions, or habit.

1 INTRODUCTION

1.1 *Background*

Information systems are a very large part of the use of technology in running a business. Many service companies customize their products with applications that are easily accessible via smartphones. This makes it easy for customers to make various transactions from anywhere and anytime.

One application that is widely used by smartphone users is food delivery applications (FDAs). People prefer buying ready-to-eat food through digital applications to cooking themselves or buying their own food (Valenta, 2019).

The results of the Nielsen Singapore survey show that 58% of Indonesians buy fast food via an online application from smartphones. People aged 18–45 years buy ready-to-eat food through an online food delivery application from smartphones as many as 2.6 times per week (Jayani, 2019). The main reasons consumers choose online applications are reduction of time and effort (39% of respondents), attractive promotions or offers (33% of respondents), practical payment options and cash discounts (21% of respondents), and a variety of food menu choices (17% of respondents).

Meanwhile, Lee et al. (2019) employed the unified theory of acceptance and use of technology model to increase the quality of information in identifying the elements of continuous intentions of FDAs for 340 respondents in South Korea. Habit has the sturdiest impact on continuous intention, as shown by performance expectancy and social influence. Information quality does not directly influence intentions of continuing use through performance expectations.

This work emphasizes the importance of information quality, performance expectancy, habit, and social influence as driving factors for users' intention to keep using FDAs. The findings build on previous research in online business in the food service sector. This study was conducted by taking case studies of the FDA in Indonesia, namely GoFood and GrabFood.

2 LITERATURE REVIEW

2.1 *Information quality*

In the era of information technology, all the businesses in the world require accurate information. Information quality is the gradation to which evidence has pleased, form, and time characteristics, which give it value for certain end users (Rai et al. 2002; O'Brien, 2005).

DOI 10.1201/9781003196013-3

Information quality also pertains to the degree, soundness, and usefulness of information, which is the yield of an information system and the eminence of the output (Ranganathan & Ganapathy, 2017; Lee et al. 2019). Good quality information creates pleasure when using and the intention to behave positively (Ahn et al. 2007). Consumers' positive perceptions about the quality of information are formed when evidence encounters opportunities during the pronouncement-making procedure and sufficient information is available (Corbitt et al. 2003; Kim & Park, 2013).

A literature review on technology receiving demonstrations that conviction in information is a feature influencing behavioral meaning (Yadav et al. 2016; Sharma et al. 2016). These arguments are the basis for formulating the following hypothesis.

H1: Information quality drive significantly impacts continuous intention.
H2: The information quality of FDA will significantly impact expectancy.
H3: The information quality of FDA will significantly impact behavioral intention.

2.2 *Unified theory of acceptance and use of technology*

Many studies use TAM (Technology Acceptance Model) to explain technology acceptance. This model has a weakness, namely, it cannot properly explain the effect of various exogenous variables on the TAM variable. Venkatesh et al. (2003) refined the TAM model by proposing a comprehensive model, namely, the intention to practice and user behavior of the information system UTAUT. Performance expectations, business expectations, social influences, and facility conditions are direct determinants of behavioral intention and use.

Performance expectations also refer to the extent to which individuals believe that the use of the system will help in improving their occupation performance (Yadav et al. 2016). Business expectations denote the affluence of practice of the structure. Positive perceptions of ease of use will lead to a greater intention to use technology (Lawan & Dahalin, 2011; San Martin & Herrero, 2012). Social influence shows that the practice of systems or technology is influenced by peer views (Yeo et al. 2017). Facility conditions are demarcated as the level of individual confidence in the existence of prearranged practical sustenance for system use (Jati & Laksito, 2012).

This study places these seven variables as factors that influence continuous intention for using a food delivery application. The hypotheses proposed are as follows:

H4: Expectation of FDA performance will significantly impact continuous intention.
H5: Attempted FDA expectations will significantly influence continuous intention.
H6: The social influence associated with FDA will significantly impact continuous intention.
H7: The facilitation conditions regarding FDA will significantly impact continuous intention.
H8: The hedonic motivation regarding application delivery will significantly impact continuous intention.
H9: The value of the delivery application price will significantly impact continuous intention.
H10: Customs with respect to delivery applications will significantly impact continuous intention.

3 METHODS

This research is a causal research that explains cause and effect. The data source is primary data obtained by distributing questionnaires to food application delivery users. The validity and reliability of the questionnaire were tested using a sample of 30 respondents. They were all users of food application delivery in Surabaya, Indonesia. The respondents are users of food application delivery in the last 6 months with a do-it-yourself process, both male and female, and at least 17 years old. The model measurement uses the confirmatory factor analysis. The structural model was analyzed using the AMOS program.

4 RESULTS

This study used a sample of 150 respondents in accordance with predetermined characteristics. 64.7% of the respondents were women, and 35.3% were men. Respondents aged 17–24 years comprised 91.3% of the sample, and those with senior high school education comprised 74%, followed by bachelors degree (25.3%). 82% of the respondents were students.

4.1 *Measurement model*

The results of confirmatory factor analysis show that each indicator has a standardized loadings value greater than 0.5 and the resulting AVE value also greater than 0.5. The indicators that make up each research variable, namely, information quality, habit, effort expectancy, performance expectancy, price value, continuance intention, facilitating conditions, hedonic motivation, and social influence, have shown good measurements. The value of construct reliability in each variable is above 0.6. This means that these indicators have good reliability. The criteria for the validity and reliability of the measurement model are good. This is also supported by the goodness-of-fit criteria for the measurement model.

4.2 *Structural model*

The structural model of the research is the effect of information quality, effort expectancy, social influence, price value, facilitating conditions, hedonic motivation, performance expectancy, and habit regarding the continuous intention of users of GoFood and GrabFood applications. The structural model was analyzed using the AMOS program (Figure 1).

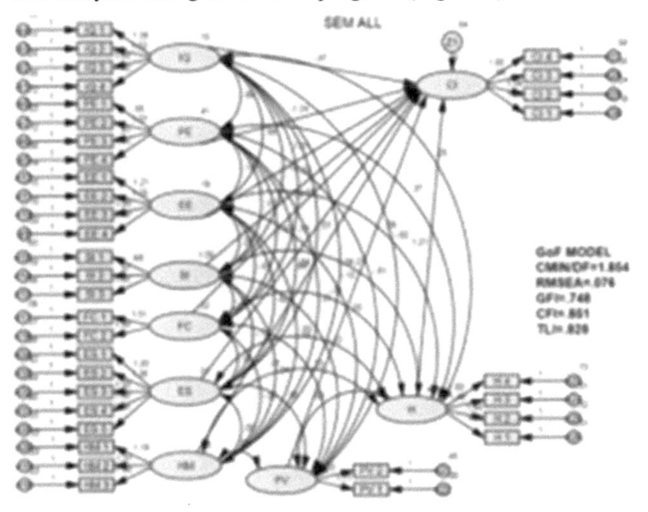

Figure 1. Structural model.

The structural model evaluation has some criteria. The goodness-of-fit value has met the established criteria. Only the TLI and CFI values of 0.828 and 0.851 are included in the Marginal Fit category. Thus, testing the research hypothesis can use the results of the structural model.

4.3 *Hypothesis testing and discussion*

Inter-variables have a significant effect if the significance value is <0.05 ($\alpha = 5\%$) or CR > 1.96. This study shows that performance expectancy and effort expectancy have no positive effect on continuous intention of FDA users. FDA use is commonplace. Regardless of the conditions, the user will continue to use them because of the demands of convenience, practicality, and other things related to the user's lifestyle.

Price value has no positive effect on continuous intention by FDA users. Users will continue to use the FDA for other reasons. The costs incurred are often not worth the effort if they have to leave the house to buy food.

Facilitating conditions and habits have no positive effect on continuance intention by FDA users. This is because the facilities made available by FDAs are commonplace and it has become a daily habit for users to use various applications, so they will continue to be used in the future.

Hedonic motivation has a positive effect on continuance intention of FDA users. This is because FDAs facilitate the desire to enjoy various trends in the culinary world easily.

Information quality has no positive effect on continuance intention of FDA users. This is because almost all application-based services provide complete services, so that no matter what information is available, the user will still use FDAs.

Social influence has no positive effect on the continuance intention of FDA users. FDAs have become a trend, so their use is not due to the social influence of the people around the users.

5 CONCLUSION

Continuous intention is positively influenced only by hedonic motivation and not other factors such as information quality, performance and effort expectancy, price value, facilitating conditions, or habit.

REFERENCES

Ahn, T.; Ryu, S.; & Han, I. 2007. The impact of web quality and playfulness on user acceptance of online retailing. *Information Management*. 44: 263–275.

Corbitt, B.J.; Thanasankita, T.; & Yi, H. 2003. Trust and e-commerce: A study of consumer perceptions. *Electronic Commerce Research and Applications*. 2: 203–215.

Jati, N.J. & Laksito, H. 2012. Analysis of the factors that influence interest in the use of the e-ticket system: Empirical study of travel agents in Semarang city. Diponegoro Journal of Accounting. 1(1): 512–524.

Jayani, D. H. 2019. Ordering Food Online is More and More Popular. *https://databoks.katadata.co.id/datapublish/2019/09/19*. Downloaded at 1 August 2020.

Kim, S. & Park, H. 2013. Effects of various characteristics of social commerce (s-commerce) on consumers' trust and trust performance. *International Journal of Information Management*. 33: 318–332.

Lawan, A. & Dahalin Z. M. 2011. Effectiveness of Telecentre using a Model of Unified Theory of Acceptance and Use of Technology (UTAUT): SEM Approach. *CIS Journal*. 2(9): 402–412.

Lee, S. W., Hye J. S. & Hyeon M. J. 2019. Determinants of Continuous Intention on Food Delivery Apps: Extending UTAUT2 with Information Quality. *Sustainability*. II: 3141.

O'Brien, J. A. 2005. *Introduction to Information Systems*. Salemba Empat. Jakarta

Rai, A., Lang, S.S. & Welker, R.B. 2002. Assessing the Validity of IS Success Models: An Empirical Test and Theoretical Analysis. *Information System Research*. 13(1): 29–34.

Ranganathan, C. & Ganapathy, S. 2017. Key dimensions of business-to-consumer websites. *Information Management*. 39: 457–465.

San Martin, H. & Herrero, A. 2012. Influence of the user's psychological factors on the online purchase intention in rural tourism, integrating innovativeness to the UTAUT framework. *Tourism Management*. 33: 341–350.

Sharma, S.K.; Joshi, A.; & Sharma, H. 2016. A multi-analytical approach to predict the Facebook usage in higher education. *Computers in Human Behavior*. 55: 340–353.

Valenta, E. 2019. The food ordering application changes the behavior of Indonesian consumers. *https://beritagar.id/artikel/berita*. Downloaded at 1 August 2020

Venkatesh, V.; Morris, M.G.; Davis, G.B.; & Davis, F.D. 2003. User acceptance of information technology, toward a uni?ed view. *MIS Quarterly*. 27: 425–478.

Yadav, R.; Sharma, S.K.; & Tarhini, A. 2016. A multi-analytical approach to understand and predict the mobile commerce adoption. *Journal of Enterprise Information Management*. 29: 222–237.

Yeo, Vincent; See, K.G.; & Sajad, R.; 2017. Consumer experiences, attitude and behavioral intention toward online food delivery (OFD) services. *Journal of Retailing and Consumer Services*. 35:150–162.

Contemporary Research on Business and Management – Noviaristanti (Ed.)

Family governance and Indonesian firm performance

N. Erisma, I. Sadalia & B.K. Fawzea
Universitas Sumatera Utara, Medan, Sumatera Utara, Indonesia

ABSTRACT: Performance of family firms in Indonesia is an interesting topic of study. Some studies had been conducted in an attempt to understand the performance of family firms, especially the ones that went public. This study aims at examining the effect of family governance on firm performance moderated by debt, firm age, and firm size. The effect of various factors relating to family governance on firm performance was tested through board size, board independence, director's degree, director's expertise, leadership structure, debt, firm age, and firm size. The moderating variables of this study include earnings per share, operating cash flow, and Tobin's Q of the company. The success of the family firm plays an important role in regulating leadership duality for the survival of their firms and protecting their legacy for future generations. The sample study comprised 58 family companies listed on the Indonesia Stock Exchange in 2016–2018. This study explored family governance and firm performance, which serve as the basis for defining the significant increase in the company's revenue. Further studies will be helpful for future business research and policymaking practice.

1 INTRODUCTION

A company's performance depends on its management's ability to utilize the company resources to increase the company value and shareholder welfare. The management's optimal productivity, efficiency, and effective use of resources and profits are required to increase the company value (Gaur et al., 2015). Maximum performance can be seen in companies that make effective use of company assets to meet the shareholders' economic interests (Haniffa & Hudaib, 2006). Using the agency theory is an approach that is closely discussed when evaluating company performance, which considers the company's board of commissioners and directors as the main mechanism for the company's internal control. Recent economic developments have shown that the family companies are a part of the governance structures of business organizations in both developed and developing countries. Indonesian corporate governance is characterized by a concentrated, family-owned, and controlled ownership structure (Setiawan et al., 2016; Prabowo & Simpson, 2011). The agency theory states that if the ownership is concentrated in the management of family companies, it triggers agency conflicts between major shareholders and minor shareholders (Azoury & Bouri, 2015; Cheng, 2014; and Yen et al., 2015).

Family companies also provide many jobs for the community. In general, family companies are built to last for generations. The amount of family ownership is the company's equity/capital owned by the family (Beuren et al., 2016). At least one family representative is involved in the company management, such as the person who acquires the company, parents, children, spouses, or heirs appointed by the company founder (Kirana & Ermawati, 2017). Family companies that go public usually have the majority voting rights controlling 25% of the company's shares, where two or more family members who are involved in company management activities hold positions such as commissioners or directors (Dussault, 2008). Family ownership has an inverse U-like relationship with company performance (Poutziouris et al., 2015). The presence of family CEOs

DOI 10.1201/9781003196013-4

has a positive relationship with company performance in Indonesia and Taiwan. However, it has a negative relationship with company performance in Hong Kong (Jiang & Peng, 2011). The increase in the placement of family members on the board of commissioners or directors resulted in the expropriation of minor shareholders, thereby reducing company performance.

2 LITERATURE REVIEW

Company performance is a measure of its manager's success in running a company. For shareholders, information about company performance is very useful to determine the effectiveness and efficiency of company management. Therefore, company performance can be measured using the formula of Tobin's Q, earnings per share, and operating cash flow (Enriques & Volphin, 2007). Tobin's Q is the formula for calculating the company market value. To replace the asset value, it is commonly calculated as the book value of the company's assets. Moreover, a company's performance can also be calculated using operating cash flow and earnings per share due to the difference in cash inflows and outflows originating from the company's operating activities. According to Aristya & Suryana (2013), the level of profit generated per share owned by investors will affect investors' assessment of a company's performance.

Family governance is a system of joint decision-making, most commonly by the board of directors and the family board, which helps the owner family regulate the relationship between wealth and its company (Hadiprajitno, 2013). So far, family governance has had a significant influence on company performance related to corporate strategic management (Marashdeh, 2014). Family control is indicated by the existence of a minimum 10% share ownership in the company owned by the founding family (Maury, 2006). A family-controlled company is an example of the most developed forms of concentrated ownership in the world, which is characterized by the large number of shares held and family participation in managing the company and maintaining the existence of corporate defenses by planning the regeneration process of the family company. This study uses the percentage of family ownership in a company structure such as board size, board independence, director's degree, director's expertise, and leadership structure.

Debt can be used as a corporate governance mechanism (Yen et al., 2015). Debt can encourage managers to work in accordance with the interests of shareholders, try their best to prevent bankruptcy, and reduce free cash flow for less valuable investments (Villalonga et al., 2015). In the context of family companies, the average family company in Australia has a higher level of debt than non-family companies (Halili et al., 2014). Debt can be calculated by dividing total debt by total assets. Firm age is the length of time a company takes to generate profits (Loderer & Waelchli, 2009). Firm age is closely related to the establishment of a company; the longer the company has been established, the better the constructive evaluation in each period is for building the company. The amount of experience and expertise in minimizing the risks experienced, high customer satisfaction, customer loyalty, and channel links are very influential in firm age (Mahajan & Singh, 2013). Firm size is the size of a company represented by its market capitalization. The firm size ratio is the number of assets owned by the company measured using the logarithm of total assets (Hartono, 2014). Firm size can also be shown through the total asset value, net sales, or equity value (Hartono, 2014). Firm size in this study used the logarithmic value of total assets, where it can be seen from the total assets owned.

3 RESEARCH METHOD

This study analyzed the effect of the board size, board independence, director's degree, director's expertise, leadership structure, debt, firm age, and firm size on earnings per share and operating cash flow. Debt, firm age, and firm size are the moderating variables. A secondary data source was used in this study. The population of this study was the 79 family companies in all sectors listed on the Indonesia Stock Exchange (IDX) in 2016–2018. The sampling technique was carried out using purposive sampling, and 58 companies fulfilled the criteria. The sample criteria were as follows:

1. Family companies listed on IDX in 2016–2018.
2. Family companies, most of whose shares are owned by domestic investors (domestic investment).
3. A family company that has consistent financial report data and annual reports for the period of 2016–2018.

This study employed a quantitative research design using panel data regression data analysis techniques, namely, for regression with more than one independent variable and the dependent variable in time series and cross-section. This study used the structural equation modeling partial least squares (SEM-PLS) analysis method using the SmartPLS software.

4 DISCUSSION AND IMPLICATION

A hypothesis was made in the significance test of the direct effect. The hypothesis testing used SmartPLS software to determine the direct effect. The results of the path coefficient and the direct effect significance test are presented in Table 1.

Table 1. Path coefficient value and P-value.

	Original Sample (O)	Sample Mean (M)	Standard Deviation (STDEV)	T Statistics (|O/STDEV|)	P Values
Debt → Firm Performance	0.253	0.236	0.113	2.242	0.025
FamGov*Debt → Firm Performance	0.196	0.192	0.082	2.405	0.017
FamGov*FirmAge → Firm Performance	−0.106	−0.103	0.034	3.115	0.002
FamGov*FirmSize → Firm Performance	0.001	−0.001	0.074	0.015	0.988
Family Governance → Firm Performance	0.083	0.080	0.043	1.905	0.057
FirmAge → Firm Performance	−0.142	−0.134	0.042	3.387	0.001
FirmSize → Firm Performance	0.262	0.283	0.097	2.692	0.007

Table 1 shows that family governance had a positive effect on firm performance (path coefficient = 0.083), but it was not significant (p-value = 0.057 > 0.05). Firm age is a significant moderator of the relationship between family governance and firm performance, with a p-value of 0.002 < 0.05. In addition, the firm size was not a significant moderator of the relationship between family governance and firm performance, with a p-value of 0.998 > 0.05.

Table 2. Validity and reliability testing.

	Average Variance Extracted (AVE)	Composite Reliability (CR)	Coefficient Determination (R2)
Debt	1.000	1.000	
Family Governance	1.000	1.000	
Firm Performance	1.000	1.000	0.332
Firm Age	1.000	1.000	
Firm Size	1.000	1.000	

The recommended AVE value is above 0.5 (Mahfud & Ratmono, 2013: 67). Table 2 shows that all AVE values are below 0.5, which indicates that the validity requirements based on AVE were met. Furthermore, the reliability test was carried out based on the composite reliability (CR) value. The recommended CR value is above 0.7 (Mahfud & Ratmono, 2013: 67). Table 2 shows that all CR values are greater than 0.7, which indicates that the recommended CR values were achieved. Furthermore, the reliability test was carried out based on the Cronbach's alpha (CA) value.

Therefore, the coefficient of determination for the firm performance variable was 0.332. This value shows that family governance, debt, firm age, and firm size were able to influence firm performance by 33.2%.

5 CONCLUSION

Overall, this study found that family governance with the percentage of variables, namely, board size, board independence, director's degree, director's expertise, leadership structure, debt, firm age, firm size had a moderating effect on firm performance. In other words, a larger board size makes monitoring management more effective. This study explains that family governance with a number of the company's board of directors who are experts in their fields and have high levels of education and the duality of leadership influences the performance of family companies in a period that reflects the health level of the company. Family businesses in Indonesia believe that the success of a company is generally controlled by members of the founding family or the company's successors. Therefore, investors and regulators need to understand that family companies have different characteristics.

Further studies are required in the future. The limitation of this study is that it considered only family companies with domestic investment as the sample. Thus, further studies can consider family and non-family businesses with foreign investment or add more variables. Although obtaining the data is more difficult, the results will be clearer in the IDX regulations.

ACKNOWLEDGEMENT

We are very fortunate to have the opportunity to conduct this research with the support of the Directorate of Research and Community Service (DRPM) of the Ministry of Research and Technology of Higher Education, who provided financial assistance in 2020 for this research.

REFERENCES

Azoury, N., Bouri, E., 2015. Principal–Principal Conflicts in Lebanese Unlisted Family Firms. J Manag Gov. Vol.19, 461–493.

Beuren, I.M., Politelo, L., Martins, J.A.S., 2016. Influence of Family Ownership on Company Performance. International Journal of Managerial Finance. Vol. 12 (5), 654-672.

Cheng, Q., 2014. Family firm research – A review. China Journal of Accounting Research. Vol. 7, 149-163.

Dussault, M. 2008. Family Business Suicide: Prevention Guide. Strategic Book Publishing and Rights Agency.

Gaur, S.S., Bathula, A., Singh, D., 2015. Ownership Concentration, Board Characteristics, and Firm Performance. Management Decision. Vol. 53 (5), 911–931.

Haniffa, R., Hudaib, M., 2006. Corporate Governance Structure and Performance of Malaysian Listed Companies, Journal of Business Finance & Accounting. Vol. 33(7) & (8), 1034–1062. Hsiao, Y.C., 2019. Outside and Family Governance Power for Firm Performance: Why Organizational Capabilities Matter? J Account Mark. Vol. 8 (1).

Jiang, Y dan Peng, M.W., 2011. Are Family Ownership and Control in Large Firms Good, Bad, or Irrelevant? Asia Pacific Journal of Management. Vol. 28, 15–39.

Kirana, J.D., Ermawati. 2017. The Role of Corporate Governance in Improving the Performance of Family Ownership Companies in Indonesia. Prosiding Seminar Nasional Penelitian dan PKM Sosial, Ekonomi dan Humaniora. Vol. 7 (2).

Marashdeh, Z. M. S. 2014. The effect of corporate governance on firm performance in Jordan.
Thesis. The University of Central Lancashire.

Poutziouris, P., Savva, C.S., Hadjielias, E. 2015. Family Involvement and Firm Performance: Evidence from UK Listed firms. Journal of Family Business Strategy. Vol. 6, 14–32.

Prabowo, M.A., Simpson, J., 2011. Independent directors and firm performance in family-controlled firms: evidence from Indonesia, Asian- Pacific Economic Literature. Vol. 25(1), 121–132.

Setiawan, D,. Bandi, B., Phua, L.K., Trinugroho, I., 2016. Ownership structure and dividend policy in Indonesia. Journal of Asia Business Studies. Vol. 10(3). 230–252.

Villalonga, B., Amit, R., Trujillo, M.A., dan Guzman, A., 2015. Governance of Family Firms.

The Annual Review of Financial Economics. Vol. 7, 635–54.

Yen, J.F., Lin, C.Y., Chen, Y.S., Huang, Y.C., 2015. Founding Family Firms and Bank Loan Contracts, J Financ Serv Res. Vol. 48, 53–82.

Contemporary Research on Business and Management – Noviaristanti (Ed.)

The influence of service quality, product quality, and lifestyle on repurchase intention mediated by customer satisfaction at Starbucks in Surabaya

D.A. Dwiyanty & Ronny
STIE Perbanas Surabaya, Indonesia

ABSTRACT: The development of the food and beverage business in the modern era is so rapid that it is not surprising to find coffee shops on many streets or in malls. Starbucks is a modern coffee shop that can often be found in malls and is spread across Indonesia. As a popular coffee shop, Starbucks needs to provide the best services and products for its consumers to maintain its existence.

1 INTRODUCTION

1.1 *Background*

Starbucks is one of the modern coffee shops that can often be found in malls and is spread across Indonesia. Starbucks is popular with the general public, as seen from the many consumers who visit Starbucks for buying coffee or for just hanging out. Starbucks visitors are predominantly students; they do their assignments and hang out with their friends at the place.

Customer repurchase intention is a stage where the choice of brand is formed. High repurchase intention reflects a high level of consumer satisfaction when deciding to choose a product. This is also affected by the quality of service.

In this study, service quality indicates a measure of the service level provided by Starbucks in fulfilling customer expectations. Customer satisfaction after purchase is highly dependent on offering performance in relation to consumer expectations. However, in Indonesia, the society and technological advances have a tendency to change people's lifestyles and perspectives, especially in the urban society with a modern lifestyle.

2 METHODOLOGY

The population in this study is Starbucks' customers. The sample of 100 respondents was taken by a non-probability sampling technique using the purposive sampling method. The primary data source was used in this study. Variables in this study included independent variables, namely, quality of service, quality of product, and lifestyle, and a dependent variable, namely, repurchase intention.

2.1 *Data analysis technique*

This study took measurements to obtain the mode and average value of respondents' responses. The mode was used to create different groups of subjects or simply to determine the characteristics

 DOI 10.1201/9781003196013-5

of the respondents. This study employed a partial least squares structural equation modeling (PLS-SEM) as a test tool to find out if there is a relationship between variables in a linear and nonlinear relationship and various indicators.

3 DISCUSSION

Quality of service is a customer perception after purchasing the related service (Demir & Aydinli, 2016). The service quality level depends significantly on consumer acceptance of the services provided in relation to their expectations. The need for a good relationship between Starbucks and its customers creates an opportunity for Starbucks to know and understand the needs and expectations of its customers. Therefore, Starbucks can provide good quality products for customers with satisfaction and create pleasant experiences and minimize experiences that tend to disappoint customers in using the product. Kotler and Armstrong (2016:11) state that product quality includes factors such as durability, reliability, accuracy, ease of operation, repair of product, and other attributes.

Nowadays, the customers tend to pay more attention to service details. In the event of a service failure, customers generally stop buying the product and spread the word to their closest people, which is detrimental to the company. Therefore, handling service failures needs to be a priority for business owners (Kumar et al., 2016). According to Sumarwan (2003), lifestyle is defined as the people's behavior on how to live, spend money, and take advantage of their time. Currently, there is a tendency in the Indonesian society to adapt a modern lifestyle; they do that by choosing, for example, a modern coffee shop such as Starbucks.

After customers make a purchase of services and products, they will evaluate whether the services and products fulfilled their expectations. This phase is known as post-purchase or post-consumption evaluation. The post-consumption evaluation of the Starbucks products will include whether or not consumers were happy with the products that they had consumed. According to Tjiptono (2012), customer satisfaction is when they realize that their needs and desires are well fulfilled. If the performance of the services and products exceeds their expectations, consumers will be satisfied; therefore, repurchase intention can occur.

Repurchase intention is a post-purchase action that is due to the consumer's satisfaction with the purchased product. Repurchase intention has two forms, namely, the intention to make a repurchase and the intention to recommend it to the closest person (Hilal and Top, 2019).

4 CONCLUSION

Based on the aforementioned discussion, if the service quality is good and the product quality is in line with expectations, then the consumers' repurchase intention level is higher. Also, if a person's lifestyle reflected in the purchased product is in line with expectations, the consumers' repurchase intention level is higher. This can occur because of customer satisfaction, which can affect repurchase intentions. In other words, customers' repurchase intention can increase in accordance with their satisfaction, that is, their experience obtained from the quality of the service provided, the quality of the product received, and the suitability of their lifestyle.

REFERENCES

Demir, A., & Aydinli, C. 2016. Exploring the Quality Dimensions of Mobile Instant Messaging Applications and Effects of Them on Customer Satisfaction. International Journal of Computer Theory and Applications, 9(22), 1–15.

Hilal, D. & Top, C. 2019. Impact of Product and Service Quality of Gated Communities on the Purchasing Intentions: Case Study in Kurdistan Region of Irak. International Journal of Economics, Commerce and Management. 7(6).

Ibzan, E., Balarabe, F., & Jakada, B. 2016. Consumer Satisfaction and Repurchase Intentions. Developing Country Studies. 6(2), 96–100.
Tarofder, A. K., Nikhashemi, S. R., Azam, S. M. F., Selvantharan, P., & Haque, A. 2016. The Mediating Influence of Service Failure Explanation on Customer Repurchase Intention through Customers Satisfaction. International Journal of Quality and Service Sciences. 8(4), 516–535.
Kotler, P., and G. Amstrong. 2016. Prinsip-prinsip Pemasaran [13th Ed.]. Jakarta: Erlangga.
Sumarwan, U. 2003. Perilaku Konsumen. Jakarta: Galia Indonesia.
Tjiptono, F. & Chandra G. 2012. Pemasaran Strategik. Yogyakarta: Andi.

Contemporary Research on Business and Management – Noviaristanti (Ed.)

The effect of good corporate governance and financial performance on the health score of regional development banks in Indonesia

W. Perwitasari & E. Kritijadi
STIE Perbanas Surabaya, Indonesia

ABSTRACT: Bank health reflects the condition and performance that need to be considered. A bank has to be healthy to get customers' trust. The indicator of a healthy bank is showing good performance or achievement during a certain period. It can be influenced by factors such as GCG, PBBA, PNBA, ALR, KKR, and MI. This study aims to find out the simultaneous and partial effects of GCG, PBBA, PNBA, ALR, KKR, and MI on bank health. The population of the study was regional development banks in Indonesia in 2014–2018. The technique used in this study was purposive sampling involving six regional development banks as the samples of the study. The data were analyzed using multiple linear regression. The results showed that GCG and MI had no positive effect on bank health; while PBBA, PNBA, and ALR had no significant negative effect. Moreover, KKR had a negative significant effect on bank health.

1 INTRODUCTION

The health scores of 26 regional development banks (RDBs) in Indonesia from 2014 to 2018 showed a downward trend. This means that there are still problems with the RDBs, especially with their health. Hence, research is needed to find out the factors influencing the health score decline. This study aims to determine the significance of the simultaneous effects of the composite scores of GCG, PBBA, PNBA, ALR, KKR, and core capital on the soundness level of RDBs in Indonesia.

Research on health by Aryanti and Balafir (2007) concluded that the NPL ratio had a significant effect on the probability of a bank being healthy and unhealthy, while the CAR, ROA, ROE, and NIM ratios showed insignificant effects on bank health. Haryati explained that of the 27 analysis variables, only 16 were significantly discriminant against the health level of national public and private banks. Sukarno (2011) explained that the performance of PT Bank DKI, based on its financial ratios, was classified as a liquid, quite solvable, profitable, and efficient commercial bank; but it was unstable in handling business risks. Using CAMELS, PT Bank DKI gained the soundness level of "healthy," which means that the bank was classified as a good bank and able to overcome the negative effects of economic conditions and the financial industry. However, the bank still had minor weaknesses that could be resolved immediately with routine action. Lutfi et al. (2014) explained the role of the board of commissioners and transparency in financial and non-financial conditions in operational efficiency. In terms of efficiency, only a properly functioning board of commissioners is able to increase the bank operation efficiency. As for profitability, the boards of commissioners and public transparency are also able to increase the profitability of banking operations in Indonesia. Sri Haryati and Kristijadi (2014) explained that the risk profile did not have a significant and positive effect on financial performance. Among the four risk profiles, liquidity risk had the best discriminatory validity. However, GCG had a significant and positive effect on financial performance, and only financial and non-financial transparency had the best convergent validity. Five indicators of bank financial performance had good validity. In addition, ROA, NIM, and CAR had good validity, in which ROA had the highest estimated loading. Mongid et al. (2012) used three stages of analysis in their research. In the first stage of analysis, INEFF was

DOI 10.1201/9781003196013-6

regressed against CAP, RISK, SIZE, and OBSTA. In the second stage, RISK was regressed against CAP, INEFF, SIZE, and NLTA. In the third stage, CAP was regressed against RISK, INEFF, SIZE, ROA, and IRC. The results showed that CAP and SIZE were negatively associated with inefficiency. However, surprisingly, RISK was insignificant. In the risk equation, the results showed that CAP and INEFF were negatively associated with risk. In the modal equation, there was a negative relationship between CAP and RISK but not with INEFF. Sufian and Noor (2012) explained that empirically, credit risk, network attachment, operating costs, liquidity, and size statistically had a significant impact on bank profitability in India. Yung (2009) found that banks with larger board sizes with lower related party borrowing rates tended to perform well. In addition, bank size is positively related to bank performance, which indicates that the bigger the bank, the better is its performance. Other findings indicated that banks listed on the stock exchange showed better performance than unlisted banks.

2 THEORETICAL FRAMEWORK AND HYPOTHESIS

The hypothesis used in this study is that the composite scores of GCG, PBBA, PNBA, ALR, KKR, and MI have a simultaneous significant effect on the health score of RDBs in Indonesia. The GCG composite score, PBBA, PNBA, ALR, KKR, and the adequacy of MI have a partially significant positive effect on the health score of RDBs in Indonesia. The bank soundness level is the result of the assessment of the bank's condition with respect to the risk and performance of the bank. Banks are required to assess the soundness level using a risk-based approach (risk-based bank rating), either individually or on a consolidated basis (Regulation of the Financial Services Authority Number 4 /POJK.03/2016).

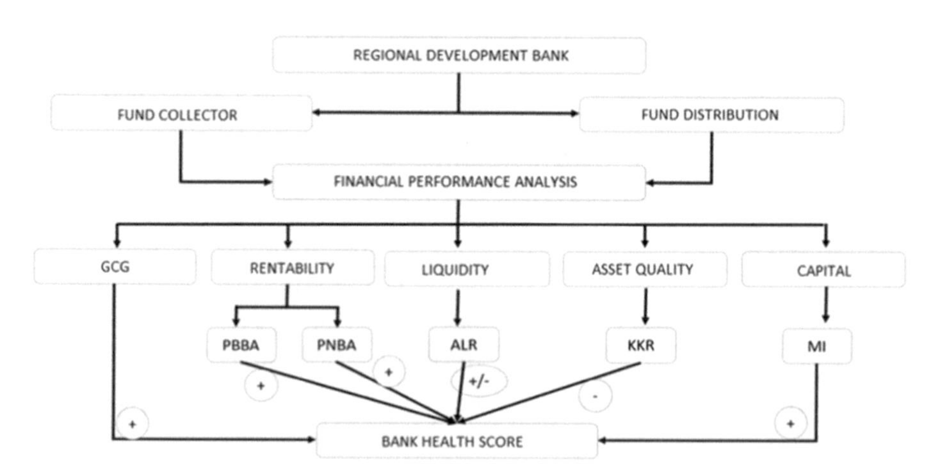

Figure 1. Research framework

3 RESEARCH METHOD

The population in this study is 26 regional development banks in Indonesia. The technique used in this study is the purposive sampling method. The sample selection criteria were as follows: banks in the Business Group II, having a core capital of over two trillion rupiahs, and having experienced a decline in health scores. Based on the criteria, six RDBs were selected as samples, namely, Bank Pembangunan Daerah (BPD) Kalimantan Barat, BPD Sumatera Barat, BPD Riau and Kepulauan Riau, BPD Bali, BPD Sulawesi Selatan dan Sulawesi Barat, and BPD Papua.

The analysis technique used is multiple linear regression with the following regression equation:

$$Y = \alpha + \beta_1 GCG + \beta_2 PBBA + \beta_3 PNBA + \beta_4 ALR + \beta_5 KKR + \beta_6 MI + ei$$

4 RESEARCH RESULTS

Table 1. Regression Coefficient

					Conclusion	
Variable	t_{count}	t_{table}	r	r^2	H_0	H_1
GCG	0.760	1.714	0.156	0.024	H0 accepted	H1 rejected
PBBA	−0.816	−1.714	−0.168	0.028	H0 accepted	H1 rejected
PNBA	−0.829	−1.714	−0.170	0.029	H0 accepted	H1 rejected
ALR	−0.553	−1.714	−0.115	0.013	H0 accepted	H1 rejected
KKR	−5.090	−1.714	−0.728	0.530	H0 rejected	H1 accepted
MI	0.255	1.714	0.053	0.003	H0 accepted	H1 rejected

GCG had a partially insignificant effect on bank health. The insignificance is because the banking sector had a low level of good corporate governance. Banks implement GCG not merely to fulfill their necessity, but also because of compliance with existing regulations. Indeed, good corporate governance will have a good impact on companies, so that it can indirectly improve the earnings quality or financial performance, which in turn can improve the company value in the eyes of investors and parties who lend money to the company. Therefore, companies can get a loan easily if they need funds to maintain their operations and reduce risks for shareholders, and it can also increase competitiveness in the global market.

PPBA had a partially insignificant effect. This is because during the period of 2014 to 2018, there was a very small change in the PPBA of the research sample banks. This condition showed that the better the PPBA obtained by the banks, the lower is the banks' health score. This result is certainly not in line with the theory stating that the higher the PPBA value, the greater is the banks' ability to generate net interest income using their assets so that the banks' health score also increases. In addition, PNBA partially had no significant effect on bank health. This is because in general, the non-interest income earned by regional development banks is very low.

ALR had a partially insignificant effect on bank health. This is due to the low average ALR level of these banks. This reflects the banks' low ability to meet their maturing obligations using primary liquid assets and secondary liquid assets, while the banks' level is generally healthy. A high ALR level indicates a higher ability to meet maturing obligations by relying on primary liquid assets and secondary liquid assets as sources of liquidity. This high ALR can result in a large amount of unused capital, so that the RDBs' opportunity to gain profits is also lost and the health score decreases.

KKR had a partially significant effect on bank health. This means that the higher the KKR level, the lower is the banks' health score. This relationship is in line with the theory stating that a higher KKR level indicates the lower ability of banks to manage credit quality, and it will have an impact on the earning interest income, which will lead banks to form provision expenses, and so the bank's health score decreases.

MI had a partially insignificant effect on bank health. This result shows that the banks could not effectively manage their core capital to increase income. In addition, the banks could not reduce costs incurred during operations, so that even though the core capital increases, the resulting income could not be maximized if it is accompanied by an increase in costs.

5 CONCLUSIONS, RECOMMENDATIONS, AND LIMITATIONS

The GCG variable had a partially positive and insignificant effect. PBBA, PNBA, and ALR variables had no significant negative effect. KKR had a significant negative effect. MI had a positive and insignificant effect on bank health scores in 2014–2018.

This research on RDBs in Indonesia has several limitations: the research period used was 2014–2018; the number of variables studied was limited (GCG, PPBA, PNBA, ALR, KKR, and MI) and other variables in the Infobank research bureau, including LDR, NPL, IRR, PDN, BOPO, and FBIR, were not used. The research subjects were only limited to the six RDBs, BPD Kalimantan Barat, Sumatra Barat, BPD Riau dan Kepulauan Riau, BPD Bali, BPD Sulawesi Selatan and Sulawesi Barat, and BPD Papua. Further research can be more representative by expanding the research subjects.

REFERENCES

Abdul Mongid, Izah Mohd Tahir, Sudin Haron. 2012. The Relationship between Inefficiency, Risk and Capital Evidence from Commercial Banks in ASEAN. International Journal of Economics and Management, Vol. 6, No. 1, 2012, p. 58–74.

Anwar Sanusi. 2013. Metodologi Penelitian Bisnis. Cetakan Ketiga. Jakarta: Salemba Empat.

Fadzian Sufian, Mohamad Akbar Nor. 2012. Determinants of Bank Performace in a Developing Economy: Does Bank Origin Matters?. Global Business Review, Volume: 13 issue: 1, page(s): 1–23.

Juliansyah Noor. 2013. Metodologi Penelitian. Jakarta: PT Kencana Prenada Media Group.

Kasmir. 2010. Manajemen Perbankan. Jakarta: PT. Raja Grafindo Persada.

Kasmir. 2012. Manajemen Perbankan. Edisi Revisi. Jakarta: PT Raja Grafindo Persada.

Lutfi, Melyza Silvy, Iramani. 2014. The Role of board commissioners and transparency in improving bank operational efficiency and profitability. Journal of Economics, Business, & Accountancy Ventura. Vol. 17, Number 1, p. 81–90.

Majalah Infobank pada periode 2014 sampai dengan 2018, mengambil data pada tanggal 7 Juni 2020.

Mo Fung Yung. 2009. The Relationship Between Corporate Governance and Bank Performance in Hongkong. Master Dissertation at Auckland University of Technology, 2009.

Otoritas Jasa Keuangan dan Publikasi Bank (Online). (www.ojk.go.id). Diakses pada tangal 14 Juni 2020.

Peraturan Otoritas Jasa Keuangan nomor 4 /POJK.03/2016 tentang Penilaian Tingkat Kesehatan Bank Umum (Online). (www.ojk.go.id), diakses pada tanggal 14 Juni 2020.

Rosady Ruslan. 2010. "Metode Penelitian Public Relations dan Komunikasi". Cetakan Kelima. Jakarta : PT. Raja Grafindo Persada.

Sri Haryati, Emanuel Kristijadi. 2014. The Effect of GCG implementation and Risk Profile on Financial Performance at Go Public National Commercial Bank. Journal of Indonesian Economy and Business. Vol. 29. Number 3, 2014, P. 237–250.

Sugiono. 2012. "Metode Penelitian Kuantitaf Kualitatif dan R&B". Bandung: Alfabeta.

Contemporary Research on Business and Management – Noviaristanti (Ed.)

Do awarded companies have fewer earnings management practices?

C. Kumandang & N.S. Hendriyeni
Sekolah Tinggi Manajemen PPM, Jakarta, Indonesia

ABSTRACT: This study aims to identify the influence of corporate social responsibility (CSR) and good corporate governance (GCG) on earnings management practices. CSR is proxied by the CSR award and GCG by the GCG award. Earnings management is measured by discretionary accruals and real earnings management. The population in this study involved all manufacturing companies listed on the Indonesia Stock Exchange whose financial reports were made fully available during the study period, that is, 2015–2019. The results of this study indicate that GCG and the number of awards received by the companies have a negative, but not significant effect on accrual earnings management and real earnings management practices, while CSR has a negative but not significant effect on accrual earnings management practices

1 INTRODUCTION

These days, companies require more than large profits to survive. They also have to show that they always pay attention to their stakeholders in their operation. The relationship between good corporate governance (GCG) and corporate social responsibility (CSR) lies in the principle of "accountability," which prioritizes stakeholders. Companies realize that their daily activities have an impact on their stakeholders, and they want to prepare it for them.

Transparency, openness, and accuracy in providing information to the stakeholders on time have attracted the attention of various institutions to evaluate the awarded companies that have properly implemented GCG principles. In Indonesia, many award events are held. However, this study considered that these two awards were the most appropriate because the assessment criteria were in line with this study. In addition, the awards are held regularly and organized by credible educational institutions. Both award events also call for strict requirements for registration, and the election of the winners is also carried out strictly based on predetermined criteria.

Logically, the company that wins the award should avoid unethical practices, such as earnings management due to the strict set of criteria. Unfortunately, a study that uses an award as a variable that affects earnings management has yet to be done. Hence, it can be said that this is the first study that used CSR and GCG awards as variables and their effects on earnings management.

2 LITERATURE REVIEW

2.1 *Earnings management*

The concept of earnings management, according to Jones (1991), consists of discretionary accruals, which are not reasonable, and non-discretionary accruals, namely, the level of accruals that are reasonable. Discretionary accrual is the recognition of accrual profit or expense that is free, unregulated, and a result of management policy choice, which can be manipulated to conform to the manager's policy. On the other hand, non-discretionary accrual is the recognition of accrual earning that is reasonable and in accordance with general accounting standards that can explain the

DOI 10.1201/9781003196013-7

company's varied economic conditions such as whether a depreciation will increase if a company's asset increases.

2.2 *Corporate Social Responsibility (CSR)*

Jones (1995) argues that companies that are committed to CSR provide incentives and are honest and ethical because they are aware that such behaviour will be the company's added value. Therefore, managers involved in CSR activities will be reluctant to practice earnings management because they are responsible for presenting transparent financial reports.

Consequently, CSR prevent the manager from performing earnings management. Thus, it is expected that there will be a negative relationship between CSR and earnings management practices. Therefore. the following hypothesis was formulated:

(H1): CSR has a negative effect on corporate earnings management practices.

2.3 *Good Corporate Governance (GCG)*

GCG is a system that regulates, manages, and supervises a business control process to increase share prices and is a form of concern for stakeholders, employees, and the surrounding community (Tunggal, 2013). According to Monks & Minow (2003), GCG is a system that regulates and controls the company to create added value for all parties.

Therefore, it is expected that the implementation of GCG can curtail earnings management practices. Therefore, the following hypothesis was formulated:

(H2): The implementation of good corporate governance has a negative effect on earnings management practices.

2.4 *Awards*

An award-winning company is the one that truly pays attention to its stakeholders and does not only pursue profit. In this study, the award events that were held by the Indonesian Institute for Corporate Directorship (IICD) for GCG proxy and the Centre of Entrepreneurship, Change and Third Sector (CECT)—Trisakti University for CSR proxy were used.

The assessment criteria for the awards used in this study included the factors of GCG and transparency, which were used to determine winners, namely, companies with high-quality financial reports with low earnings management practices, constituting a negative relationship between receiving awards and earnings management practices. Therefore, the following hypothesis was formulated:

(H3): The number of awards received by the company has a negative effect on the company's earnings management practices.

3 RESEARCH METHOD

The number of samples used for this study was in accordance with the data of manufacturing companies listed on the IDX as of Sept 21, 2020. After an examination, 131 companies with complete financial data were found.

The proxies used to measure CSR and governance were the awards received by the companies from 2015 to 2019. Each award received equals 1 (one), and 0 (zero) score was given for no award. As for the "number of awards" variable, 1 (one) was given if the company received two awards and 0 (zero) was given if the company did not receive any awards or only received one award. The proxies used to measure earnings management were accrual earnings management, which is an absolute value, and real earnings management through operating cash flow.

This study also included several control variables. It included the company size that was measured by the natural logarithm of a company's total assets with formula Ln (total asset), return on asset (ROA), return on equity (ROE), and the use of the big four public accountant service

The modified Jones model described by Dechow et al. (1995) was used to measure accrual earnings management as follows:

$$DCA = CA_{i,t}/TA_{i,t-1} - [ß_0(1/TA_{i,t-1}) + ß_1((\Delta Sales_{i,t} - \Delta TR: TA_{i,t-1}) + ß_2(PPE_{i,t}/TA_{i,t-1}) + €_{i,t}$$

where DCA is discretionary current accrual, CA is current accrual, TA is a total asset, ΔSales is the difference between this year's sales and the previous year's ΔTR is the difference between this year's sales receivables and the previous year's and PPE is property plant and equipment.

The model used to measure REM in this study is as follows (Roychowdhury, 2006)

$$CFO_t/TA_{t-1} = a_0 + a_1(1/TA_{t-1}) + b_1(Sls_t/TA_{t-1}) + b_2(\Delta Sls_t/TA_{t-1}) + €_t$$

The following regressions were used to test the hypotheses (H1 and H2) and study the relationship between earnings management and CSR and GCG using several control variables:

$$\text{AEM or REM} = a0 + b1\ GCG + b2\ CSR + b3\ Big4 + b4\ ROA + b5\ ROE + b6\ Size + e$$

To test the H3 hypothesis and identify the relationship between earnings management and the number of awards received by the company, the following regression model will be used:

$$\text{AEM or REM} = a_0 + b_1\ \text{Number of } \textit{awards} + b_2\ Big4 + b_3\ ROA + b_4\ ROE + b_5\ Size + e$$

4 RESULTS AND DISCUSSION

Based on the available financial data, the relevant data was processed using the STATA-16 program. The regression results for accrual earnings management showed that hypotheses H1 and H2 were proven. It was found that the independent variables GCG and CSR were proven to have a negative effect on accrual earnings management practices. However, if viewed from its significance, both GCG and CSR had not been shown to have a significant effect on accrual earnings management practices. Meanwhile, the results of multiple regression for H3 showed that the number of awards received by the company had a negative effect on accrual earnings management practices, even though the effect was not significant. These results are similar to the study by Kumala and Siregar (2020), who found that CSR has a negative effect on accrual earnings management practices. Kamran and Shah (2014) found that the GCG mechanism has a negative effect on accrual earnings management practices. The regression results for all control variabels have a significant effect on accrual earnings management practices for all hypotheses. Size, ROE, and the big four have a negative effect, but ROA has a positive effect on accrual earnings management practices.

The regression results for real earnings management showed that the hypotheses H2 and H3 were proven. It was found that the independent variable GCG and the number of awards were proven to have a negative effect on real earnings management practices through operating cash flows, and this is similar to the study by Calvo (2015), who found that GCG has a negative effect on real earnings management practices through operating cash flows. However, H1 was not proven, since CSR had a positive effect on real earnings management practices through operating cash flows. This is similar to the study by M. Liu et al. (2017), who found that CSR does not affect real earnings management practices. However, if viewed from its significance, both GCG and CSR had not been shown to significantly influence real earnings management practices through operating cash flows. The regression results for all control variables have a positive but not significant effect on real earnings management practices through operating cash flows

5 CONCLUSION

The results of this study suggest that awarding of GCG to companies and the number of awards received by a company had not been able to reduce the practice of accrual earnings management (AEM) and real earnings management through operating cash flows, even though the effect was negative. The awarding of CSR to companies also had not been able to reduce the practice of AEM, although the effect was negative. Meanwhile, it did not affect the practice of real earnings management through operating cash flows

These results are similar to the study by Subekti et al. (2010), who found that companies listed on the IDX generally practiced earnings management based on the company's operational activities. This also supports the study by Leuz et al. (2002), who succeeded in proving that Indonesia (an East Asian country) was one of countries that had weak protection against earnings management practices

Subsequently, the results mentioned above can be used as an input for investors and other stakeholders, which indicated that companies that receive awards from credible institutions can still be considered as investment options because they are proven to have better GCG and CSR practices than other companies.

REFERENCES

Calvo, S. G. (2015). Analysing The Relationship Between Corporate Social Responsibility, Discretionary Accruals And Real Earnings Management. Universiteit Van Amsterdam.

Dechow, P. M. (1995). Detecting Earnings Management Author(s): Detecting Earnings Management. The Accounting Review.

Jones, J. J. (1991). Earnings Management During Import Relief Investigations. Journal of Accounting Research.

Jones, T. M. (1995). Instrumental Stakeholder Theory: A Synthesis of Ethics and Economics. Academy of Management Review.

Kamran, & Shah, A. (2014). The Impact Of Corporate Governance and Ownership Structure On Earnings Management Practices: Evidence From Listed Company in Pakistan. The Lahore Journal of Economics.

Kumala, R., & Siregar, S. V. (2020). Corporate Social Responsibility, Family Ownership and Earnings Management: The Case of Indonesia. Social Responsibility Journal.

Leuz, C. N. (2002). Investor Protection and Earnings Management: An International Comparison.

Liu, M., Shi, Y., Wilson, C., & Wu, Z. (2017). Does family involvement explain why corporate social responsibility affects earnings management? Journal of Business Research,

Monks, R. A. (2003). Corporate Governance. Blackwell Publishing.

Roychowdhury, S. (2006). Earnings management through real activities manipulation. Journal of Accounting and Economics.

Subekti, I. W. (2010). The Real and Accruals Earnings Management: Satu Perspektif Dari Teori Prospek. Simposium Nasional Akuntansi XIII. Purwokerto.

Tunggal, A. W. (2013). Internal Audit dan Good Corporate Governance. Erlangga.

Contemporary Research on Business and Management – Noviaristanti (Ed.)

Financial inclusion and its impact on performance of MSMEs with efficiency and financial flexibility as mediation: Empirical evidence from Malang City

K. Ratnawati
Faculty of Economics and Business, Brawijaya University, Malang, Indonesia

ABSTRACT: This study analyzes the effect of financial inclusion on the performance of MSMEs and examines the role of mediation efficiency and financial flexibility. The simple random sampling technique was used with a total of 100 respondents and analyzed using partial least square (PLS). The results showed that financial inclusion had an effect on the performance of MSMEs, both directly and through mediation, namely, financial efficiency and flexibility. Increased financial inclusion has a major impact on improving the performance of MSMEs through efficiency, rather than through financial flexibility. Increasing access to financial services for MSMEs, which increases with increased efficiency in the form of a financial service approach to MSMEs, will improve the performance of MSMEs.

1 INTRODUCTION

MSMEs are a vital sector for the Indonesian economy (Zain, 2010); the number of MSME units in Indonesia comprises 99.99% of the total business actors (Kementrian Koperasi dan UMKM RI, 2017). Abdmoulah and Jelili (2013), Chauvet and Jacolin (2017), Lee et al. (2019), Khan (2011), and Morgan and Pontines (2014) examined the impact of financial inclusion on the growth of MSMEs in developed and developing countries, which shows that financial inclusion has a positive and significant effect on company performance and growth. According to Khan (2011) and Morgan and Pontines (2014), financial inclusion reduces liquidity barriers and encourages investment, thereby increasing output and employment opportunities. The proper implementation of financial inclusion can increase the economic activities of MSMEs (Egbetunde, 2012; Martinez, 2011; Mbuotor and Uba, 2013; Okafor, 2012; Onaolapo, 2015; Yaron et al, 2013). However, the latest research conducted by Ejiofor et al. (2020) firmly concludes that financial inclusion has no effect on the growth of MSMEs. The results of the research by Ejiofor et al. (2020) are in accordance with the findings of Awoyemi, Ogunyikamni, and Akamolafe (2015), which also state that financial inclusion has no effect on the performance of MSMEs. This shows the inconsistency of findings related to the effect of financial inclusion on the performance of MSMEs. According to Rifa'i (2017), it is difficult for MSMEs to get financed flexibly from formal financial institutions because their business background can be categorized as unbankable.

Research related to the effect of financial inclusion on the growth of MSMEs has been conducted by several researchers, namely, Abdmoulah and Jelili (2013), Banarjee (2014), Chauvet and Jacolin (2017), Lee et al. (2019), Khan (2011), and Morgan and Pontines (2018). Several other studies have also concluded that proper implementation of financial inclusion can increase economic activity, including improving the performance of MSMEs (Egbetunde, 2012; Goodland, Onumah, and Amadi, 2012; Khan, 2011; Martinez, 2011; Mbotor and Uba, 2013; Okafor, 2012; Onaolapo, 2015; Yaron, Benjamin, and Piprek, 2013).

DOI 10.1201/9781003196013-8

Hypothesis 1: Financial inclusion has a positive and significant effect on the performance of MSMEs.

According to Mardiasmo (2009), efficiency is closely related to the concept of productivity. Efficiency indicators describe the relationship between resource inputs by an organizational unit (e.g., staff, wages, and administrative costs) and the resulting output. The greater the output compared to the input, the higher will be the level of efficiency of an organization (Mardiasmo, 2009). Ergungor (2006); Kempson et al., (2004); and Grohman et al. (2017) have examined the relationship between financial inclusion and efficiency in developing countries.

Hypothesis 2: Financial inclusion has a positive and significant effect on efficiency.

Financial flexibility is one of the most interesting themes these days. This is because of a survey conducted by Graham and Harvey (2001), and the results obtained from 392 chief financial officers (CFOs) from various companies in America showed that financial flexibility is the most important factor in determining the composition of the capital structure. According to Byoun (2008), financial flexibility shows the level of capacity and the speed of a company in mobilizing its financial resources or taking preventive, reactive, and exploitative actions in order to maximize the firm value.

Hypothesis 3: Financial inclusion has a positive and significant effect on financial flexibility.
Hypothesis 4: Efficiency has a positive and significant effect on the performance of MSMEs.

Kevane and Wydick (2001) also stated that providing credit to MSMEs encourages economic growth through increase of business capitalization, creation of job opportunities, and long-term income growth. More credit means more company formation and hence economic growth (Aghion and Bolton, 1997). On the other hand Brown et al. (2005) and Chauvet and Ehrhart (2018) found that access to external credit will substantially increase sales growth and company growth (Banarjee, 2014). Dimitrov and Tice (2006) reveal that credit constraints, either due to bad credit or a recession, can encourage companies to experience lower sales growth rates.

Hypothesis 5: Financial flexibility has a positive and significant effect on the performance of MSMEs.

According to Okafor (2012) and Nurjannah (2017), financial inclusion can accelerate the flow of credit to MSMEs and support the growth of MSMEs. MSMEs without access to affordable credit will experience difficulties in expanding their business (Donati, 2016). Calcagnini et al. (2014) show that collateral easily allows companies to access credit, especially to reduce payment costs. Mishkin (2007) noted that efficiency through opening of many bank branches and the entry of new financial service providers in the financial market can pave the way for the provision of varieties of financial products and services that are in accordance with the needs of MSMEs, which was confirmed by Chandan and Mishra (2010), who observed that the presence of an institutional structure for financial services such as offices, branches, and personnel resulted in increased access to financial services by MSMEs. Fafchamps and Schündeln (2013) found evidence that there is a significant relationship between the availability of local banks and faster growth for MSMEs.

Hypothesis 6: Financial inclusion has a positive and significant effect on the performance of MSMEs with efficiency as a mediator.
Hypothesis 7: Financial inclusion has a positive and significant effect on the performance of MSMEs with financial flexibility as a mediator.

2 RESEARCH METHODS

Determination of the sample in this study was done using Roscoe's theory reported by Saidani, Rachman, et al. (2013) by determining a sample of 100 MSMEs in Malang City. The data were collected using the simple random sampling method and analyzed using PLS.

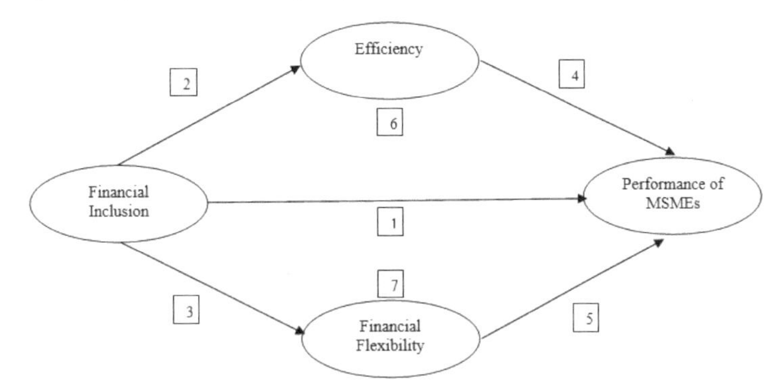

Figure 1. Research framework

3 RESULT AND DISCUSSION

3.1 *Research result*

The outer model test results show that all variables can be said to be valid and reliable and can be used for further analysis. Hypothesis testing is presented in Table 1.

Table 1. Hypothesis testing

	Original Sample (O)	**Sample Mean (M)**	**Standard Deviation (STDEV)**	**T Statistics (|O/STDEV|)**	**P Values**
Financial inclusion → performance of MSMEs	0.870	0.876	0.012	91.660	0.000
Financial inclusion → Efficiency	0.876	0.868	0.017	76.281	0.000
Financial inclusion → Financial Flexibility	0.862	0.875	0.015	82.779	0.000
Efficiency → Performance of MSMEs	0.841	0.847	0.014	69.011	0.000
Financial flexibility → Performance of MSMEs	0.882	0.895	0.010	36.892	0.000
Financial inclusion → Efficiency → Performance of MSMEs	0.795	0.876	0.022	95.008	0.000
Financial inclusion → financial flexibility → Performance of MSMEs	0.836	0.937	0.013	78.683	0.000

Source: *Output* PLS, 2020

3.2 *Discussion*

Financial inclusion has a significant positive effect on the performance of MSMEs, which is in line with several previous studies conducted by Egbetunde (2012); Goodland et al. (2012); Khan (2011); Martinez (2011); Mbuotor and Uba (2013); Okafor (2012); Onaolapo (2015); and Yaron et al. (2013). Considered as one of the most important dimensions of financial development, financial efficiency is defined as the extent to which the financial system fulfills its functions (Olgu, 2014). This finding regarding a positive relationship between financial inclusion and efficiency is in line with that reported by García and José (2016), who argued that increased financial inclusion due to intensive participation in the financial system by low-income clients can lead to high transaction and information costs. Prasad (2010) also argued that at the country level, financial inclusion can increase efficiency in financial intermediation by increasing domestic savings and investment, thereby promoting economic stability.

The Encyclopedia of Business (2015) states that financial flexibility is the ability of a company to adapt to problems and opportunities and borrow money if needed (Brigham et al., 1999). Agha (2014) and Denis and McKeon (2011) considered financial flexibility as the reserve capacity for leasing credit and loans. Bergant (2015) defined financial flexibility as the ability (capacity and speed) of a business system to finance the exploitation of business opportunities without disturbing its financial stability. When faced with financing constraints, companies should prioritize debt financing at lower costs when obtaining external financing and use short-term investment and untapped loan capacity to maintain financial flexibility in order to maintain constant growth in their firm value (Yang and Zhang, 2020).

Efficiency and financial flexibility have a significant positive effect on the performance of MSMEs in Malang City, which is in line with previous research studies conducted by Mishkin (2007), Primus (2015), and Ahmad and Ali (2015). The mediation efficiency and financial flexibility have an effect on financial inclusion and the performance of MSMEs, and the results of the analysis show a significant positive effect. The results of the variable distribution analysis show that the financial inclusion is in a fairly good category, and the performance of MSMEs in Malang city also shows that Malang UMKM actors have high MSME performance.

4 CONCLUSIONS

Increasing financial inclusion also improves efficiency by simplifying formal credit procedures and reducing bank credit costs. Increasing efforts to build customer satisfaction and loyalty as an indicator of financial flexibility will improve the performance of MSMEs. Efforts to increase financial flexibility have the strongest influence on improving the performance of MSMEs, if it is through financial intermediation by opening banking branches, so as to improve lender and borrower relations. Increasing financial inclusion will improve the performance of MSMEs through financial flexibility and access to finance, but the impact is not as strong as through financial flexibility alone.

REFERENCES

Abdmoulah, W., & Jelili, R. Ben. (2013). Access to Finance Thresholds and the Finance-Growth Nexus. *Economic Papers: A Journal of Applied Economics and Policy*, *32*(4), 522–534.

Agha, Mahmoud.(2014): *The Effect of Financial Flexibility and Credit Transitions on Corporate Investment and Financing Decisions*. Found on internet: March 2015 http://scholargoogleusercontent.com/scholar?q=cache:L2qyOphUXCAJ:scholar.google.com/+agha+mahmoud+%22The+effect+of+financial+flexibility+and+credit+rating+transition%22&hl=s1&as_sdt=0,5&as_vis=1.

Aghion, P. and Bolton, P. (1997). A Theory Of Trickle-Down Growth and Development. *Review of Economic Studies*. Vol. 64, pp. 151–172.

Ahmad, Syed Z. dan Arif, Afida M.M. (2015). Strengthening Access to Finance for Women-Owned SMEs in Developing Countries. *Equality, Diversity and Inclusion: An International Journal*, Vol. 34 Iss 7 pp. 634–639.

Bergant, Ž. (2015): *Financna prožnost in naložbena sposobnost*. Poslovodno racunovodstvo, Letnik , Števìlka 1, 38–61.

Brigham F. E., Gapenski C. L., Daves R. P. (1999): *Intermediate Financial Management*. 6th ed. Orlando: The Dryden Press.

Brown, J. D., Earle, J. S., dan Lup, D. (2005). What Makes Small Firms Grow? Finance, Human Capital, Technical Assistance, and the Business Environment in Romania. *Economic Development and Cultural Change*, Vol. 54 No. 1, pp. 33–70.

Byoun, S. 2008. How and When do Firms Adjust their Capital Structures Toward Targets? Journal of Finance. 63(6): 3069–3096.

Calcagnini, G., Farabullini, F., dan Giombini, G. (2014). The impact of guarantees on bank loan interest rates. *Applied Financial Economics*, Vol. 24 No. 6, pp. 397–412.
Chauvet, L., & Jacolin, L. (2017). Financial Inclusion, Bank Concentration, and Firm Performance. *World Development*, *97*, 1–13.
Denis, J.D., McKeon, S.B. (2011): *Debt Financing and Financial Flexibility: Evidence from Pro-active Leverage Increases.* http://papers.ssrn.com//sol3/papers.cfm?abstract_id=1361171.
Dimitrov, V., dan Tice, S. (2006). Corporate diversification and credit constraints: Real effects across the business cycle. *The Review of Financial Studies*, Vol. 19 No. 4, pp. 1465–1498.
Donati, C. (2016). Firm Growth and Liquidity Constraints: Evidence from the Manufacturing and Service Sectors in Italy. *Applied Economics*, Vol. 48 No. 20, pp. 1881–1892.
Egbetunde, T. (2012). Bank credits and rural development in Nigeria (1982–2009). *International Journal of Finance and Accounting*, *1*(3), 45–52.
Ejiofor, E., Camillus, Ok. N., & Ubogu, F. E. (2020). Effects Of Financial Inclusion On The Growth Of Cottage Firms In Nigeria. *American Research Journal of Humanities & Social Science (ARJHSS)*, *3*(1), 06–14.
Encyclopedia of Business. (2015): *Income statements.* http://www.swlearning.com/finance/maness/short-term2e/powerpoint/ch02/sld008.htm.
Ergungor, OE., (2001). Theories of Bank Lon Commitments. Economic Review-Federal Reserve Bank of Cleveland, p.1.
Fafchamps, Marcel dan Schündeln, Matthias. (2013). Local Financial Development and Firm Performance: Evidence from Morocco. *Journal of Development Economics,* Vol 103 pp. 15–28.
García, M. J.R and José, M. (2016). Can financial inclusion and financial stability go hand in hand? Economic Issues, 21 (2) (2016), pp. 81–103.
Graham, J., Harvey, C. (2001). The Theory and Practice of Corporate Finance: Evidence from The Field. Journal of Financial Economics. 60(2):187–243.
Goodland, A., Onumah, G., & Amadi, J. (2012). *Rural Finance* (No. 1). Chatham, U.K.
Kempson, Ruth M. (1995). Semantics. Cambridge University Press. Jakarta: Airlangga Press.
Kevane, M., Wydick, B. (2001). Microenterprise Lending to Female Entrepreneurs: Sacrificing Economic growth for Poverty Alleviation. *World Development*, Vol. 29, No. 7, pp. 1225–1236.
Khan, S. H. R. (2011). Financial inclusion and financial stability: are they two sides of the same coin? *Indian Bankers Association & Indian Overseas Bank, Chennai*, (November), 1–12.
Lee C.-C, Wang C.-W, & Ho S.-J. (2019). Financial Inclusion, Financial Innovation, and Firms' Sales Growth. *International Review of Economics and Finance*.
Mardiasmo. 2004. Efisiensi dan Efektifitas. Penerbit: Andy Jakarta
Martinez, M. V. (2011). *The Political Economy of Increased Financial Access*. Georgetown University.
Mishkin, Frederic S. (2008). Ekonomi, Uang, Perbankan dan Pasar Keuangan. Edisi 8.Buku 2. Jakarta: Salemba Empat.
Morgan, P. J., & Pontines, V. (2014). Financial Stability and Financial Inclusion. In *ADBI Working Paper Series* (No. 488). Tokyo: East Asian Bureau of Economic Research.
Nurjannah, Laila. (2017). *Peran Inklusi Keuangan Terhadap Perkembangan UMKM di Yogyakarta*. UIN Sunan Kalijaga Yogyakarta.
Okafor, F. O. (2012). Financial Inclusion: An Instrument for Economic Growth and Balanced Development in Rural Areas. *Journal of the Chartered Institute of Bankers of Nigeria*, *6*(8), 38–45.
Olgu, O. (2014). Handbook of Research on Strategic Development and Regulatory Practice in Global Finance. IGI Global.
Onaolapo, A. R. (2015). Effects of financial inclusion on the economic growth of Nigeria. *International Journal of Business and Management Review*, *3*(8), 11–28.
Prasad E. 2010. Financial Sector Regulation and Reforms in Emerging Markets: An Overview. Working Paper Cambridge: National Bureau of Economic Research.
Saidani, Basrah, M. Aulia Rachman, and Mohamad Rizan. (2013). Pengaruh Kualitas Produk Dan Desain Produk Terhadap Keputusan Pembelian Sepatu Olahraga Futsal Adidas Di Jakarta Timur. *Jurnal Riset Manajemen Sains Indonesia*, Vol. 4, No. 2.
Yang, Liu dan Zhang, Youtang. (2020). Digital Financial Inclusion and Sustainable Growth of Small and Micro Enterprises—Evidence Based on China's New Third Board Market Listed Companies. Sustainability Journal, 12, 3733; doi:10.3390/su12093733.
Yaron, J., Benjamin, M. P., & Piprek, G. L. (2013). *Rural Finance: Issue, Design, and Best Practices.* Washington DC.
Zain, N. H. (2010). Keterkaitan Perbankan Syariah dengan Usaha Mikro Kecil Menengah (UMKM) di Indonesia. *Dikta Ekonomi*, *7*(2), 146–158.

Contemporary Research on Business and Management – Noviaristanti (Ed.)

The relationship among job demand, job satisfaction, and job burnout of bank employees in Malang

E.Z. Firdaus, N. Noermijati & K. Ratnawati
Management Department, Faculty of Economic and Business, Universitas Brawijaya, Indonesia

ABSTRACT: The purpose of this study was to examine the direct effect of job demand and job burnout on job satisfaction and to see the mediating role of job burnout in the effect of job demand on job satisfaction of bank employees in Malang. The hypotheses were tested using the analysis tool partial least square (PLS) with the WarpPLS.7.0 program, based on data from 131 bank employees in Malang. As expected, the results of this study indicate that job demand has a significant negative direct effect on job satisfaction and significant positive direct effect on job burnout, but job burnout has no significant direct effect on job satisfaction. This research also shows that job burnout has no role in mediating the effect of job demand on job satisfaction.

1 INTRODUCTION

Technological developments require banks to continue to innovate by following the existing developments to continue to compete. One of the most important strategies is the digitization of human resource management. Employees are considered an important asset to the company. One of the most important things that employees should consider is the level of job satisfaction. Job satisfaction is considered hugely important within the field of human resources and organizational behavior. The level of job satisfaction is influenced by intrinsic and extrinsic motivational factors, the quality of supervision, social relationships with work groups, success rates, and individual failures in their work. (Armstroang, M. & Taylor, S., 2014). Individual job failures can occur when employees are unable to meet the job demands and when they are experiencing burnout.

In the JDCS Model (Job demand/Control/Support Model), reduced job satisfaction is one of the negative contributions that result from high job demands, which then produce long-term psychological stimulation and discomfort, as well as long-term activation (Karasek, R., 1979; Karasek, R., 1998; Karasek, R. & Theorell, T., 1990). Burned out employees are more likely to show greater absenteeism, higher turnover intention, as well as actual turnover, lower job satisfaction, and lower organizational commitment (Bakker et al., 2003; Maslach & Leiter, 2008). The purpose of this study was to examine the direct effect of job demand and job burnout on job satisfaction and to see the mediating role of job burnout in the effect of job demand on job satisfaction of bank employees in Malang.

Job satisfaction is a feeling of pleasure or positive emotions that arise from assessing a job or experience at work (Luthans, F., (2011). In the JDCS Model, reduced job satisfaction is one of the negative contributions that result from high job demands, which then produce long-term psychological stimulation and discomfort, as well as long-term activation ((Karasek, R., 1979; Karasek, R., 1998; Karasek, R. & Theorell, T., 1990). Supporting this opinion, the results of research by Lu, L. *et al.* (2015) and Hernandez, W. *et al.* (2018) show that job demand, as measured by the work constraints indicator, has a significant negative effect on job satisfaction. Based on the theoretical study and the results of the research, the following hypothesis can be formulated:

H1: Job demand has a significant negative effect on job satisfaction.

DOI 10.1201/9781003196013-9

The job demands–resources model extends the previous work–stress model to include demand and resources (Hakanen, J. *et al.*, (2017). The integration between the job demand process and the job resource process (resources) can finally achieve a balance where an employee is challenged enough to stay involved in the workplace and is supported to avoid fatigue (Demerouti, E. & Bakker, A., 2011). According to Hobfoll, S. E. and Freedy, J. (1993), job burnout develops from job demands and work environment. Several studies have shown that job demand has a significant positive effect on various job burnout indicators, as research results from the studies by Bakker, A. *et al.* (2004); Montgomery, A. *et al.* (2015); Alarcon, G. (2011); Hernandez, W. *et al.* (2018); Yener, M. and Coskun, O. (2013) indicate. Based on the theory and the research results, the following hypothesis can be formulated:

H2: Job demand has a significant and positive direct effect on job burnout.

Job burnout is a chronic, work-related affective state that develops gradually over time as a consequence of prolonged stress at work. (Sokka, L. *et al.*, 2016). High job burnout was found to have a direct impact on employee job satisfaction levels. Supporting this opinion, the results of research by Chen, S. and Chen, C. (2018) and Muhammad, A. and Hamdy, H. (2005) show that job burnout has a significant negative direct effect on job satisfaction. In line with this, in the research by Kim, W. *et al.* (2017), the results of the direct effect analysis show that job burnout as measured by indicator exhaustion, cynicism, and professional inefficacy has a significant negative effect on job satisfaction. Based on the theoretical study and the results of the research, the following hypothesis can be formulated:

H3: Job burnout has a significant negative effect on job satisfaction.

The term job burnout was coined to describe a psychological syndrome characterized by negative emotional reactions to work as a result of prolonged exposure to a stressful work environment. (Maslach, C. *et al.*, 2001). Hernandez, W. *et al.* (2018) found that job burnout has a significant direct effect on job demand and job satisfaction. The results of this study also indicate that job burnout moderates the effect of job demand on job satisfaction. Based on the theoretical study and the results of the research, the following hypothesis can be formulated:

H4: Burnout mediates the direct effect of job demand on job satisfaction.

2 RESEARCH METHOD, DATA ANALYSIS, AND MEASUREMENT

Quantitative research with survey methods was conducted in this study to obtain information from respondents. This research is a causal associative study aiming to explain the variables studied and the causal relationship between one variable and another. The study population was permanent employees with a minimum working period of 1 year. Questionnaires were distributed to 145 bank employees in Malang. The validity and reliability tests were conducted to measure questionnaire feasibility. Data analysis in this study was done using descriptive statistical analysis by analysis tool partial least square (PLS) with the WarpPLS.7.0 program. The mediation test was carried out using the Sobel test.

This study's job demand variables are measured by three indicators: qualitative demand, quantitative demand, and organizational demand. This study's indicators and job demand items refer to the questionnaire's measurement scale on the Experience and Evaluation of Work (QEEW) (Schaufeli, W., 2015; Schaufeli, W. 2017). This study's indicator for measuring job burnout variables refers to the Chinese version of the Maslach Burnout Inventory-General Survey (MBI-GS) (Hu & Schaufeli, 2011; Schaufeli, W. et al., 1996), which consists of emotional exhaustion, depersonalization, and cynicism. Job satisfaction variables are measured in this study using overall job satisfaction indicators, including indicators of pride in work, loyalty to work, love of work, and satisfaction with work. Overall job satisfaction indicators in this study are taken from the works by Quarstein et al. (1992) and Crossman, A. and Abou-Zaki, B. (2003).

3 RESULT AND DISCUSSION

3.1 *Research result*

Of the questionnaires that were distributed to bank employees in Malang, 131 were collected (88%); 56.5% of them were female employees, with tenure levels in the range of 1–5 (47.4%) years. Eighty-four employees were married (64.1%), with ages ranging from 26 to 30 (47.4%), and 112 employees had a bachelor's degree (85.5%).

This hypothesis testing is done by looking at the results of the p-value. The criterion is significant if the p-value is less than 0.05; then H0 is accepted or significant. However, if the p-value is greater than 0.05, then H0 is rejected or insignificant. The results of the analysis of hypothesis testing using the WarpPLS software application are as follows.

The p-value for testing hypothesis 1 is 0.01, and the path coefficient for direct influence is −0.39. So hypothesis 1 is accepted, and it is concluded that job demand has a significant negative direct effect on job satisfaction. The p-value of the testing hypothesis 2 is 0.01, and the path coefficient for direct influence is 0.44. So hypothesis H2 is accepted, and it is concluded that job demand has a significant positive effect on job burnout. The coefficient value of the pathway for testing hypothesis 3 is −0.04, and the p-value is 0.35. So hypothesis 3 is rejected, and it can be concluded that job burnout has no significant direct effect on job satisfaction.

The fourth hypothesis in this study was conducted to examine the mediating role of job burnout. Testing the role of mediation is carried out using the Sobel test. The Sobel test results show that hypothesis 4 has a p-value of 0.309 and an estimated value of −0.497. So hypothesis 4 is rejected, and it can be concluded that job burnout has no role in mediating the effect of job demand on job satisfaction.

4 DISCUSSION

Hypothesis 1 in this study is accepted, meaning that the job demand variable has a significant negative effect on the job satisfaction variable. Job demand negatively affects job satisfaction, meaning that the more job demands the employees must deal with, the less will be the level of job satisfaction felt by employees. These results support the results of previous research conducted by Karanika-Murray et al. (2017), who showed that job demands have a significant negative direct effect on job satisfaction. However, this study's results do not support any of the results of previous studies conducted by Nauman, S. et al. (2019), who showed that job demand does not have a significant direct effect on job satisfaction. On the other hand, in the research by Nauman, S. et al. (2019), it was also shown that emotional labor has a positive significant direct effect on job satisfaction. Many other factors can affect the level of employee job satisfaction. Armstrong, M. (2009) stated that the level of job satisfaction is influenced by intrinsic and extrinsic motivation factors, quality of supervision, social relations with workgroups and stages of success, and individual failure in their work.

Testing the direct effect of job demand on job burnout shows that there is sufficient evidence to accept hypothesis 2, which states that job demand has a significant positive effect on job burnout. The positive path coefficient sign indicates a direct relationship between job demand and job burnout. These results suggest that the higher the job demands that the employees must fulfil, the higher the job burnout that employees will experience. The results of this study are similar to those of the research by Hakanen, J. *et al.* (2017); Montgomery, A. *et al.* (2015); and Hu, Q. *et al.* (2011). Schaufeli, W. et al. (2017) also showed that job demand has a significant positive effect on job burnout. The results of this study support the results of previous studies such as the research studies of Babakus, E. et al. (2009); Brauchli, R. et al. (2013); Chen, S. and Chen, C. (2018); Evers, A. *et al.* (2017); Wang, Y. *et al.* (2016); and Yener, M. and Coskun, O. (2013), which showed that job demand has no significant positive effect on job burnout.

This study indicates that job burnout has no significant effect on job satisfaction, or in other words, hypothesis 3 is rejected. These results illustrate that job burnout does not have a significant direct effect on job satisfaction. These results do not support the results of previous studies conducted by Chen, S. and Chen, C. (2018); Kim, W. *et al.* (2017); and Muhammad, A. and Hamdy, H. (2005), which indicated that job burnout has a negative effect on job satisfaction. Many other factors can affect the level of job satisfaction of employees. Armstrong, M. (2009) stated that the level of job satisfaction is influenced by intrinsic and extrinsic motivational factors, the quality of supervision, social relationships with work groups, and levels of individuals' success and failure in their work.

The results of the Sobel test show that job burnout has no role in mediating the effect of job demand on job satisfaction. In other words, hypothesis 4 is rejected. The results of this study refute several previous studies that claim that high levels of burnout lead to lower job satisfaction, increased intentions for change and poor sleep quality, or even depression and suicide (Lu, A. & Gursoy, D., 2016).

5 CONCLUSIONS

Research shows that job satisfaction is not directly influenced by job burnout; however, job satisfaction is directly influenced by job demand. Job demand also directly affects job burnout rates. Therefore, it is very important for companies, especially when considering the level of job demand that is given to employees. It is not easy to reduce the impact of job demand on employee burnout. Companies can try to change employee perceptions of job demands and, at the same time, make improvements in job resources. This action should significantly reduce work wear and tear and lead directly to higher performance.

REFERENCES

Alarcon, G. M. (2011). A meta-analysis of burnout with job demands, resources, and attitudes. *Journal of Vocational Behavior*, *79*(2), 549–562. https://doi.org/10.1016/j.jvb.2011.03.007

Armstrong, M. (2009). *Armstrong 'S Handbook Handbook of Human Resource Management Practice*. 288. http://196.29.172.66:8080/jspui/bitstream/123456789/2178/1/152.pdf

Armstrong, M., & Taylor, S. (2014). *A Handbook of Human Resource Management Practice*. Kogan Page Publishers. http://books.google.com/books?id=NVnd4s6JtioC&pgis=1

Babakus, E., Yavas, U., & Ashill, N. J. (2009). The Role of Customer Orientation as a Moderator of the Job Demand-Burnout-Performance Relationship: A Surface-Level Trait Perspective. *Journal of Retailing*, *85*(4), 480–492. https://doi.org/10.1016/j.jretai.2009.07.001

Bakker, A. B., Demerouti, E., de Boer, E., & Schaufeli, W. B. (2003). Job demand and job resources as predictors of absence duration and frequency. *Journal of Vocational Behavior*, *62*(2), 341–356. https://doi.org/10.1016/S0001-8791(02)00030-1

Bakker, A. B., Demerouti, E., & Verbeke, W. (2004). Using the job demands-resources model to predict burnout and performance. *Human Resource Management*, *43*(1), 83–104. https://doi.org/10.1002/hrm.20004

Chen, S. C., & Chen, C. F. (2018). Antecedents and consequences of nurses' burnout: Leadership effectiveness and emotional intelligence as moderators. *Management Decision*, *56*(4), 777–792. https://doi.org/10.1108/MD-10-2016-0694

Crossman, A., & Abou-Zaki, B. (2003). Job satisfaction and employee performance of Lebanese banking staff. *Journal of Managerial Psychology*, *18*(4), 368–376. https://doi.org/10.1108/02683940310473118

Demerouti, E., & Bakker, A. B. (2011). The Job Demands?Resources model: Challenges for future research. *SA Journal of Industrial Psychology*, *37*(2), 1–9. https://doi.org/10.4102/sajip.v37i2.974

Evers, A. T., Yamkovenko, B., & Van Amersfoort, D. (2017). How to keep teachers healthy and growing: the influence of job demands and resources. *European Journal of Training and Development*, *41*(8), 670–686. https://doi.org/10.1108/EJTD-03-2017-0018

Hakanen, J. J., Seppälä, P., & Peeters, M. C. W. (2017). High Job Demands, Still Engaged and Not Burned Out? The Role of Job Crafting. *International Journal of Behavioral Medicine*, *24*(4), 619–627. https://doi.org/10.1007/s12529-017-9638-3

Hernandez, W., Yanchus, N. J., & Osatuke, K. (2018). Evolving the JD-R model: The moderating effects of job resources and burnout taxonomies. *Organization Development Journal*, *36*(1), 31–53.
Hobfoll, S. E., & Freedy, J. (1993). Conservation of resources: A general stress theory applied to burnout. In *Professional burnout: Recent developments in theory and practice* (pp. 115–133).
Hu, Q., Schaufeli, W. B., & Taris, T. W. (2011). The Job Demands-Resources model: An analysis of additive and joint effects of demands and resources. *Journal of Vocational Behavior*, *79*(1), 181–190. https://doi.org/10.1016/j.jvb.2010.12.009
Karanika-Murray, M., Michaelides, G., & Wood, S. J. (2017). Job demands, job control, psychological climate, and job satisfaction: A cognitive dissonance perspective. *Journal of Organizational Effectiveness*, *4*(3), 238–255. https://doi.org/10.1108/JOEPP-02-2017-0012
Karasek, R. (1998). Demand/Control model: a social, emotional, and physiological approach to stress risk and active behaviour development. In *Encyclopaedia of Occupational Health and Safety, International Labour Office* (4th ed, pp. 34.6–34.14). International Labour Office.
Karasek, R., & Theorell, T. (1990). *Healthy work: Stress, productivity and the reconstruction of working life*. Basic Books, Inc.
Karasek Robert A., J. (1979). Job Demands, Job Decision Latitude, and Mental Strain: Implications for Job Redesign. *Administrative Science Quarterly*, *24*(2), 285–308. http://www.jstor.org/stable/2392498
Kim, W. H., Ra, Y. A., Park, J. G., & Kwon, B. (2017). Role of burnout on job level, job satisfaction, and task performance. *Leadership and Organization Development Journal*, *38*(5), 630–645. https://doi.org/10.1108/LODJ-11-2015-0249
Lu, A. C. C., & Gursoy, D. (2016). Impact of Job Burnout on Satisfaction and Turnover Intention: Do Generational Differences Matter? *Journal of Hospitality and Tourism Research*, *40*(2), 210–235. https://doi.org/10.1177/1096348013495696
Lu, L., Lin, H. Y., Lu, C. Q., & Siu, O. L. (2015). The moderating role of intrinsic work value orientation on the dual-process of job demands and resources among Chinese employees. *International Journal of Workplace Health Management*, *8*(2), 78–91. https://doi.org/10.1108/IJWHM-11-2013-0045
Luthans, F. (2011). *ORGANIZATIONAL BEHAVIOR: An Evidence-Based Approach* (12th ed.). McGraw-Hill/Irwin.
Maslach, C., & Leiter, M. P. (2008). Early Predictors of Job Burnout and Engagement. *Journal of Applied Psychology*, *93*(3), 498–512. https://doi.org/10.1037/0021-9010.93.3.498
Maslach, C., Schaufeli, W. B., & Leiter, M. P. (2001). Job Burnout. *Encyclopedia of Mental Health: Second Edition*, *52*, 397–422. https://doi.org/10.1016/B978-0-12-397045-9.00149-X
Montgomery, A., Spânu, F., B?ban, A., & Panagopoulou, E. (2015). Job demands, burnout, and engagement among nurses: A multi-level analysis of ORCAB data investigating the moderating effect of teamwork. *Burnout Research*, *2*(2–3), 71–79. https://doi.org/10.1016/j.burn.2015.06.001
Muhammad, A. H., & Hamdy, H. I. (2005). Burnout, supervisory support, and work outcomes: A study from an Arabic cultural perspective. *International Journal of Commerce and Management*, *15*(3–4), 230–243. https://doi.org/10.1108/10569210580000199
Nauman, S., Raja, U., Haq, I. U., & Bilal, W. (2019). Job demand and employee well-being: A moderated mediation model of emotional intelligence and surface acting. *Personnel Review*, *48*(5), 1150–1168. https://doi.org/10.1108/PR-04-2018-0127
Quarstein, V. A., McAfee, R. B., & Glassman, M. (1992). The situational occurrences theory of job satisfaction. *Human Relations*, *45*(8), 859–873. https://doi.org/10.1177/07399863870092005
Schaufeli, W. B. (2015). Engaging leadership in the job demands-resources model. *Career Development International*, *20*(5), 446–463. https://doi.org/10.1108/CDI-02-2015-0025
Schaufeli, W. B. (2017). Applying the Job Demands-Resources model: A 'how to' guide to measuring and tackling work engagement and burnout. *Organizational Dynamics*, *46*(2), 120–132. https://doi.org/10.1016/j.orgdyn.2017.04.008
Sokka, L., Leinikka, M., Korpela, J., Henelius, A., Ahonen, L., Alain, C., Alho, K., & Huotilainen, M. (2016). Job burnout is associated with dysfunctions in brain mechanisms of voluntary and involuntary attention. *Biological Psychology*, *117*, 56–66. https://doi.org/10.1016/j.biopsycho.2016.02.010
Wang, Y., Huang, J., & You, X. (2016). Personal resources influence job demands, resources, and burnout: A one-year, three-wave longitudinal study. *Social Behavior and Personality*, *44*(2), 247–258. https://doi.org/10.2224/sbp.2016.44.2.247
Yener, M., & Coskun, Ö. (2013). Using Job Resources and Job Demands in Predicting Burnout. *Procedia – Social and Behavioral Sciences*, *99*, 869–876. https://doi.org/10.1016/j.sbspro.2013.10.559

Contemporary Research on Business and Management – Noviaristanti (Ed.)

Does adaptive marketing capabilities mediate the association between market orientation and marketing performance?

Maryanti, A.S. Hussein & K. Ratnawati
Brawijaya University, Malang, Indonesia

ABSTRACT: SMEs, which play a very important role in the improvement of the Indonesian economy, have been rapidly growing every year. One of the industrial sectors that is growing rapidly is the creative industry. Therefore, the purpose of this research is to examine the link between market orientation and marketing performance through adaptive marketing capabilities. A self-administered survey was conducted targeting SMEs in the creative industry sector in Malang, East Java. Two hypotheses were tested using SmartPLS 3.0. The results demonstrate that adaptive marketing capabilities did not mediate the relationship between market orientation and marketing performance.

1 INTRODUCTION

1.1 *Background*

The growth of small and medium enterprises (UKM), which has been occurring every year, has been able to improve the Indonesian economy. The creative industry also cannot be separated from this, which has also been growing rapidly recently. The 2018 World Conference Creative Economy stated that the creative industry sector in Indonesia has contributed 852 trillion Rupiah to the gross domestic product (GDP) or equivalent to 7.3 percent of Indonesia's total GDP over the last three years.

Based on presidential regulation number 72 of 2015, there are 16 creative industry sub-sectors, out of which there are three sub-sectors that dominate with a percentage above 10 percent: culinary with a percentage of 41.69%, fashion at 18.15%, and crafts at 15.07%. Meanwhile, there are sub-sectors that are growing rapidly. They are visual communication design, music, video animation, and architecture.

Through the R-A theory, market orientation is identified as a source of competitive advantage and innovation as a supporting tool in maintaining competitiveness. Market orientation basically requires innovative steps in accordance with market conditions and consumer expectations; therefore, market orientation is seen as an innovative action. The most important thing in marketing is to meet the needs of customers, which can be achieved through innovative practices in the creation of new goods and services, so as to achieve superior performance as directed by organizations in the market. (Remli et al., 2013; Pardi et al., 2014).

To always be able to contribute to the regional economy, SMEs must have a good performance. With a good performance, SMEs will be able to survive, grow, and develop. Marketing performance is one of the keys to the success of SMEs with intense business competition. The high marketing performance of a business will provide a large advantage for SMEs. Therefore, it is important for an SME to improve its marketing performance.

In the midst of rapidly changing business conditions, SMEs are required to be able to accommodate these changes. Therefore, the concept of marketing capability cannot be fully applied. It is necessary to develop a concept of adaptive marketing capabilities. From the point of view of satisfying customer needs, traditional marketing management has usually focused exclusively

DOI 10.1201/9781003196013-10

on reactively meeting explicit customer needs, while ignoring latent or implicit demand (Miles & Darroch, 2004; Guo, H. et al, 2018). The novelty in this paper is the developing concept of adaptive marketing capabilities as stated by Guo, H. et al, who observed that instead of being reactive to change, the new function of marketing capabilities (adaptive marketing capabilities) is to anticipate, identify, and explore customers' explicit and implicit needs and sometimes even create change in the external environment (Guo, H. et al, 2018). The purpose of this research is to investigate the effect of market orientation on marketing performance and to investigate the effect of market orientation on marketing performance through adaptive marketing capabilities.

2 LITERATURE REVIEW

2.1 *RA theory and industrial competition*

Competition can be seen as a process that focuses on the market position with regard to competitive advantage. Continuous advantage will be achieved if the company continues to invest in and accumulate resources and competitors fail to replicate these resources because they are protected by patents, are socially complex, and/or exhibit a decline in economic time. Innovation is crucial to competition and growth (O'Keeffe et al., 1998).

2.2 *Market orientation*

An organization's market orientation helps the management to identify consumers whose value requirements best serve the organization's unique capabilities. Market-oriented organizations continue to gather information on consumers, competitors, and markets; view information from an overall business perspective; decide how to deliver superior consumer value; and take action to provide value to consumers (Craven, 2006).

Wang (2009) defines market orientation as a set of behaviors that refers to the collection of intelligence on customer needs and external forces that shape those needs (intelligence generation), the extent to which the external intelligence obtained is disseminated within the company (intelligence dissemination), and actions that are taken in response to external intelligence generated and disseminated (responsiveness).

2.3 *Adaptive marketing capabilities*

Adaptive marketing capabilities can be defined as capabilities that can be extended proactively to sense and act on market signals, continue to learn from market experiments and integrate and coordinate social network resources to adapt to market changes, and predict industry trends (Guo et al., 2018). Adaptive marketing capabilities are the company's ability to identify and take advantage of the emerging market opportunities, and the development of adaptive capabilities is often accompanied by the evolution of organizational forms (Polat and Akgün, 2015).

2.4 *Marketing performance*

Voss and Voss (2000) state that marketing performance can be defined as a business measure of the company's success rate, including sales turnover, number of customers, profits, and sales growth. Companies want to achieve better marketing performances to make them more effective and increase market share and profitability (Utaminingsih, 2016).

3 RESEARCH METHODOLOGY

3.1 *Data collection*

The relationship between market orientation and marketing performance and the mediating effect of adaptive marketing capabilities were investigated. A self-administered survey was conducted targeting SMEs in the creative industry sector in Malang, East Java. The population in this study is SMEs engaged in the creative industry sector in Malang. Non-probability techniques with a purposive method were applied in this study because there were no official records related to the exact number of creative industry actors in Malang. The criteria used in this study are as follows:

1. Businesses that are classified as small and medium-sized business.
2. Businesses that have been running for at least two years continuously.

A total of 100 respondents were involved in the quantitative inquiry, but only 68 could be used for further analysis. From 16 subsectors of the creative industry, only 9 subsectors could be involved in this research. They are advertising, culinary, fashion, handicraft, photography, product design, visual arts, application and game developer, and visual communication design. There were more females than males, and about 66.2 percent of the respondents were young adults. Almost half of the respondents were high school graduates. 41.2 percent of the respondents had 2–5 years of business experience, and nearly 50 percent of the respondents had a turnover of under 10 million Rupiah per month.

4 FINDINGS

4.1 *Convergent validity*

The convergent value is the value found in each loading factor in the relationship of latent variables with indicators, which is expected to be greater than 0.6.

Table 1. Convergent validity

Indicator	Convergent validity	Conclusion
MO01	0.662	Valid
MO02	0.717	Valid
MO03	0.805	Valid
MO04	0.653	Valid
MO06	0.668	Valid
AMC01	0.654	Valid
AMC06	0.661	Valid
AMC07	0.728	Valid
AMC08	0.742	Valid
MP01	0.842	Valid
MP02	0.932	Valid
MP03	0.880	Valid

Source: Primary Data Processed, 2020

Based on Table 1, it is revealed that all the indicators have a good validity value, and thus it is implied that these indicators can measure their variables.

4.2 *Average Variance Extracted (AVE)*

Average variance extracted (AVE) on the variable market orientation (0.513 > 0.5), adaptive marketing capabilities (0.550 > 0.5) and marketing performance variables (0.784 > 0.5) is obtained, and it can be seen that all constructs have met the criteria (Table 2).

Table 2. Average Variance Extracted (AVE)

Variables	Average Variance Extracted (AVE)
Market Orientation	0.513
Adaptive Marketing Capabilities	0.550
Marketing Performance	0.784

Source: Primary Data Processed, 2020

4.3 *Cronbach's Alpha*

The reliability test can also be strengthened with Cronbach's Alpha. Table 3 shows that the Cronbach's Alpha value for all constructs is above 0.7; for market orientation, it is 0.758; for adaptive marketing capabilities, it is 0.720; and for marketing performance, it is about 0.863. The results show that all variables are reliable.

Table 3. Cronbach's Alpha

Variables	Cronbach's Alpha
Market Orientation	0.513
Adaptive Marketing Capabilities	0.550
Marketing Performance	0.784

Source: Primary Data Processed, 2020

4.4 *Coefficient of determination*

The result of this study shows that market orientation and adaptive marketing capabilities influence marketing performance with $R2 = 0.317$. This means that market orientation and adaptive marketing capabilities have an effect on marketing performance.

4.5 *Result of hypotheses*

Table 4 shows that the relationship between market orientation and marketing performance is significant with a T statistic of 2.187 (>1.96). The p-values are smaller than 0.05 (0.029). Thus, H1, which states that market orientation affects marketing performance, is accepted. Adaptive marketing capabilities as the mediating variable cannot affect the relationship between market orientation and marketing performance because the score of the T statistic is 0.584 (<1.96). The p-values are greater than 0.05 (0.560). Thus, H2, which states that market orientation affects marketing performance through adaptive marketing performance, is rejected.

Table 4. Result of hypotheses

	T Statistics	P-Values	Result
MO → MP	2.187	0.029	Accepted
MO → AMC → MP	0.584	0.560	Rejected

Source: Primary Data Processed, 2020

5 CONCLUSION

This research has shown that an increase in marketing performance can occur when there is an increase in market orientation, but adaptive marketing capabilities could not increase marketing performance. As with any studies, there are some drawbacks in this research such as the samples chosen were limited to the SMEs in the creative industry sector and not all the subsectors were involved in this research. Thus, future studies are suggested to cover all the subsectors in the creative industry and select more sectors and also other variables that are related to marketing performance.

REFERENCES

Guo, H., Xu, H., Tang, C., Liu-Thompkins, Y., Guo, Z., & Dong, B. (2018). Comparing the impact of different marketing capabilities: Empirical evidence from B2B firms in China. *Journal of Business Research*, 93(May 2017), 79–89. https://doi.org/10.1016/j.jbusres.2018.04.010

Merrilees, B., Rundle-Thiele, S., & Lye, A. (2011). "Marketing capabilities: Antecedents and implications for B2B SME performance". Industrial Marketing Management, 40(3),368–375. https://doi.org/https://doi.org/10.1016/j.indmarman.2010.08 005

O'Keeffe, M., Mavondo, F., & Schroder, B. (1998). the Resource-Advantage Theory of Competition: Implications for Australian Agribusiness. *Agribusiness Perspective Papers, 2*, 1–13.

Utaminingsih, A. (2016). MEDIA EKONOMI DAN MANAJEMEN Vol. 30 No. 2 Juli 2015. Media Ekonomi Dan Manajemen, 30(2), 161–177.

(2019, Oktober 19). Retrieved from BEKRAF: https://www.bekraf.go.id/pustaka/page/data-statistik-dan-hasil-survei-khusus-ekonomi-kreatif

(2019, Oktober 19). Retrieved from Departemen Koperasi: http://www.depkop.go.id/uploads/tx_rtgfiles/SANDINGAN_DATA_UMKM_2012-2017_.pdf

(2019, Nopember 14). Retrieved from Malang Merdeka: https://malang.merdeka.com/kabar-malang/kota-malang-terpilih-sebagai-kota-kreatif-indonesia-190626w.html

(2019, Nopember 25). Retrieved from Badan Pusat Statistik Malang: https://malangkota.bps.go.id/statictable/2017/06/14/537/luas-wilayah-dan-persentase-luas-wilayah-di-kota-malang-terhadap-luas-kota-malang.html

(2019, Desember 11). Retrieved from PMK3I: https://www.kotakreatif.id/16-subsektor

Contemporary Research on Business and Management – Noviaristanti (Ed.)

Peer-monitoring, credit discipline training, and ultra-microcredit loan repayment performance: The case of Mekaar program in Indonesia

R. Rokhim, I.A.A. Faradynawati, & A.D. Yonathan
Department of Management, Faculty of Economics and Business, Universitas Indonesia, Indonesia

W.A. Perdana
Investment Banking, Mandiri Sekuritas, Indonesia

P.G.L. Natih
University of Oxford, Oxford, UK

ABSTRACT: This research seeks to investigate the impact of peer-monitoring and credit discipline training on ultra-microcredit loan repayment performance. It was performed by analyzing primary and secondary data obtained from Mekaar, an ultra-microcredit program for women, introduced by a state-owned company, PT Permodalan Nasional Madani (PNM), as a part of the national poverty alleviation program. The findings reveal that peer-monitoring and credit discipline training have significant impacts on loan repayment rates. The implementation of a joint liability system enforces Mekaar clients to monitor their peers actively. Participants attended weekly meetings and paid the instalment on time, mainly because of the fear of being judged and scolded by other members. Respondents also realized the importance of credit discipline to determine their creditworthiness in their upcoming credit period.

1 INTRODUCTION

After being severely hit by the Asian Financial Crisis in 1998, Indonesia experienced significant economic growth and became the largest economy in Southeast Asia. Despite this miraculous GDP growth, poverty and inequality still remain the country's biggest challenges. The Indonesian poverty rate was estimated to be 8%–8.5% by the end of 2019. The government formulated various strategies to develop access to financing and capacity building to address this issue. Hence, the ministry established PT Permodalan Nasional Madani (PNM)—a state-owned company—as the implementing agency to channel government loans to micro, small and medium enterprises (MSMEs). PNM provides two main products, UlaMM for small and medium firms and Mekaar, a women-only ultra-microcredit program.

1.1 *Microfinance and access to finance*

Roughly half of adults in Indonesia remain un-banked. The main barrier to establishing an account with a bank is distant inaccessible formal financial services. However, unlike the global statistics on financial inclusion, Indonesia has a higher percentage of women having an account than men. One of the biggest contributors to this improvement is microfinance programs, particularly ones targeting women. Microfinance emergence is considered a solution to eradicate poverty and reduce inequality (Rokhim 2017). In emerging markets, female ownership of SMEs represents 30 to 37 percent of all SMEs (i.e., 8 million to 10 million women-owned firms). In particular, Indonesia had 11 million, or 43 percent of the 26 million SMEs led or owned by women (World Bank 2014). The women-owned or women-led SMEs contributed ten percent to Indonesia's GDP in 2014. Despite the significant contribution, female entrepreneurs and employers face significantly greater challenges than male counterparts in accessing financial services. According to the Global Findex, women

 DOI 10.1201/9781003196013-11

are less likely than men to have formal bank accounts. In developing economies, women are 20 percent less likely than men to have an account at a formal financial institution and 17 percent less likely to borrow money formally in the past year. Therefore, unbanked women often have a reliance on non-formal loan providers, such as loan sharks, which are very risky and often very expensive (Rokhim et al. 2016). Following the 'Grameen Bank' model, many companies—Micro Finance Institutions (MFIs)—were set up worldwide to help the poor access financial services. The broader term embraces efforts to collect savings from low-income households and, in some places, to help distributing and market clients' output (Grameen Bank 2011).

1.2 *Microfinance in Indonesia*

The microfinance industry in Indonesia has long existed, even since the Dutch colonialization period. Bank Rakyat Indonesia (BRI) is the most well-known player in this industry with its Micro Business Division (referred as BRI Units) (Charitonenko & Afwan 2003; BRI 2017). BRI Units were established in 1970 and have provided various financial products to serve the micro-business segment. In 2018, the Central Bank of Indonesia issued a regulation that requires banks to have at least 20% of their loan portfolio allocated to microloans. Moreover, the Republic of Indonesia's President Joko Widodo encourages banks to provide loans more evenly, especially to micro, small, and medium enterprise segments. Initiated in 2015, PT Permodalan Nasional Madani (PNM), through the Mekaar program, aims to improve Indonesia's financial inclusion by providing small loans to pre-prosperous women; thus, they have capital to open or develop a business. The program targets women as their clients as women are frequently marginalized in accessing formal financial institutions. Mekaar is considered the most reachable microcredit provider for the poor since it is available in all provinces in Indonesia. This program also provides very small amounts of money for loans (IDR2 million, equivalent to US$150, on average per person), far below the lowest loan limit provided by banks (Rokhim 2017). Mekaar offers distinctive products compared to other microcredit products. One example is the very small amount of loans compared to conventional banks, known as ultra-microcredit loans. Furthermore, the program also obliges clients of Mekaar to attend a weekly group meeting and several rituals before the collection of weekly installment payments by officers. In two years after its establishment, Mekaar demonstrated a significant growth level in the number of clients. At the end of 2018, Mekaar had 4 million active clients from across Indonesia, with only 0.0023% non-performing loans in its lending portfolio.

1.3 *Joint liability lending, peer-monitoring, and credit discipline training*

Joint liability lending works through prompting borrowers working on independent projects to self-select into groups to receive loans. If the group fails to meet its obligations, the micro-lender will cut off all members from future credit until the debt is repaid. The lending group members are responsible for their repayment and their partners' repayments. This lending scheme can improve social welfare if the group members are willing to cover for each other when they cannot repay their loans (Bayer & Shatragom 2013). In the absence of collateral, joint liability lending relies on peer-monitoring, which manifests itself quite often as a peer pressure mechanism to enforce loan repayment. The mechanism is a subjective experience of feeling depressed, pressed, and coerced to take certain actions or actions caused by others' pressure, urge, and challenge to a group member (Brown et al. 1986). In other words, peer pressure is a subjective experience of feeling depressed, pressed, and daring to take certain actions. They distinguish between internal pressure (or guilt) and external pressure (or shame). Some previous studies (see, for example, Wang, Lu, and Malhotra 2011; Liberati & Camillo 2018) found that personal behaviour and attitudes could substantially influence financial behaviour. Furthermore, Treynor (2009) defines peer pressure as the influence exerted by a peer group or individual, encouraging other individuals to change their attitudes, values, or behaviours to conform to group norms. Peer pressure is usually considered negative. However, in the case of micro-loan, group financing can positively influence an individual's behaviour. Peer pressure impacts individuals differently. It can be categorized as positive if the results make the person feel better, healthier, or happier, but negative if it makes people unhappy, unwell, or uncomfortable. This pressure can generally move people as they want to be accepted and fit in a group, leading to conformity.

This paper's remainder is organized as follows: Section 2 presents the literature review related to microfinance and the peer-monitoring scheme. Section 3 explains the methodology employed in this research. Section 4 discusses the research findings, and section 5 elaborates the conclusions and suggestions for policy implications.

2 DATA AND METHODOLOGY

This research collected and analyzed primary and secondary data obtained by the account officers of Mekaar through online and paper-based questionnaires, which were collected from April to May 2017. The questionnaire was distributed across 14 Provinces in Indonesia, including Aceh, West Sumatera, South Sumatera, Riau, Lampung, Jakarta, Banten, West Java, Central Java, East Java, Yogyakarta, West Nusa Tenggara, East Nusa Tenggara, and South Celebes. The secondary data were the number of customers and the distribution of these customers in each district. This research employs cluster sampling, which assumes the sub-populations to represent the diversity of respondents in the population. The questions used in this research were closed-ended multiple choice and yes or no questions.

3 RESULTS

In this research, 1,563 respondents were disaggregated into four age groups: (i) between 17–25 years old, (ii) between 26–40 years old, (iii) between 41–55 years old, and (iv) over 50 years old. The 26–40 years old group was the largest respondent group in this research, with 814 respondents or 52 percent of the total respondents sampled. Mekaar prioritizes channelling their loans to women in the productive age group. The second largest group is the 41–55 years old group. There were 660 respondents from this age group, or 42 percent of the total respondents. The third-largest group is 17–25 years old, consisting of 85 respondents or five percent of the total respondents. The over 50 is the smallest group that only consists of four respondents. This distribution of respondent ages was based on the rules applied to receiving the ultra-microcredit loans, which restrict recipients to be adult women. If the women are already married, they must get permission from their husbands to obtain Mekaar ultra-microcredit. Most of the respondents are from Java Island, the most populous area in Indonesia. West Java and Central Java are the respondents' biggest contributors, with 467 and 452 respondents, respectively. The number of respondents from Java comprises 84.7% of the total respondents, while non-Java respondents consist of only 238 people or equal to 15.2% of the total respondents.

QUESTION 1 TO QUESTION 6 indicate the peer-monitoring impact on the group members. The vast majority of the respondents answered yes to these questions, indicating the importance of peer pressure in effecting discipline and other necessary behaviours among members. Most of them were afraid of being judged and scolded by the group members. The finding is in line with Brown et al. (1986) stating that members would feel afraid, depressed, and ashamed because of peer pressure when skipping the weekly meeting, coming late for the meeting, or not paying the weekly instalment. Hence, the number of absenteeism and late payment rates were relatively low. Peer pressure was considered effective, especially when clients live in the same neighbourhood, as in the Mekaar program. Clients were grouped based on their geographical closeness, making the potential social pressure higher. This finding is in accordance with Gould (2020), stating that, concerning debt repayment, individuals in a collectivist country are obliged to repay their loans and encourage other group members to act in the same manner.

QUESTION 7 AND 8 indicate the obedience of the members to their religion and God. Most of the members believe that if they pray together, they will be able to repay the loan. In a religious society like Indonesia, people tend to believe in God's power to make everything possible. During the weekly meeting, praying sessions were believed to give clients moral support to strengthen their ability to earn money and repay their loans. According to McGregor et al. (2010) and Soenke et al. (2013), the belief in divine control supports goal pursuit by easing anxiety and feelings of uncertainty. QUESTION 9 shows that most of the members are afraid of not being invited for the next loan scheme period when violating the group loan regulations. If a client failed to repay debts or disrespect group regulations, expulsion and blacklist from the group for the next round of the loan period would await. The expelled client will be difficult to join other groups commonly living in the same neighbourhood and doing peer screening before selecting group members. QUESTION 10 shows that most of the members were afraid of the account officers who would not give the clients a loan for the next period when the violation of rules is determined. Hence, clients consider account officers the highest authorities that keep or kick a client out from the group. This belief was reinforced by members of the group and is manifested in the high level of conformity within loan recipients.

4 CONCLUSION

Mekaar has proven to be very successful since its establishment. It can increase the number of its clients exponentially while maintaining its low default rate at the same time. As a joint liability credit program, Mekaar achieved its success by implementing a peer-monitoring scheme and providing credit discipline training that enforces group members to comply with the group's regulations. The compliance is reinforced mainly by the fear of being judged or scolded by other group members. These mechanisms minimize default rates and teach members about group values like punctuality and compliance. Borrowers frequently comply with the rules because they want to be included in the next loan period. Failure to obey the rules can cause clients to be blacklisted from the next round of the loan period in all neighbourhood groups. Account officers of Mekaar also play a crucial role in managing members and achieving program objectives. The account officer acts as a monitoring agent and a role model providing credit discipline training to clients during the continuous weekly meeting.

REFERENCES

Bank Rakyat Indonesia. (2017). *Annual Report*, Indonesia.

Bayer, R., Shatragom, S. (2013). *Is joint liability lending more efficient than individual lending? A theoretical and experimental analysis*.

Brown, B.B., Clasen, D.R., & Eicher, S.A. (1986). Perceptions of peer pressure, peer conformity dispositions, and self-reported behaviour among adolescents. *Developmental psychology*. 22:521–530.

Charitonenko, S., & Afwan, I. (2003). *Commercialization of microfinance in Indonesia*. Jakarta: Asian Development Bank.

Gould CL. 2010. Grameencredit: One solution for poverty, but maybe not in every country. *Pacific Basin Law Journal* 28 (1): 1–24.

Grameen Bank Report (2011)

Liberati, C. & Camillo, F., 2018. Personal values and credit scoring: new insights in the financial prediction. *Journal of the Operational Research Society, 69*(12), pp.1994–2005.

McGregor, I., Nash, K., & Prenctice, M. (2010). Reactive approach motivation for religion. *Journal of Personality and Social Psychology*. 99, 148–161.

Ministry of State Owned Enterprises Republic of Indonesia. (2014). *PT Permodalam Nasional Madani (Persero)*.

Rokhim, R., Sikatan, G., Lubis, A., & Setyawan, M. (2016). Does microcredit improve wellbeing? Evidence from Indonesia. *Humanomics*, Vol. 32 Iss 3 pp. 258–274.

Rokhim, R., (2017). Why has Ultra Microfinance been a Policy Success in Indonesia? Working paper *2017 IAFICO Global Financial Consumers: Financial Consumer, Society and Global Development,* Fudan University, Shanghai, Tiongkok, 3–4 November 2017, ISBN 979-11-962446-0-6.

Soenke, M., Landau, M.J., & Greenberg, J. (2013). Sacred amor. Religions' role as a buffer against the anxieties of life and the fear of death. *APA Handbook of psychology, religion, and spirituality, Vol. 1 (pp. 105–122)*. Washington DC, APA Press.

Wang, L., Lu, W. & Malhotra, N.K., 2011. Demographics, attitude, personality, and credit card features correlate with credit card debt: A view from China. *Journal of economic psychology, 32*(1), pp. 179–193.

Contemporary Research on Business and Management – Noviaristanti (Ed.)

Stock movement prediction

V.A.W. Hapsari & R. Rokhim
University of Indonesia, Indonesia

ABSTRACT: Economic growth in the 21st century relies on the participation of the people in economic activities (e.g., trade and commerce and investing). In Indonesia, one particularly common economic activity is to invest in the stock market due to the wide array of companies that investors could choose to invest in. Therefore, this paper will discuss the prediction of stock prices using the Gauss–Newton representation-based algorithm (GNRBA). The proposed method provides users with a more effective algorithm. Investors and potential investors could use the methods discussed in this paper to make an informed decision in investing.

1 INTRODUCTION

The integration of international economies creates an interdependency between all the countries in the world. As economies across the world integrate more, there will be more transfer of goods, services, and capital. In the context of macroeconomy, capital investment is a component of a country's national income, better known as gross domestic product (GDP), based on the following equation: $Y = C + I + G + (x - m)$, where Y refers to GDP, C refers to public consumption, I refers to the total investment, G refers to government spending, x refers to export, and m refers to import. GDP is directly in proportion with investment, which means that as investment increases, GDP will also increase. Investment is one of the major drivers of Indonesia's economic growth because the inflow of capital through investments can be used to produce goods and services for the general public. The amount of investment in Indonesia has been growing quite significantly. A report from the United Nations Conference on Trade and Development (UNCTAD) titled "World Investment Report 2019" states that Indonesia is ranked 18th in FDI inflows. The collective FDI inflow of Asian countries grew by 4% to $512 billion. This growth is due to the increase of FDI inflows to China, Hong Kong, Singapore, and Indonesia (United Nations Conference on Trade and Development (UNCTAD), 2019).

The recorded investment in the Indonesian stock market grew year on year. One factor that stimulated this growth is a regulation issued on 6 January 2014 by Otoritas Jasa Keuangan (OJK), which changed the minimum amount for purchasing stocks in the stock market. This new regulation stated that the minimum amount of shares that can be bought is 100 shares, and it allowed more Indonesians to buy shares from the stock market with less capital required. According to Badan Koordinasi Penanaman Modal (BKPM), the amount of investment in the Indonesian Stock Exchange in 2015 to 2019 has been steadily increasing. There is a 30% growth in investors (1.1 million investors in 2019) based on the Single Investor Identification (SID) compared to previous years (Badan Koordinasi Penanaman Modal (BKPM), 2020).

The stock market has a noisy characteristic, which refers to the lack of holistic information regarding the behavior of the stock market. This lack of information is known as noise. The non-stationary characteristic of the stock market demonstrates the ever-changing distribution of the financial time series. The financial time series is random in the short term but is rather predictable in the longer term (Kumar & Thenmozhi, 2006). Unpredictable factors and occurrences (e.g., sudden political changes, traders' expectations, and a global pandemic) could alter the financial time series, such as the composite stock index. Additionally, the correlation between the financial

DOI 10.1201/9781003196013-12

time series with other data series could also shift in time. Taking all these factors into account, reading the movement of the stock market is challenging.

To minimize risk in the stock market, multiple research studies have been done to predict the movement of the stock market using machine learning to help investors in the stock market, such as support vector regression (Bao, Liu, Guo, & Wang, 2005), random forest (Kumar & Thenmozhi, 2006), decision tree (Tiwari, Pandit, & Richhariya, 2018), and artificial neural networks (Patel, Shah, Thakkar, & Kotecha, 2015). This research will utilize the Gauss–Newton representation-based algorithm (GNRBA). The writer chose this method of machine learning because there has not been any research studies using the GNRBA method to predict the movement of the stock market. In previous research, the GNRBA method is proven to have an accuracy rate of 98% when used to study breast cancer (Dora, Agrawal, Panda, & Abraham, 2017) and cervix cancer (Rustam, Hapsari, & Solihin, 2019).

2 LITERATURE REVIEW

2.1 *Technical analysis indicators*

Various methods can be used to predict the movement of stock prices. One such method is to use technical analysis, which examines patterns in past stock price movements. This historical data is the independent variable (explanatory) in examining technical analysis indicators. These indicators are mathematical formulas, which use stock prices and transaction volumes to find a pattern to the movement of the corresponding stock price (Achelis, 2001). In this research, there will be nine technical indicators used: moving average (MA) comprised of simple moving average (SMA) and exponential moving average (EMA), moving average convergence/divergence (MACD), momentum (MOM), relative strength index (RSI), stochastic oscillator (SO) %K and %D, rate of change (ROC), and commodity channel index (CCI).

3 METHODOLOGY

3.1 *Data validation*

The raw data supplied to the algorithm will be separated into two categories: training data, for model creation purposes in machine learning, and data testing, for analyzing the data model and to measure the accuracy of the model. The data is separated by utilizing a data cross-validation method, namely, stratified shuffle split (Pedregosa, 2011).

The data will randomly be categorized randomly into k group(s). Each data group has a balanced class proportion, which will be randomized with each iteration of the algorithm. This results in the algorithm getting better chances to learn from different data compositions to achieve the highest accuracy rate possible.

3.2 *Gauss–Newton representation-based algorithm*

The Gauss–Newton algorithm started from a data model paired to the actual data, data (S_i, t_i), where $i = 1,2, \ldots, n$ will be adjusted according to the model $p(\beta, s_i)$ which is nonlinear to $\mathbf{s}$, while t_i is the result of the actual data. The purpose of utilizing this model is to ensure a fit, which reduces the residue between the data models and the data (Dennis & Schnabel, 1996).

In this research, $g_i(\mathbf{s}) = p(\beta, s_i) - t_i$ is a function of the smallest nonlinear quadratic problem

$$\min_{\beta} \sum_{i=1}^{n} (p(\beta, s_i) - t_i)^2 \tag{1}$$

where β is the parameter, $p(\beta, \mathbf{s})$ is the data-fitted model, n is the amount of data and (s_i, t_i), and $s_i \in \mathbb{R}^d$, $t_i \in \mathbb{R}$, $n \geq 1$, $d \geq 1$.

4 EXPERIMENTS

The stock price data used as the input to the GNRBA program is a data set of 1,264 rows of daily stock prices (with adjusted closing) and 17 columns of technical analysis indicators consisting of SMA10, SMA12, SMA20, SMA26, SMA50, EMA10, EMA12, EMA20, EMA26, EMA50, %K,%D, MOM9, ROC12, MACD, RSI, and CCI; one weekly average price column, one column to indicate whether the stock price is higher or lower compared to the weekly average price column, one holistic average price column, one column to indicate whether the stock price is higher or lower compared to the holistic average price column, and one column to indicate whether the stock price is higher or lower compared to the previous trading day (daily). If the current stock price is higher compared to the previous day's price, then the stock price movement is defined as increasing. Conversely, if the current stock price is lower compared to the previous day's price, then the stock price movement is defined as decreasing. However, if the current stock price is similar to the previous day's price, the author would define this as increasing (considering its stability).

Table 1. A Comparison of the GNRBA Method's Accuracy to the Daily Stock Price of UNVR, MYOR, and ICBP from 2015 to 2019.

Issuer	Highest Accuracy	Lowest Accuracy
UNVR	95.35%	92.76%
MYOR	90.70%	86.90%
ICBP	93.02%	89.51%
Overall	95.35%	86.90%

Table 1 shows the result of the GNRBA method used on UNVR, MYOR, and ICBP, which shows that the highest result accuracy is recorded when using the UNVR stock price as its input, where the data composition used is divided into 30% data testing and 70% data training. This composition yielded an accuracy of 95.35%. Meanwhile, the result of the GNRBA method used on UNVR, MYOR, and ICBP shows that the lowest result accuracy is recorded when using the MYOR stock price as its input, where the data composition used is divided into 20% data testing and 80% data training, yielding an accuracy of 86.90%.

5 CONCLUSION

The computation of the GNRBA is executed by calculating the Euclidean distance between testing samples and class contribution. GNRBA yields an accuracy of higher than 90%. With this, the GNRBA can be an alternative solution to aid investors in predicting stock price movements.

REFERENCES

Achelis. (2001). Technical Analysis from A to Z (2 ed.). New York: McGraw-Hill.

Bao, Y., Liu, Z., Guo, L., & Wang, W. (2005). Forecasting Stock Composite Index by Fuzzy Support Vector Machines Regression. Fourth International Conference on Machine Learning and Cybernetics, (pp. 3535–3540). doi:10.1109/ICMLC.2005.1527554

BKPM. (2020, Januari 29). Realisasi Investasi Indonesia 2019 Naik 48,4% dalam 5 Tahun. Jakarta, DKI Jakarta, Indonesia: Badan Koordinasi Penanaman Modal. Retrieved Juni 17, 2020, from https://databoks.katadata.co.id/datapublish/2020/01/29/realisasi-investasi-indonesia-2019-naik-484-dalam-5-tahun#

Dennis, J., & Schnabel, R. (1996). Numerical Methods for Unconstrained Optimization and Nonlinear Equations. New Jersey: Prentice-Hall.

Dora, L., Agrawal, S., Panda, R., & Abraham, A. (2017). Optimal Breast Cancer Classification using Gauss- Newton Representation Based Algorithm. Expert Systems with Applications, 85, 134–145. doi:10.1016/j.eswa.2017.05.035
Kumar, M., & Thenmozhi, M. (2006). Forecasting Stock Index Movement: A Comparison of Support Vector Machines and Random Forest. 9th Capital Markets Conference Paper. Indian Institute of Capital Markets. doi:10.2139/ssrn.876544
Patel, J., Shah, S., Thakkar, P., & Kotecha, K. (2015). Predicting Stock and Stock Price Index Movement using Trend Deterministic Data Preparation and Machine Learning Techniques. Expert System Application, 42(1), 259–268.
Pedregosa, e. a. (2011). Scikit-learn: Machine Learning in Python. Journal of Machine Learning Research, 12, 2825–2830.
Rustam, Z., Hapsari, V., & Solihin, M. (2019). Optimal Cervical Cancer Classification using Gauss-Newton Representation Based Algorithm. 4th International Symposium on Current Progress in Mathematics and Sciences. 2168(1), pp. 020045-1–020045-6. Jakarta: AIP Conference Proceedings. doi:10.1063/1.5132472
Tiwari, S., Pandit, P., & Richhariya, P. (2018). Predicting Future Trends in Stock Market by Decision Tree Rough- set Based Hybrid System with HHMM. International Journal of Electronics and Computer Science Engineering, 1(3), 1578–1587. Retrieved from https://pdfs.seman-ticscholar.org/29a7/e1e1386db50bbb22710d63eb820f9c86ae1b.pdf
UNCTAD. (2019). World Investment Report. Switzerland: United Nations Conference on Trade and Development. Retrieved June 17, 2020, from https://unctad.org/en/PublicationsLibrary/wir2019_overview_en.pdf

Contemporary Research on Business and Management – Noviaristanti (Ed.)

The influence of football players' credibility on consumers' loyalty commitment to a football club

A.Z. Qashri & Y. Alversia
Master of Management, Faculty of Economics and Business, Universitas Indonesia, Indonesia

ABSTRACT: The purpose of this research is to examine the impact of football players' credibility on consumers' loyalty commitment to a football club, particularly in Indonesia. The research involved a specific football club, Madura United FC. Data were collected by distributing questionnaires (n = 1,388) to Madura United FC fans and were analyzed using SEM-PLS. The findings indicated that trustworthiness in a football player plays a great role in increasing both brand credibility and attitude of a football club. Brand credibility of a football club proved to be a significant factor affecting fans' loyalty commitment to a football club. This research provides insights into football clubs, particularly on the importance of achieving high brand credibility for achieving fans' loyalty commitment toward a football club.

1 INTRODUCTION

Football is one of the most well-known sports in the world. According to a study conducted by Terrel et al. in 2005, the number of football fans around the world had reached billions and the numbers were steadily increasing over time (Terrel et al., 2005). Moreover, sports are product-led, which means that what happens in a game or a match is what gives strength to the product (Chadwick, 2017), and what happens in a game relies completely on the team. Hence, the main asset of a football club is its players. Yu (2005) argued that athlete's good performance and positive image has a significant effect on building a successful brand and an endorsement, thus making athletes an important asset for a club, not only because of their technical skills and contribution to the club's performance in a competition, but also as an advantage to use their image for marketing purposes (Castillo, 2007).

This research focuses on the Indonesian football market, specifically Madura United FC. Madura United FC is a relatively young club that was established back in 2016, but has gained many followers and fans, since the city had never experienced having a professional football club representing its area. Considering that there has been currently minimum research on the marketing value that an athlete brings to the stakeholders of a football club, this research is significant for the football industry because it aims at investigating which characteristic in an athlete that will bring the most benefit to football clubs. Moreover, this research also aims at finding out the influence of a football club's credibility and attitude to fans' loyalty commitment.

2 LITERATURE REVIEW

2.1 *Football players as an endorser of a football club*

An identification of a target group is often influenced by an individual's identification with an associated group, when the target group is perceived to represent the group (Heere et al., 2011). If a consumer is attached to a human brand, it offers a significant potential for endorsers (Thomson,

 DOI 10.1201/9781003196013-13

2006). Thus, professional athletes in sports are direct endorsers of their respective teams (Carlson & Donovan, 2013). Athletes also represent an important brand association for their team (Gladden & Funk, 2002), where sports teams are strongly linked to their personalities inside the team compared to other brands (Smith et al., 2006).

2.2 *Football club as a brand*

Like retail brands, football clubs brand themselves the way they would like to be perceived. Branding in a football club creates a point of differentiation and individuality, provides a motive for people to buy or consume the products, and provides a chance for consumers to be more loyal to a brand. The most important outcome of branding for football clubs is that it leads to brand equity; in which, equity in football's case may be built by team-related factors such as quality and credibility of players that a football team possess (Chadwick, 2017).

2.3 *The source credibility model*

The source credibility model contends that a message's effectiveness depends on the perceived level of expertise, trustworthiness, and attractiveness in an endorser (Hovland & Weiss, 1951). Expertise is defined as the skills and the historical performance of the players. Trustworthiness is the attitude of dignity, believability, and honesty possessed by the endorser as observed by the consumer (Erdogan, 1999). Attractiveness is something that creates a positive effect on the brand credibility and brand attitude (Erdogan, 1999; Wang et al., 2016), where brand credibility is defined as believability of the information produced by a brand (Erdem & Swait, 2001), and brand attitude is the defined as summary of evaluation of a particular brand that influences behavior (Spears & Singh, 2004).

H1: Consumer's attitude of a club's players, which includes attractiveness (H1a), trustworthiness (H1b), and expertise (H1c), positively influences the club's brand credibility.
H2: Consumer's attitude of a club's players, which includes attractiveness (H2a), trustworthiness (H2b), and expertise (H2c), positively influences the club's brand attitude.

Sweeney and Swait (2008) stated that there is a positive effect of brand credibility on loyalty commitment. Loyalty commitment reflects an emotional and positive sentiment of the customer toward the idea of staying in the firm's customer base (Sweeney & Swait, 2008). Moreover, brand attitude has been proven to positively affect loyalty, and brand loyalty is essentially based on positive brand attitude, brand trust, and brand affective commitment (Gomez & Rubio, 2010).

H3: Brand credibility positively influences loyalty commitment.
H4: Brand attitude positively influences brand attitude.

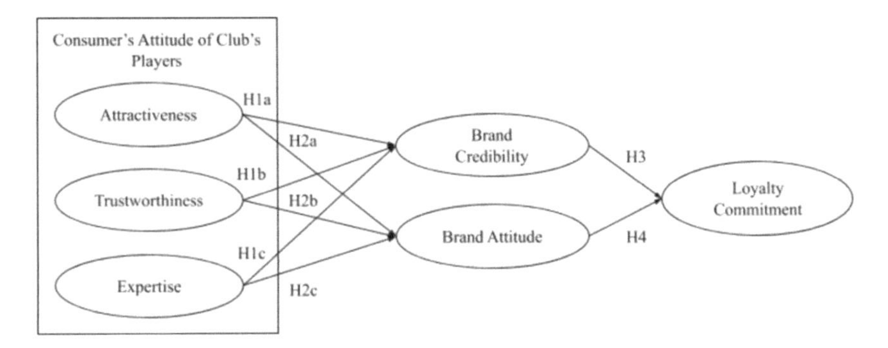

Figure 1. Conceptual model.

3 METHOD

Data were collected through questionnaires that were distributed on social media. One screening question was placed in the beginning of the questionnaire to prevent non-fans to fill in the questionnaire. Among 1,388 respondents, 1,379 respondents were Madura United FC fans, and thus, 1,379 answers were further analyzed. Most of the respondents (47.3%) were 10–20 years old, and 73.2% of them resided in East Java, Indonesia. Moreover, most of the respondents (79.2%) have been a fan of Madura United FC for more than 2 years, and 50.6% of respondents have bought an official merchandise from Madura United FC at some point. 85.8% of respondents have watched a live match featuring Madura United FC in a stadium, and 38.7% of them have watched a live match for less than 10 times.

Each of the consumer's attitude of club players' dimension used three indicators and were formed using a six-point scale to avoid neutral answers (Wang et al., 2017). Brand credibility, brand attitude, and loyalty commitment also used three indicators formed using a six-point scale (Sweeney & Swait, 2008).

4 RESULTS

This research used PLS-SEM to examine the causal relationships between variables (Garson, 2016). Most of the hypotheses were supported, except for H1c (O = 0.002, T Statistics = 0.066, P-Value = 0.471) and H2c (O = −0.026, T Statistics = 0.721, P-Value = 0.471), and both explain the relationship of expertise with both brand credibility (H1c) and brand attitude (H2c). Attractiveness was concluded to have a significant effect on both brand credibility (H1a) (O = 0.142, T Statistics = 4.122, P-Value = 0.00) and brand attitude (H2a) (O = 0.145, T Statistics = 3.720, P-Value = 0.00). Moreover, both brand credibility (H3) (O = 0.436, T Statistics = 8.293, P-Value = 0.00) and brand attitude (H4) (O = 0.193, T Statistics = 3.518, P-Value = 0.00) were concluded to have a positive and significant relationship with loyalty commitment.

5 DISCUSSION

Attractiveness and trustworthiness had a positive and significant effect on brand credibility and brand attitude, and trustworthiness most aptly explains both brand credibility and brand attitude compared to the two other variables of attractiveness and expertise. Expertise was the variable that least affected both brand credibility and brand attitude. To explain this result, a football team is considered as an icon and representation of a local area (Van Houtum & Van Dam, 2002). A fan is defined as an individual with a lasting connection to a particular subject (Thorne & Bruner, 2006). It means that a football fan's demographic location might influence fan fanaticism. Furthermore, fan loyalty creates a constant patronage even when the team fails to produce solid results on the field (Gladden & Funk, 2002). This means that even in the absence of expertise, fans might still show a positive behavior toward the team they support.

Both brand credibility and brand attitude have a positive and significant effect to loyalty commitment. This is in accordance with previous studies that successfully proved that high brand credibility and positive brand attitude will lead to an increase in loyalty commitment to a brand (Sweeney & Swait, 2008; Gomez & Rubio, 2010)

6 CONCLUSION AND IMPLICATION

This study aims at examining the relationship of a consumer's attitude of a club's player (which includes attractiveness, expertise, and trustworthiness of a player) with fans' loyalty commitment to a football club. The results indicated that in order for football clubs to reach a higher level of brand attitude and brand credibility, the management of a football club should consider choosing a trustworthy and an attractive football player. Interestingly, expertise of a player has an insignificant relationship with both brand credibility and brand attitude. This result should be interpreted with extra caution, since team performance, which is directly influenced by individual expertise, is one of the most determining factors for the team's success (Chappelet et al., 2017).

In order to increase fans' loyalty commitment toward a club, its management should focus on increasing the club's credibility by being consistent in their service quality, pricing, promotion, and their channels, including communication with customers (Sweeney & Swait, 2008).

REFERENCES

Carlson, Brad D. 2013. Human brands in sport: Athlete brand personality and identification. *Journal of Sport Management*. 27(3): 193–206.

Castillo, J.C. 2007. The concept of loyalty and the challenge of internationalization in post-modern Spanish football. *International Journal of Iberian Studies*, 20(1): 23–40.

Chadwick, S. 2017. *Handbook of Football Association Management* (3): 117–138. Nyon: UEFA Education Programme.

Chappelet, J. & Aquilina, D. 2017. *Handbook of Football Association Management*. Nyon: UEFA Education Programme.

Erdem, T., Swait, J. 2001. Brand equity as a signalling. *J. Consumer Psychol*. 7(2): 131–157.

Erdogan, B.Z. 1999. Celebrity endorsement: a literature review. *J. Mark. Manag*. 15(4).

Garson, G. D. 2016. *Partial Least Squares: Regression and Structural Equation Models*. Statistical Associates Publishers.

Gladden, J.M. & Funk, D.C. 2002. Understanding brand loyalty in professional sport: examining the link between brand associations and brand loyalty. *International Journal of Sports Marketing and Sponsorship*. 3(1).

Gómez, M. & Rubio, N. 2010.Re-thinking the relationship between store brand attitude and store brand loyalty: a simultaneous approach. *The International Review of Retail, Distribution and Consumer Research*. 20(5): 515–534.

Heere, B., James, J.D., Yoshida, M., & Scremin, G. 2011. The effect of associated group identities on team identity. *Journal of Sport Management* 25: 1–44.

Hovland, C.I. & Wiess, W. 1951. The influence of source credibility on communication effectiveness. *Public Opin. Q.* 15 (4).

Smith, A., Graetz, B.R. & Westerbeek, H.M. 2006. Brand personality in a membership-based organization. *International Journal of Nonprofit and Voluntary Sector Marketing*, 11(3): 251–266.

Spears, N. & Singh, S. N. 2004. Measuring Attitude toward the Brand and Purchase Intentions. *Journal of Current Issues & Research in Advertising*. 26(2): 53–66.

Sweeney, J. & Swait, J. 2008. The Effects of Brand Credibility on Customer Loyalty. *Journal of Retailing and Consumer Services*. 15(3).

Terrel, E., Herd, J., Marcus, L., Sams, J. & Sullivan, J. 2005. *The Business of Soccer*. Retrieved 16 May 2020 from: http://www.loc.gov/rr/business/BERA/issue3/soccer.html

Thomson, M. 2006. Human brands: Investigating antecedents to consumers' strong attachments to celebrities. *Journal of Marketing* (70): 104–119.

Thorne, S. & Bruner, G.C. 2006. An exploratory investigation of the characteristics of consumer fanaticism. *Qualitative Market Research: An International Journal*. 9(1): 51–72

Van Houtum, H. & Van Dam, F. 2002. Topophilia or topoporno? Patriotic place attachment in international football derbies. *International Social Science Review*. 3 (2): 231–248.

Wang, S. W, Kao, G.H.S. & Ngamsiriudom, W. (2017). Consumers' Attitude of Endorser Credibility, Brand and Intention with Respect to Celebrity Endorsement of the Airline Sector. *Journal of Air Transport Management*. 60: 10–17.

Yu, C. 2005. Athlete endorsement in the international sports industry: a case study of David Beckham. *International Journal of Sports Marketing and Sponsorship*, 6(3): 189–199.

Contemporary Research on Business and Management – Noviaristanti (Ed.)

Comparative analysis of internal and external factors on profits in conventional and Islamic bank systems for the long and short terms in the 2013–2019 period

E.P.D. Anggoro & A.K. Wardini
Universitas Terbuka, South Tangerang, Indonesia

M.E. Siregar
Indonesian Banking Development Institute, Jakarta, Indonesia

ABSTRACT: Commercial banks comprise conventional commercial banks (CCB) and Islamic commercial banks (ICB). They use different principles; ICB uses Islamic principles, while CCB uses conventional principles. Therefore, there are also differences in how they earn profits and the internal and external factors that influence it. This research employs the error correction model and difference test. The samples of CCB and ICB were recorded in the 2013–2019 Indonesian Banking Survey and the Islamic Banking Survey. The research results reveal that CCB did not make provisions, so that profits remained high, but it had an impact on financing expenses from non-performing loan (NPL) for the next year. Meanwhile, ICB made backups, and so the profit earned reduced, but it had a lighter financing expenses from non-performing financing (NPF), so that the continuity/sustainability of ICB was better maintained.

1 INTRODUCTION

1.1 *Research background*

According to Kasmir (2008, p.25), a bank is a financial institution that serves the public by collecting funds, channeling the funds back to the public, and providing other bank services. Based on Law No. 10 of 1998 regarding banking, there are two types of banks, namely, commercial banks and rural banks. Commercial banks comprise conventional commercial banks (CCB) and Islamic commercial banks (ICB). In collecting, storing, and distributing funds, banks need operational funds. Bank operational funds are drawn from the money deposited by customers, funds from investors, and profits.

According to Horne and Wachowicz (2005, p. 235), return on assets (ROA) measures the overall effectiveness in generating profits through the available assets; in other words, it is the power of generating returns on invested capital. In 2015, the commercial banks' ROA reached their lowest point due to the crisis experienced by Greece in 2015, which was a continuation of the 2008 crisis. This was caused by debts that matured in June 2015 to the IMF amounting to 243 billion Euros (Pane, 2016). As a result of this failure of payments, Greece went into bankruptcy, while currently carrying out its business using borrowed money. This incident forced the Greek government to tighten capital controls in order to prevent a rush of cash outflow (Wirabrata, 2015). In addition, this crisis was exacerbated by high levels of non-performing loans (NPL) at banks in Greece in 2015, where the NPL value reached 35.6% (Aggelopoulos and Georgepoulus, 2016).

For the Indonesian economy, this crisis caused a weakening of the exchange rate in 2015; it reached 10.89% (Wirabrata, 2015). Moreover, this crisis was also affected by inflation (Wirabrata, 2015). In June and July 2015, the inflation experienced by Indonesia was 7.26%, and this inflation rate was the highest throughout 2015. Decline in gross domestic product (GDP) also led to

 DOI 10.1201/9781003196013-14

the decline in the commercial banks' ROA. The decline in average GDP growth per quarter in 2014 reached 0.12%, which caused a decrease in the commercial banks' ROA by 1.15% in 2014. According to Macroeconomic Dashboard (2014), it was caused by slowing GDP growth in the mining and quarrying sector as well as industry. The slowdown in GDP growth indicates a decline in economic activity, which affects the ability of debtors to repay loans (bad credit), and it can affect bank profits.

In addition to external factors, ROA fluctuations are also influenced by internal factors. One of the internal factors affecting the ROA of CCB and ICB is the soundness of the bank. In Indonesia, to find out the health level of the bank, CCB refers to the Bank Indonesia Regulation No. 13/1/PBI/2011 regarding the Rating System for Commercial Banks, while for ICB refers to the Financial Services Authority Regulation No. 8/POJK.03/2014 regarding Assessment of Soundness Level of Sharia Commercial Banks and Business Units. In both regulations, the indicators used in assessing the soundness level of a bank consist of risk profile, good corporate governance (GCG), profitability (earnings), and capital. The better the management of risk management, GCG, profitability, and capital owned by the bank, the better placed the bank in generating profits. In addition, there are several other variables linked to bank profits, such as the bank size in terms of total assets. According to Kosmidou (2008), banks with large total assets are more profitable than banks with small total assets due to a large level of efficiency in generating profits.

1.2 *Literature review*

The grand theory in this research is the productive theory of credit. Halem (in Taswan, 2006) argued that bank liquidity will be guaranteed if productive assets are obtained from short-term loans that are easier to disburse during the normal business condition. In this theory, the bank focuses on providing short-term loans through repayment (installments) of the credit as a source of liquidity. This installment payment is also used as capital. This theory was chosen as the main theory, considering that banks need capital to generate profits.

1.2.1 *Differences between CCB and ICB*

ICB has a difference in the interest system used compared to CCB. CCB uses the interest rate system, while the ICB uses the profit-sharing system (Ascarya, 2005). ICB adopts this system because interest rates contain elements of usury, which are prohibited by Islam. According to Islamic laws, the interest system has an element of injustice because the borrower is required to pay more than the loan without paying attention to the advantages/disadvantages of the borrower. On the other hand, the profit-sharing system used by Islamic banks is a system where the borrower and lender share the risks and benefits in the agreement, so that both parties do not feel disadvantaged in this case. This difference shows how CCB and ICB generate profits.

1.3 *Method*

The sample of this research was the aggregate data of CCB and ICB available in Indonesian Banking Statistics (SPI, *Statistik Perbankan Indonesia*, and SPS, *Statistik Perbankan Syariah*), which were taken from January 2013 to December 2019. The independent variables consist of internal factors (credit risk, liquidity risk, profitability, capital, and total assets) and external factors (inflation, exchange rates, and GDP). The analysis technique used the error correction model (ECM) and difference tests.

2 RESULT

2.1 *CCB analysis*

The variables that significantly affected the ROA of CCBs in the long-term equation are capital adequacy ratio (CAR), net interest margin (NIM), total assets, and middle exchange rates. The

CAR negatively affected the ROA of CCBs; the CAR increased in conditions where capital growth was faster than the growth in the risk-weighted asset (RWA). With the increase in RWAs, even if they are lower than the capital increase, banks experience an increase in risk; so they need to add allowance for impairment losses (CKPN, *Cadangan Kerugian Penurunan Nilai*) to deal with these risks. The addition of CKPN by the bank shows that the bank incurs a fee, causing a decrease in profit/return. However, after correcting the imbalance of the CAR variable in this long-term equation, the short-term equation for the CAR variable became insignificant.

Furthermore, NIM negatively affected the ROA of CCBs when the increased NIM caused by the growth of net interest income was faster than the growth of productive assets. However, an increase in RWAs was also followed by the increase/growth of productive assets. With the increase in RWAs, there was an increase in risk, which led banks to make an allowance in the form of CKPN. Therefore, the issuance of fees for the CKPN could reduce bank profits/returns.

Furthermore, total assets positively affected the ROA of CCBs. This result is in accordance with that reported by Kosmidou (2008), who observed that a bank with large total assets has a larger economy of scale than a bank with small total assets, and thus, the bank will operate efficiently to generate higher and better profits.

Finally, the middle rate negatively affected the ROA when the depreciation of the Rupiah exchange rate affected the increase in the price of imported goods. This could cause an increase in production costs, which affected the ability of customers to repay loans/credits, and hence, this affected bank profits.

2.2 *ICB analysis*

The variables that significantly affected the ROA of ICBs in the long-term equation are financing-to-deposit ratio (FDR), non-performing financing (NPF), and total assets. FDR positively affected the ROA of ICBs when the FDR increased, which indicated that the bank was expanding to finance funds to customers in the form of productive and consumptive financing. With the expansion of financing channeled carefully, the profit or return of the bank will automatically also be higher.

Furthermore, NPF negatively affected the ROA of ICBs. When the NPF was high, the bank added the allowance in the form of CKPN. In other words, the bank incurred additional costs for CKPN to not reduce the profit earned for financing provisions.

Furthermore, total assets positively affected the ROA of ICBs. This result is in accordance with that reported by Kosmidou (2008), who observed that a bank with large total assets has a larger economy of scale than a bank with small total assets, and thus the bank will operate efficiently to generate higher and better profits.

Based on the short term, the variables that significantly affected the ROA of ICBs were non-performing financing (NPF) and the middle exchange rate. NPF still negatively affected the ROA of ICBs, so the cause was still the same. Meanwhile, the middle exchange rate negatively affected the ROA of ICBs when the exchange rate weakened (depreciation). This was caused by an increase in the price of imported goods, because of which production costs increased and the ability of customers to repay loans decreased, leading to the decline in bank profits. The difficulty of repaying loans could be exacerbated if the debtor borrowed funds in the form of US$ because the exchange rate of Rp/US$ was getting weaker, so that the value of credit repayments increased.

2.3 *Differences between CCBs and ICBs*

Overall, the difference between CCBs and ICBs lies in the interest system used by CCBs, which is the interest rate system, while ICBs adopt a profit sharing system. In other words, the methods of obtaining profits used by these two banks are different; CCB relies more on interest and ICB relies more on profit sharing (Ascraya, 2005, p.1). Moreover, the underlying asset is not followed by the customer. For example, a customer borrows funds from CCBs to buy a house; then the loan provided is in the form of cash. However, the customer can later use the funds for other purposes, even if the house purchase is not realized. Unlike the CCB, when a customer applies for financing to buy a house, the ICB will find a house for the customers, so they will receive a house instead of cash. Based on the Wilcoxon test, it is concluded that H0 was rejected; thus, there was a difference between the average ROA of CCBs and ICBs. This result also shows that CCBs and ICBs were different in managing profits. CCBs tended not to make provisions in the form of CKPN, so that profits remained high but affected the financing expenses from NPL in the further year. Meanwhile, ICBs would have a backup in the form of CKPN to reduce the profit earned by them, but they had less financing expenses from NPF, which affected the continuity/sustainability of ICBs.

3 CONCLUSION AND SUGGESTION

3.1 *Conclusion*

The internal and external factors had influences, both long-term and short-term, on the ROA of CCBs and ICBs. However, several different factors influenced it partially; the ROA of CCBs was influenced by CAR, NPL, NIM, exchange rate, and total assets, while the ROA of ICBs was influenced by FDR, NPF, exchange rate, and total assets. Based on these results, it can be concluded that a different handling is needed in managing earnings.

In addition, this research revealed that there was a difference in the average value of ROA between CCBs and ICBs based on the results of different tests, which were supported by the research results, in which CCBs and ICBs were different in managing profits. CCBs tended not to make provisions in the form of CKPN, so that profits remained high but affected the financing expenses from NPL in the further year. Meanwhile, ICBs would have a backup in the form of CKPN to reduce the profit earned by them, but they had less financial expenses from NPF, which affected the continuity/sustainability of ICBs.

3.2 *Suggestion*

For further research, it is necessary to find a replacement variable that is not partially significant between the two banks, namely, GDP, which may be replaced by per capita income, or inflation, which is replaced by the consumer price index or to create a new composite index. Furthermore, the policymakers at the national level do not make rules that equalize between CCBs and ICBs because these two types of banks have different ways of generating profits.

REFERENCES

Aggelopoulos, Eleftherios, and Antonios Georgopoulos. (2016). Bank Branch Efficiency under Environmental Change: a Bootstrap DEA on Monthly Profit and Loss Accounting Statements of Greek Retail Branches. European Journal of Operational Research, Volume 261, Issue 3: p.1170–1188.

Ascarya, and Diana Yumanita. (2005). Seri Kebanksentralan No. 14 Bank Syariah: Gambaran Umum. Jakarta: Bank Indonesia

Fajari, Slamet, and Sunarto. (2017). Pengaruh CAR, LDR, NPL, BOPO Terhadap Profitabilitas Bank (Studi Kasus Perusahaan Perbankan Yang Tercatat di Bursa Efek Indonesia Periode Tahun 2011 Sampai 2015). Proceeding at Seminar Nasional Multi Disiplin Ilmu & Call for Papers Unisbank Ke-3:.853–862.

Horne V. James, and John M Wachowicz. (2005). Prinsip-prinsip Manajemen Keuangan (Fundamental of Financial Management). 12th eds. Translated by Dewi Fitriasari. Jakarta: Salemba Empat.

Kosmidou, K. (2008). The Determinants of Banks' Profits in Greece during The Period Of EU Financial Integration. Managerial Finance, 34(3): 146–159.

Pane, Boy Yusuf. (2016). Kegagalan Yunani Memanfaatkan Bailout dalam Upaya Mengatasi Krisis Ekonomi Tahun 2018. Jurnal Online Mahasiswa Fakultas Ilmu Sosial dan Ilmu Politik, Vol 3, No. 1: 1–12.

Taswan. (2006). Manajemen Perbankan. Yogyakarta: UPP STIM YPKP

Wirabrata, Achmad. (2015). Krisis Yunani dan Turbulensi Ekonomi Indonesia. Info Singkat Ekonomi dan Kebijakan Publik Vol VII No. 13/I/P3DI/Juli/2016: 13–16.

Contemporary Research on Business and Management – Noviaristanti (Ed.)

Analysis of synergy of the capital budgeting method and intellectual capital disclosure

N. Astrini, A.K. Wardini & Z. Hidayah
Universitas Terbuka, South Tangerang, Indonesia

ABSTRACT: The diminishing fossil energy reserve has forced Indonesia to prioritize renewable energy (RE) development through its energy resources. However, RE power plant companies in Indonesia are challenged by the doubtfulness of potential investors and developers concerning their profitability and sustainability. To address this issue, this study aims to analyze the synergy between the capital budgeting method and the disclosure of intellectual capital (IC) of a wind farm expansion project in increasing their investment viability. The data were collected through interviews and documentation. The interactive analysis was used in this study. It was found that both the capital budgeting method and IC disclosure mutually supported the analysis of investors by helping them take better investment decisions. However, future studies are required to analyze the presentation of IC disclosure in other RE plants to increase Indonesia's RE development funding opportunities.

Keywords: capital budgeting methods, intellectual capital disclosure, investment decisions, project funding, renewable energy.

1 INTRODUCTION

Energy crises arise when the available fuel reserves show a deficit compared to consumption needs. In such a case, the government will prevent the scarcity of energy by developing new renewable energy sources.

Funding for renewable energy (RE) power plants in Indonesia is one of the major factors that has hindered the development of the project. Adequate funding would overcome RE project obstacles such as inadequacy of technological mastery, human resources, and raw materials procurement. The investment decision of potential investors is strongly related to the success of obtaining funding. It was mentioned that capital budgeting and investor behavior influence the investor decision-making process (Ekawati, 2016). Investor behavior includes their psychological factors in dealing with investment risk, which is part of behavioral finance (Baker and Ricciardi, 2014). It was also explained that investment decisions are influenced by several financial drivers, namely, a rational evaluation of the economic opportunities of a project, and non-financial drivers related to the investor's background in terms of education and experience (Masini and Menichetti, 2013). These aspects have a chance to positively influence investor perceptions of a prospect of a project through non-financial information on its true value. In this case, it can improve the project funding opportunities and generate positive idealism, especially for RE power plant development in Indonesia.

Non-financial factors of RE projects in Indonesia, regarding their influence on investment decisions, should include several aspects related to the specific advantages of the project that are not revealed in financial factors. It can be shown in the intellectual capital (IC) of the project. IC is an intangible device that will provide value creation for a project with adequate management. Razafindrambinina and Anggreni (2011) explained the three components of IC, which are human capital, structural capital, and customer or relational capital.

 DOI 10.1201/9781003196013-15

This study is focused on analyzing the synergy between the capital budgeting method and IC disclosure to obtain funding for the RE power plant in Indonesia, specifically funding for the windfarm expansion project. The problem formulation is about how the synergy between the capital budgeting method and the IC disclosure of windfarm expansion projects can increase investment viability for the project.

2 METHOD

This study employed a qualitative approach, and its framework is shown in Figure 1. The figure shows the interference of IC in investment decision making. IC disclosure is a non-financial factor that influences the ultimate goal of obtaining project funding.

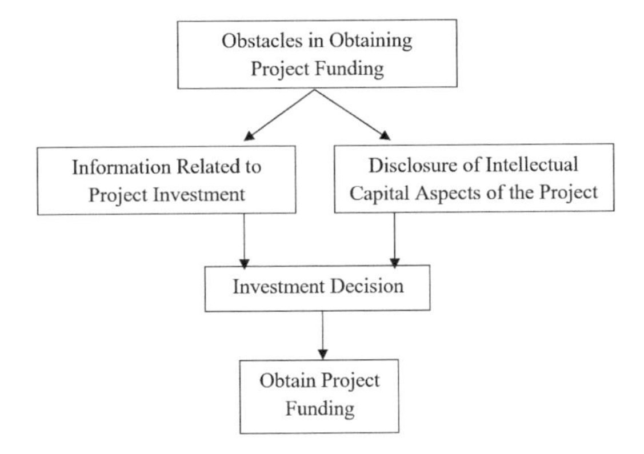

Figure 1. Framework.

The data were collected using semi-structured interviews and documentation after the key informants were selected using the purposive sampling method. The informants covered four different professions, namely, investor, head of development of the wind farm expansion projects, head of project finance, and lecturer.

Interactive analysis, which was developed by Miles and Huberman (2014), was adopted in this study. In this study, data reduction, presentation, and verification cycle would be repeated if the conclusion is still considered to be deficient, which requires additional data. It was conducted until the objectivity and validity of the data was reached. Thus, conclusions that can be accounted for were obtained.

3 RESULTS AND DISCUSSIONS

This study explained the mutually supporting relationship between the capital budgeting method and the disclosure of IC in increasing the investment viability. The interviews and documentation process revealed that the capital budgeting method played a huge role and was perceived to be effective in supporting the investment feasibility analysis of a project in certain industries and certain project characteristics. However, other factors were required as a broader perspective about investment opportunities for a project. The non-financial factors influence investment decisions but are not reflected in financial statements, especially the income statement. It could consist of environment impact management, employment, and others. The interview excerpts related to the abovementioned explanation were taken from informants who were investors, heads of project development, heads of project finance, and lecturers in the faculty of economics.

JS (2019) said that:

Actually, some investors also see 'additionality', meaning how it impacts, for example, social aspects, development, and the number of jobs created; for example, when we build power plants, how many people will be affected? CO_2 reduction is also taken into account. The 'additionality' is more about the positive points of the project, which are not measured in profit. For example, CO_2 is reduced, which is not measured in profit. Another example is the development of the local economy by carrying out a project. There may be a measure, but it is beyond economic returns.

In the financial model, an investor certainly wants to see how much the return is, how solid the model is, whether the model uses assumptions that make sense and are correct, its sensitivity, its tariff, and its selling price.

NP (2019) said that:

Starting from the FS, so we analyze the availability of resources, availability of technology, access, as well as eligibility for construction, then the risks associated with other things such as permitting, regulation, and then we determine its value in monetary terms, what it will be like. Then after that, a financial model is made that is projected for the next 30 years or during the operation of the project, to determine whether the project is profitable, or not to be developed, financed, and built, and it seems that the main parameter is still IRR.

We have Info Memo … We display this Info Memo to relevant investors asking for funding after our FS is received.

IS (2019) said that:

There must be a financial model as a basis for later talking with PLN as a customer. The financial model must be attached because it will later reflect the ability of the project's payback period. The purpose is not only for investors but also for lenders.

The power purchase agreement (PPA) is one of the triggers to get financial close with investors. Whereas with PLN, PPA is important as a determining factor that the project can go ahead. The PPA stated how much should be invested in the project.

ME (2019) said that:

To attract investors … hmmm … feasibility study, because capital budgeting is only part of the feasibility study of a project, right? Budgeting is only a financial aspect, whereas there are other aspects that we must look at if we want to adopt a project. FS in terms of work culture is suitable. Tools to know whether a project is profitable or not must be seen from all aspects, there is a balanced scorecard, also financial models.

After reviewing several previous studies and interviewing several informants, the analysis result is shown in Figure 2.

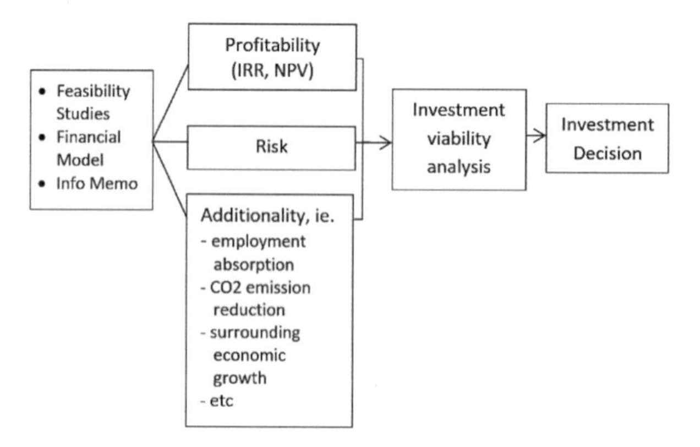

Figure 2. A synergy of the capital budgeting method and intellectual capital disclosure in increasing the investment feasibility of windfarm expansion projects.

In Figure 2, information obtained from feasibility studies (FS), financial model (FM), and information memorandum (IM) is in the form of financial and non-financial information. In this case, non-financial factors significantly reduced investment risk perception, which was not reflected in the financial statements. Both financial and non-financial information was valuable as an analysis material for investors in assessing the long-term investment feasibility of a project. In Figure 2, it is shown that the capital budgeting method represents the level of profitability. On the other hand, IC disclosure is presented as an aspect that influences the project's risk level preference and the project's positive impact information in 'additionality'.

In terms of investment decisions of the wind farm expansion project, FM, FS, and IM have a broader scope regarding financial and non-financial information. In FS and IM, several aspects of the IC were presented. There was information concerning the solutions to mitigate the project risks, the use of more efficient information technology systems, the relationship between project management and the central and the local government, and the method to approach the surrounding community. The results of the analysis are expected to help investors make better investment decisions.

4 CONCLUSIONS AND RECOMMENDATIONS

The financial and non-financial information obtained from FS, FM, and IN was valuable for investors in assessing the feasibility of project investment. The capital budgeting method is a financial aspect that represents the profitability or return on investment. On the other hand, IC disclosure is a non-financial aspect that influences the project risk level preference. The results of the analysis of financial and non-financial factors were mutually supportive and could help investors make better investment decisions.

It is suggested that further studies that analyze the IC disclosure on the RE industry project in Indonesia be conducted. Therefore, it could increase funding opportunities and improve RE development in Indonesia.

REFERENCES

Baker, H. Kent and Ricciardi, Victor. (2014). *Investor Behaviour: The Phsycology if Financial Planning and Investing*. New Jersey: John Wiley & Sons, Inc.

Bruggen, Alexander, Vergauwen, Philipe and Dao, Mai. (2009). Determinants of Intellectual Capital Disclosure: Evidence from Australia. *Journal of Management Decision*, vol. 47 No.2, 2009. pp. 233–234.

Droj, L. (2016). Premises of Behavioral Finance in Rational Decision Making.*The Annals of The University of Oradea Economic Sciences Journal*, vol. 1 No. 1, pp. 671–681

Ekawati, Erni. (2016). *Manajemen Keuangan*. Jakarta: Universitas Terbuka.

Masini, Andrea and Menichetti, Emanuela. (2012). Investment Decisions in the Renewable Energy Sector: An Analysis of Non-Financial Drivers. *Technological Forecasting and Social Change*, vol. 80, No. 3, pp: 510–524.

Miles, M. B., Huberman, A. M., & Saldana, J. (2014). *Qualitative Data Analysis: A Methods Sourcebook* (3rd ed.). Thousand Oaks, CA: Sage.

Razafindrambinina, D. dan Anggreni, T. (2011). IC and Corporate Financial Performance of Selected Listed Companies in Indonesia. *Malaysian Journal of Economic Studies*, vol. 48 No. 1, pp. 61–77.

Sofian, Saudah dan Tayles, Mike E. dan Pike, Richard H. (2004). Intellectual Capital: An Evolutionary Change in Management Accounting Practices. Presentation at The Fourth Asia Pacific Interdiciplinary Research: Singapore.

Contemporary Research on Business and Management – Noviaristanti (Ed.)

The relevance of efficiency market theory to changes in valuation in food and beverages companies listed on the Indonesia stock exchange

P.A. Sebastian, D.P. Abdulrachman & N.S. Hendriyeni
PPM School of Management, Indonesia

ABSTRACT: This study aims to find out the application of the efficiency theory to changes in valuation in food and beverages (F&B) companies listed on the Indonesia Stock Exchange in 2017–2019. The efficiency theory states that changes in stock prices are influenced by changes in the fundamental factors of the company. Changes in stock prices that are contrary to fundamental changes indicate an inefficient market. The results of this study can assist investors in making investment decisions. This study gives investors a model that they can use in making investment decisions using company fundamental data with indicators such as price to selling ratio, price to earnings ratio, and price to book value ratio. The ratios were obtained from the 2017–2019 quarterly financial reports. The market efficiency data on stock prices were data on the day the financial statements were issued (t), the day before the financial statements were issued (t-1), and one day after the financial statements were issued (t+1). Then, changes in stock prices were compared with changes in the fundamental value of the company's financial statements to obtain the market efficiency value for every F&B company during the 2017–2019 time period. This study obtained the efficiency level and company valuation in three stages: first, the calculation of the company's fundamental figures and their quarterly development; second, the calculation of the share movement according to the financial statement publication date; and third, an analysis of share movement relevance to the company's fundamentals to obtain the market efficiency value of each company. Keywords: price to selling ratio, price to earnings ratio, price to book value ratio, financial statement, market efficiency, Indonesia Stock Exchange

1 INTRODUCTION

1.1 *Research background*

The food and beverages (F&B) industry is growing rapidly in Indonesia, which has the fourth largest population in the world. Thus, the consumer's need for F&B products is a profitable potential for investment. In the past 20 years, many multinational companies have conducted foreign direct investment (FDI) by establishing large factories, such as Unilever, Heinz, Frisian Flag, Perfetti Van Melle, Mondelez, Coca Cola Amatil, Danone, and Nestle.

This level of investment interest in the Indonesian fast moving consumer goods (FMCG) sector motivates this study on the investors' basis in choosing the best FMCG companies listed on the Indonesia Stock Exchange (IDX) in the 2017–2019 period.

The investors' decision-making methods such as using the market efficiency theory for investing in FMCG companies, as reported in this study, can be used by local investors as the basis for investment. Based on *Kustodian Sentral Efek Indonesia* (KSEI/Indonesia Central Securities Depository) IDX data, the number of investors who had customer fund accounts and were able to conduct stock transactions at the end of December 2019 was 2,478,243 SIDs (single investor identification), which increased 53.04% from 2018; it was only 1,619,372 then. Among them, 1,768,485 SIDs were mutual fund customers.

 DOI 10.1201/9781003196013-16

This sizeable increase was dominated by young people, as shown by KSEI IDX data; 44.62% were aged 21–30. They were mostly (59.41%) males working in private companies (53.69%) and having undergraduate education (48.23%). The 53.04% increase from 2018 indicates that there is great potential for the younger generation to invest.

Novice investors can make mistakes because they do not understand companies' fundamental factors and they lack the ability to analyze which stocks have good potential and understand the methods commonly used in making investment decisions coupled with risk analysis in investing before making decisions in buying shares on the IDX. This problem became the background of this study.

1.2 *Research problem*

Based on the background, the formulated research problems are as follows:

a. Is the market efficiency theory relevant to stock price changes in F&B companies on IDX?
b. Does the change in the valuation of F&B companies affect market efficiency?
c. Do F&B companies listed on IDX have good market returns and safe risk value?

1.3 *Research objectives*

This study aims to:

a. Determine the relevance of the market efficiency theory to changes in stock prices in F&B companies listed on IDX.
b. Determine the effect of changes in valuation on market efficiency in F&B companies listed on IDX.
c. Determine the level of return and risk on investment in F&B companies listed on IDX.

1.4 *Research object*

This study examined 17 public companies engaged in the F&B sector listed on IDX by calculating the companies' fundamental and technical values based on financial statements and changes in stock prices in 2017–2019.

1.5 *Research significance*

This study provides a method for making a purchase decision of an F&B company stock listed on IDX, thereby reducing the risk of investment loss.

2 THEORETICAL REVIEW AND HYPOTHESIS

2.1 *Market efficiency theory*

There are four definitions of efficient capital markets. The first definition is based on the intrinsic value of securities. The efficiency measure is seen from the extent to which security prices deviate from their intrinsic value (Beaver, 1989). Thus, an efficient market can be defined as a market where the values of securities do not deviate from their intrinsic value.

The second definition is based on the accuracy of security prices. Fama (1970) defined that a capital market is efficient when the prices of securities accurately reflect the existing information.

The third definition is based on the distribution of information. Beaver (1989) stated that the market is efficient only if the prices of securities act as if everyone observes the information system. This definition implies that if everyone observes one information system, then everyone is considered to have the same information. Thus, weak and semi-strong efficiency always occurs because it is assumed that the information has been distributed and made available to the public. However, Beaver also argued that there is a cost to obtaining this information. Therefore, the information disseminated may only be accepted by some investors and the market becomes inefficient.

The fourth definition is based on dynamic processes (Jones, 1995). This definition considers the asymmetric distribution of information and describes how prices will adjust because of the asymmetric information. The market is said to be efficient if the information dissemination is prompt, so that the information becomes symmetrical, that is, everyone has this information.

2.2 *Multiple valuation*

Multiple Valuation is a valuation measurement based on the sales value (price to selling ratio), net income value (price to earnings ratio), and equity value (price to book ratio) within a certain period (Damodaran, 2012). It tends to measure only the company value historically. A low ratio value means that the possibility of development still exists. However, it cannot estimate the growth of the company.

3 RESEARCH METHOD

3.1 *Research type, data source and sample collection method*

This is a basic study, for which the data were collected using a quantitative approach based on secondary data of stock prices from 17 F&B companies listed on IDX in 2017–2019. In this study, the market efficiency and company valuation were tested and recommendations for investors were given. The data were taken from 17 F&B companies listed on IDX in the 2017–2019 period. Data were collected using a sampling frame method that was conducted on 17 companies in the F&B sector listed on IDX in 2017–2019, and the samples were compared to financial reports issued in certain periods by companies or data on IDX.

3.2 *Regression test*

Multiple regression analysis was employed in this study. Multiple regression analysis is a statistical technique using parameter coefficients to determine the influence of the independent variable on the dependent variable. Hypothesis testing, both partially and simultaneously, was carried out after the regression model was free from violations of classical assumptions. The aim was that the results can be interpreted appropriately and efficiently. The regression equation is as follows (Weston & Copeland, 1992):

$$\text{Stock Price Change (Y)} = \alpha 1 + \alpha 2\ \text{PBVR} + \alpha 3\ \text{PER} + \alpha 4\ \text{PSR} + \text{e} \quad (1)$$

where
Y = stock price change
$\alpha 1$ = constant
$\alpha 2, 3, \alpha 4$ = estimation coefficient of PBV, PER, and PSR
PBVR = price to book value ratio/stock price
PER = price to earnings ratio/stock price
PSR = price to sales ratio/stock price
e = error

4 RESULTS AND DISCUSSION

The regression test analyzed the changes in stock prices in 17 consumer goods companies listed on IDX in 2017–2019 based on the stock change data influenced by the PBVR, PER, and PSR in each quarter.

The regression analysis results employed a maximum error tolerance of 5%, showing that the effect of change in stock prices on the changes in the companies' fundamental ratio was significant in PBVR, PSR, and PER. The adjusted R square value ranged from 0.75 to 0.87, which indicates that this value is valid to be presented because it is above 0.5.

It was also found that the PBVR and PSR coefficient values were below 0.01. This finding indicated that the values had a significant effect on stock price changes. Meanwhile, it was found that stock value change did not have any influence due to its PER significance above 0.01.

The results of the normality test on all 204 data revealed that the results were normally distributed because the probability was above 0.05. Therefore, the data can be used in regression calculations.

The probability level of the data significance was above the 5% confidence level. Therefore, it can be concluded that the regression model did not show heteroscedasticity.

The multicollinearity test analyzed the correlation matrix of independent variables. A high correlation (generally > 0.90) found between variables indicates multicollinearity in the data of stock price changes during H0, H-1, and H+1.

5 CONCLUSIONS AND RECOMMENDATIONS

This study concludes that changes in PER did not have a significant effect on changes in stock prices. Instead, changes in PER occasionally affected stock prices in the opposite direction. Meanwhile, changes in PSR had a significant effect on changes in stock prices. It was also found that the changes in PBVR had a strong influence on stock price, as the increase of PBVR will increase stock prices and vice versa.

The normality, heteroscedasticity, and multicollinearity tests showed that the data were normally distributed, homogeneous, and linear, except for PER data whose values tended to be different from changes in PBVR and PSR values. From the findings, it can be concluded that to investigate the effect of market efficiency changes on the companies' fundamental ratios, PBVR is an important parameter. Therefore, investors must look at PBVR before purchasing stock rather than considering the net profit value because the changes tend to be inconsistent.

Further studies on the effect of fundamental ratios on market efficiency can be done to test companies in sectors other than consumer goods listed on IDX. Moreover, this study can be developed by adding other factors that are assumed to influence stock price changes such as free cash flow to equity (FCFE), free cash flow to firm (FCFF), debt to equity ratio (DER), and other supporting factors for making investment decisions.

REFERENCES

Beaver, W.H. 1998. *Financial Reporting: An Accounting Revolution*. Third Edition. Prentice Hall International, Inc.

Damodaran, Aswath. 2012. Investment Valuation: Tools and Techniques for Determining The Value of Any Asset 3rd Edition, John Wiley & Sons, Inc, New Jersey USA.

Fama, E.F. 1970. Efficient Capital Markets: A Review of Theory and Empirical Work. *Journal of Finance*. Vol. 25. 383–417.

Jones, P. 1996. *Investments Analysis And Management*, Fifth Edition. John Wiley & Sons, Inc.. The United States of America.

Weston & Copeland. 1995. Managerial Finance 9th ed., The Dryden Press, 1992. Terjemahan A. Jaka Wasana dan Kibrandoko, Manajemen Keuangan, Jakarta: Binarupa Aksara.

Contemporary Research on Business and Management – Noviaristanti (Ed.)

Green M&A

R. Prasetiya & R. Rokhim
Faculty of Economics and Business, Universitas Indonesia, Indonesia

ABSTRACT: This research examines the impact of green mergers and acquisitions (GMA) on the company's performance (return on asset). This research tested the sample of mergers and acquisitions of firms using OLS regression and a comparative test with t-statistics. The green factor had a statistically positive and significant effect with ROA changes after three years of mergers and acquisitions. In both green and non-green sub-samples, the average ΔROA (T + 3) has a minus value in three years after mergers and acquisitions. However, the green sample has a better average ΔROA (T + 3) value, close to zero. These results indicate that green deals, despite the higher cost of green mergers and acquisitions (green premium), are expected to improve bidder companies' financial performance in the future. This research investigated the opportunities for an environmentally friendly transition to a green economy by increasing energy efficiency and renewable energy.

1 INTRODUCTION

Global warming and climate change are caused by environmental damage. Global warming is getting worse each year, as carbon dioxide (CO2) produced by burning fossil fuels keeps increasing (Hill, 2020). The increase in CO2 emission has prompted the Indonesian government to reduce environmental pollution by transitioning from conventional energy to renewable energy. This action follows the agreement in Paris on December 12, 2015, held by the UNFCCC. The Paris Agreement, attended by 197 countries, aims to combat climate change and accelerate action and future low-carbon investments (UNFCCC, 2015).

Many companies use CSR activities to show concern for the environment. Unfortunately, CSR does not always provide higher value creation than the cost of implementing it, creating an assumption that CSR activities are detrimental to some companies (Jonikas, 2014). Moreover, to instill the corporate value that positively impacts the future environment, companies begin acquiring green technology from eco-friendly companies to increase efficiency and reduce pollution levels. This action is often referred to as green mergers and acquisitions (GMA) (Salvi, Petruzzella & Giakoumelou, 2018).

Mergers and acquisitions, in general, can positively affect the company's financial performance in the future. However, many studies also show that mergers and acquisitions can damage company performance (Ferreira, Reis, & Pinto, 2016). These varying research reports lead researchers to test indicative GMA against financial performance changes.

GMAs are expensive, as they require companies to adopt green technology, adjust new green business models, and train employees. Bidder companies tend to be willing to pay highly (green premium) to acquire target companies with green businesses (Salvi, Petruzzella, & Giakoumelou 2018), even though the mergers and acquisitions, in general, can either improve or deteriorate companies.

2 LITERATURE REVIEW

Environmental management has become essential in the last few decades because it mitigates the environmental or ecological damage caused by company activities. Proper environmental

 DOI 10.1201/9781003196013-17

management also demonstrates a company's commitment to environmental sustainability; hence, it improves company image and creates moral capital that can reduce stakeholder sanctions' severity in harmful activities (Salvi, Petruzzella, & Giakoumelou 2018).

GMA refers to acquisitions, mergers, and other companies' economic activities to acquire green resources and develop green technology (Lu, 2021). GMA can be utilized as a company strategy in changing and improving environmental protection. In recent years, industries with high pollution levels have carried out green mergers and acquisitions to increase investment in environmental sustainability and adopt environment-friendly technologies (Lu, 2021).

GMAs help heavy pollution companies to regain stakeholders' trust and reduce their risk of sanctions caused by environmental damage (Li, Liu, Liu, & Liu 2020). Direct business innovation will positively impact companies' competitiveness, profits, and business sustainability. Furthermore, research on the effects of GMA on company changes based on the ROA ratio in companies in European and American regions supports the application of GMA. The findings show that GMA positively and statistically impacts ROA for the next two to three years. This tendency encourages bidder companies to obtain good CSR and environmental management, even with a high premium price (Salvi, Petruzzella, & Giakoumelou 2018).

3 METHODOLOGY

3.1 *Research data*

A sample of companies that carried out mergers and acquisitions in the 2007–2016 period and their financial data were analyzed to understand the impact of green mergers and acquisitions on companies' performance. The companies conduct business in emerging market countries and developed markets in the Asia Pacific and Europe. Eikon Thomson Reuter's database was accessed to examine the companies' transactions. The specific criteria for companies' transactions taken from the Eikon Thomson Reuters database are as follows: (1) M&A go public companies, (2) minimum takeover of 50.01%, (3) sample taken from January 1, 2007, to December 31, 2016, (4) consisting of companies in the highly polluting industrial sector (chemicals, oil and gas, metals and mining, other energy and power, paper and forest products, petrochemicals, and power), and (5) complete financial data from a year before and three years after the mergers and acquisitions.

3.2 *Research model*

Assessment of the company performance three years after (T + 3) mergers and acquisitions or GMA can be seen from the change in the value of ROA one year before (T – 1) to three years after (T + 3) mergers and acquisitions.

Figure 1. Research model

The OLS regression and the mean comparison test with t-statistics were used for the test.

3.3 *T-test*

Hypothesis testing in this research uses parametric statistics. The parametric statistical method used in this research is the mean comparison test with the t-test. This method compares the mean differences between samples of green and non-green at three years (T + 3) after mergers and acquisitions. The t-statistic value and probability value were calculated to determine the mean difference between the two samples.

3.4 *OLS regression*

OLS regression testing regression model ΔROA (T + 3) is as follows:

$$\Delta ROA(T+3) = \alpha + \beta_1 Lev + \beta_2 LnTA + \beta_3 LnDV + \beta_4 Cash + \beta_6 GA + \beta_7 Green + e \quad (1)$$

Before testing the t-statistics and OLS regression, the classical assumptions were tested to provide the best linear unbiased estimator (BLUE). This research used cross-sectional data. Hence, the classical assumption tests use the normality test, heteroscedasticity test, and multicollinearity test.

4 RESULTS

4.1 *T-test results*

Initially, the regression model is tested for its normality, heteroscedasticity is tested with the Glejser test, and a multicollinearity test is conducted with the variance inflation factor (VIF). The regression models show that data are normally distributed. The heteroscedasticity test using the Glejser test found no heteroscedasticity between each variable, and the multicollinearity test using VIF showed no multicollinearity symptoms on the regression model.

The sample is divided into two sub-samples, green and non-green samples, to test the mean ROA change caused by green mergers and acquisitions.

Table 1. Average whole sample, green sub-sample, and non-green sub-sample

Sample	Δ ROA (T + 3)
Whole Sample	−0.492 (0.479)
Green Sample (A)	−0.158 (0.444)
Non-*Green* Sample (B)	−0.603 (0.439)
Difference (A–B)	0.445***

Notes: *P < 0.1, **P < 0.05, ***P < 0.01

The findings are not in line with previous research, which shows that the green sample average ΔROA has a positive value and is better than the average value of non-green ΔROA (Salvi, Petruzzella, & Giakoumelou, 2018), whereas in this study, the result has a minus value three years after mergers and acquisitions. In previous research, a comparison was made using a sample of green mega deals and non-green mega deals in Europe and America. The findings show that companies that did mega deals and had more acquisition experience are superior in the integration processes, such as in terms of cultural alignment and establishing a vision and mission. Also, they have a better performance than the agreement in the short and long terms (Hu, Li, Li, & Wang, 2020). However, the average green sample in this research shows a smaller minus value. It is closer to zero than the average ΔROA of the non-green samples at three years after mergers and acquisitions, and the T-test confirmed that the difference is statistically significant at 1%.

4.2 *OLS regression*

A cross-sectional OLS regression analysis for the whole sample after three years of merger and acquisition deal completion was conducted to investigate the effect of green mergers and acquisitions on change ΔROA (T + 3)

The results show a significant positive impact of green mergers and acquisitions on bidder post-acquisition performance with regard to ΔROA (T + 3). The results can be seen as follows:

Table 2. OLS Regression Impact of Green Deals on ΔROA (T + 3)

ΔROA (T + 3)			
Variabel	**Coefficient**	**Standard Error**	**Probability**
Constant	−0,7662**	0,3538	0,0321
LEV	0,0203	0,1171	0,8625
LNTA	0,0288	0,0233	0,2176
LNDS	−0,0139	0,0209	0,5066
CASH	0,2064**	0,0818	0,0128
HOM	−0,2142**	0,0854	0,0134
GA	0,0655	0,0806	0,4175
GREEN	0,3499***	0,0891	0,0001
Adjusted R-Squared	0,2391		
N	144		

Notes: *P < 0.1, **P < 0.05, ***P < 0.01

OLS regression test results show that green deals have a positive and statistically significant impact on ΔROA at three years after mergers and acquisitions. The previous study results also indicate that green mergers and acquisitions have a positive and statistical impact on ΔROA (Salvi, Petruzzella, & Giakoumelou 2018). Specifically, ΔROA (T + 3) has a coefficient of 0.3499 and significance at 1%. The findings show that green has a strong impact after three years of mergers and acquisitions.

5 CONCLUSIONS

This research analyzes and determines the impact of green mergers and acquisitions and non-green merger and acquisitions on changes in public companies' performance in emerging market and developed market countries in the Asia-Pacific and European regions in 2007–2016. The research investigates the change in the ratio of return on assets (ROA) three years after mergers and acquisitions.

T-test and OLS regression indicate a positive and statistically significant impact on ROA changes in bidder companies affected by mergers and acquisitions with green indications three years after the events. Even though the research did not find that the average change in ROA has a positive value after mergers and acquisitions, the average change in the green companies' ROA is better than that of the non-green companies. Hence, the research indicates that GMA and non-green mergers and acquisitions generally negatively affect companies' value. Nevertheless, the results show that the ROA changes for the green companies are closer to zero than those for the non-green companies. The average ROA increases three years after mergers and acquisitions. We believe that GMA positively affects bidder companies in the long term, especially with regulations supporting green business development in the future.

REFERENCES

Hill, M. (2020). *Understanding Environmental Pollution.* Cambridge: Cambridge University Press.

UNFCCC. (2015, December 12). *The Paris Agreement.* Retrieved from UNFCCC International: https://unfccc.int/process-and-meetings/the-paris-agreement/the-paris-agreement

Jonikas, D. (2014). Value created through CSR measurement possibilities. *Social and Behavioral Sciences*, 189–193.

Salvi, A., Petruzzella, F., & Giakoumelou, A. (2018). Green M&A Deals and Bidders' Value Creation: The Role of Sustainability in Post-Acquisition Performance. *International Business Research*, 96–105.

Ferreira, M. P., Reis, N. R., & Pinto, C. F. (2016). Three decades of strategic management research on M&As: Citations, co-citations, and topics. *Global Economics and Management Review*, 13–24.

Lu, J. (2021). Can the green merger and acquisition strategy improve the environmental protection investment of listed company? *Environmental Impact Assessment Review*, 1–12.

Li, B., Liu, X., Liu, Y., & Liu, M. (2020). *Green M&A, Business Model Innovation and Sustainability of Heavy Polluters: Evidence from the China's "Environmental Protection Storm."* Department of Finance, The University of Sydney.

Hu, N., Li, L., Li, H., & Wang, X. (2020). Do Mega-Mergers Create Value? The Acquisition Experience and Mega-Deal Outcomes. *Journal of Empirical Finance*, 119–142.

Contemporary Research on Business and Management – Noviaristanti (Ed.)

An investigation of workload, work–life balance, and flexible working in relation to turnover intention of Information Technology (IT) workers

A.S.K. Murti & F. Martdianty
Master of Management, Universitas Indonesia, Indonesia

ABSTRACT: The turnover rate of information technology (IT) is one the highest compared to other sectors. To minimize the turnover intention, companies need to take several factors into account, including workload, work–life balance, flexible working, and job satisfaction. The aim of this study is to analyze the relationship among workload, work–life balance, flexible working, and job satisfaction in relation to turnover intention among IT workers. This study is quantitative in nature with a cross-sectional design. Data were collected by distributing questionnaires online. The data obtained covered 272 respondents with a background as IT workers and were subsequently analyzed using the structural equation modeling method in Lisrel software. The results suggest that work–life balance and flexible working can have an indirect effect on turnover intention mediated by job satisfaction. On the other hand, workload has no indirect effect on turnover intention with the mediating effect of job satisfaction.

1 INTRODUCTION

The need for information technology (IT) professionals has multiplied in almost every industrial sector in Indonesia (Puslitbang-Aptika-IKP Kominfo, 2019). On the contrary, there is no corresponding growth in the supply of IT talents, as there is a deficit of workers who have IT skills (The World Bank, 2018). The supply–demand gap of IT professionals leads to staffing problems, as technology professionals are not satisfied with their current positions and likely to leave the organization to find other employment opportunities (Lo, 2015; Moore, 2000). It is in line with prior studies that have consistently pointed job satisfaction as a powerful factor that plays a significant role in regards to individual's intention to leave organizations.

As a result of the relentless technological changes in a competitive and dynamic environment, the role of IT workers, both professional and non-professional, becomes vulnerable to an increase in workload and work stress that lead to job dissatisfaction (Li & Shani, 1991). Besides the significant effect of workload on job satisfaction, several studies have shown that there are additional factors affecting IT workers' job satisfaction, such as work–life balance and flexible working. Therefore, this study investigates whether or not workload, work–life balance, and flexible working have an indirect effect on employees' intention to leave the organization with the mediating role of job satisfaction.

2 LITERATURE REVIEW

Turnover intention differs from turnover in definition; turnover intention is a form of one's behavioural attitude to leave the organization, while turnover is the actual action of leaving the organization or company (Lu, Lu, Gursoy, & Neale, 2016). Lo (2015) suggests job dissatisfaction as one of the important factors triggering someone to leave an organization, especially IT workers. Meanwhile, job satisfaction is a form of evaluation of the work they do and their response toward

DOI 10.1201/9781003196013-18

their work experiences (Berry, 1998). It leads people to compare what they feel in their organization and what other organizations offer for their position (Mobley, 1979). The balance of work and personal life is also an important issue for organizations that positively affects job satisfaction (Noor, 2011). The employees' contented state in a working environment will help them to balance between their personal and professional needs. Flexible working is defined as an organizational plan in terms of working hours and workplaces, but without the intention to reduce the working hours of individual workers (Deshwal, 2015). In addition, Allen (2001) argues that the specific benefit of flexibility can have a positive impact on job satisfaction but can also have a negative impact on turnover intention. Azar, Khan, and Eerde (2018) in their research study explain the role of job satisfaction in mediating the relationship between the use of flexible working and turnover intention. Qureshi and Khan (2011) stated that long working hours due to workload will have an opposite impact on job satisfaction. This is due to disruption of welfare in other places outside their work. The time spent by employees involves completing tasks, interests, and responsibilities at work, both directly and indirectly (Johari, Tan, & Zulkarnain, 2018). Chung, Jung, and Sohn (2017) in their study conclude that workload is a form of work stress. This study particularly states that job stress can be related to job satisfaction, one of which is by comparing their work with other people's jobs. Job satisfaction is one of the things that is negatively related to personnel turnover or the change in personnel in an organization. Based on the explanation given above, these hypotheses are suggested:

H1: Job satisfaction mediates the effect of workload on turnover intention.
H2: Job satisfaction mediates the effect of work–life balance on turnover intention.
H3: Job satisfaction mediates the effect of flexible working on turnover intention.

3 RESEARCH METHOD

This study is quantitative in nature, using a purposive sampling technique to test the proposed model (Figure 1). Data collection was conducted by distributing online questionnaires to a total of 272 IT workers in Indonesia. The questionnaire used in this research was adapted from previous studies concerning turnover intention (Wang, Xu, Zhang, & Li, 2020). Job satisfaction was measured using the Minnesota Satisfaction Questionnaire (MSQ) (Weiss, Dawis, and England, 1967). Workload (Holland, Tham, & Sheehan, 2019) and work–life balance were divided into three dimensions: work interference–personal life (WIPL), personal life—interference work (PLIW), and work–personal life enhancement (WPLE) (Hayman, 2005). Flexible working used two main dimensions, namely, schedule flexibility and place flexibility (Pierce & Newstrom, 1983; Katherine, Mei, Eleanor, & Mary, 2005). All measures used a 6-point Likert scale (6 = strongly agree to 1 = strongly disagree). Negatively scored items were reverse-coded for the analysis. The data were analyzed using structural equation modeling (SEM) with SPSS 22 and LISREL 8.3 software. Each hypothesis has a significant relationship if the t-value and p-value ≥ 1.96 (Hair et al., 2014).

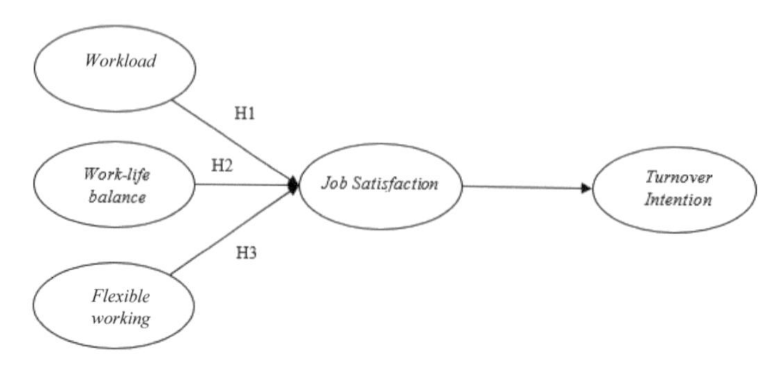

Figure 1. Research model.

4 RESULTS

Based on the survey results, the demography of the respondents is as follows: 72.8% were males, 77.8% were not married, 89% were aged 21–30, 81.3% had an undergraduate degree as the highest level of education, 80.5% worked as staff, and 49.3% had worked for 1–3 years. The research model has eight good fit indexes (RMSEA, ECVI, TLI, NFI, IFI, CFI, AIC, and CAIC). Therefore, it can be concluded that the research model meets the good-fit criteria for the measured data sample.

The hypothesis testing results show that one of the three hypotheses was accepted. By testing the indirect effect using the Sobel test, one hypothesis did not fulfill the significance criteria (t-value $\geq$ 1.96 and p-value $\geq$ 0.05) (Hair, et al., 2014). As in mediation testing, it was found that only work–life balance and flexible working have an indirect effect on turnover intention mediated by job satisfaction. Table 1 shows the results of hypothesis testing.

Table 1. Hypothesis test results

H	Path	*T-value*	*P-value*	Evaluation
1	Workload → Job Satisfaction → Turnover Intention	−1.33	0.182	Not Supported
2	Work–life balance → Job Satisfaction → Turnover Intention	−2.42	0.015	Supported
3	Flexible Working → Job Satisfaction → Turnover Intention	−2.74	0.006	Supported

Source: Result of statistical data using LISREL

5 DISCUSSION AND CONCLUSION

The result of the first hypothesis test shows that job satisfaction has no mediating effect between workload and turnover intention. The respondents stated that IT workers must complete or finish their work quickly, but they were given enough time to complete the work. Despite the relatively excessive workload, they did not find it difficult or burdensome because it was compensated by the time provided by the company. In the dualistic model of the passion theory presented by Varelland et al. (2003), there is a conceptual passion analysis based on activities that are related to job demand. One type of passion is harmonious passion, which is described by workers who are passionate about their work. So they participate in their workplace more positively, and they see more control, more support, and less excessive workload. It is argued that the workload given to IT workers is not excessive, which allows them to have passion for their work, so that the perceived workload actually forms a positive outcome for the IT workers.

Job satisfaction is influenced by work–life balance, meaning that job satisfaction can negatively affect turnover intention or reduce employee turnover intention. This finding supports Noor (2011), who explained that improving the balance between life and work can lead to higher productivity and higher job satisfaction. It is important for companies to make policies or programs to minimize the existence of work that interferes with the personal life of individuals, especially in terms of the time that workers must sacrifice which can lead to disruption of workers' personal lives.

It is proven that job satisfaction is positively and significantly able to mediate the relationship between flexible working and turnover intention. The results of this study are in line with Azar, Khan, and Eerde (2018), who suggested that the use of work flexibility and the intention to leave the organization can be mediated by job satisfaction and conflict between life and work. With the company's efforts to implement work flexibility, this policy can influence a person's desire to leave the organization when that person considers the policy as a form of company concern for workers (Azar, Khan, & Eerde, 2018). Moreover, work flexibility can contribute to workers to better adjust working hours and individual needs, and it can lead to job satisfaction of these workers (Uglanova & Dettmers, 2013).

REFERENCES

Azar, S., Khan, A., & Eerde, W. V. (2018). Modelling lingkages between flexible work arrangements' use and organizational outcomes. *Journal of Business Research, 91*, 134–143.

Allen, T. D. (2001). Family Supportive work environments: The role of organizational perceptions. *Journal of Vocational Behaviour, 58*, 414–435.

Berry, L. (1998). *Psychology at Work: An Introduction to Industrial and Organizational Psychology* (2nd ed.). Boston: McGraw Hill.

Chung, E. K., Jung, Y., & Sohn, Y. W. (2017). A Moderated mediation model of job stress, job satisfaction and turnover intention for airport security screeners. *Safety Science, 98*, 89–97.

Deshwal, P. (2015). Impact of Flexible Working Arrangements in the MNCs of Delhi. *International Journal of Applied Research, 1(13)*, 810–814.

Hair, J. F., Black, W. C., Babin, B. J., & Anderson, R. E. (2014). Multivariate data analysis. Essex, UK:Pearson Education.

Hayman, J. R. (2005). Psychometric assessment of an instrument designed to measure work/life balance. *Research and Practice in Human Resource Management*, 13(1), 85–92.

Holland, P., Tham, T. L., & Sheehan, C. (2019). The Impact of Precieved Workload on Nurse Satisfaction with Work-life Balance and Intention to Leave the Occupation. *Applied Nursing Research, 49*, 70–76.

Johari, J., Tan, F. Y., & Zulkarnain, Z. T. (2018). Autonomy, Workload, Work-life Balance and Job Performance Amon Teachers. *International Journal of Educational Management, 32 (1)*.

Katherine, M. C., Mei, L., Eleanor , H. W., & Mary, B. W. (2005). How Cirtual are we? Measuring virtuality in a Global organization. *Information Systems Journal*.

Li, E. Y., & Shani, A. B. (1991). Stress dynamics of information systems managers: A Contingency Model. *Journal of Management Information Systems, 12*(2), 107–130.

Lo, J. (2015). The information technology workforce: A review and assessment of voluntary turnover research. *Information System Frontiers, 17*(1), 387–411.

Lu, L., Lu, A. C., Gursoy, D., & Neale, N. R. (2016). Work engagement, job satisfaction and turnover intentions: a comparison between supervisors and line-level employees. *International Journal of Contemporary Hospitality Management, 28*(4), 737–761.

Mobley, W. H., Griffeth, R. W., Hand, H. H., & Meglino , B. M. (1979). Review and conceptual analysis of the employee turnover process. *Psychological Bulletin, 86*(3), 493–522.

Noor, K. (2011). work-life balance and intention to leave among academics in Malaysian public higher education institutions. *International Journal of Business and Social Science, 2*(11), 240–248.

Pierce, J. L., & Newstrom, J. W. (1983). Design of Flexible Work Schedules and Employee Responses: Relationships and Process. *Journal of Occupational Behaviour, 4*(4), 247–262.

Puslitbang-Aptika-IKP Kominfo. (2019). *Buku Saku Kebutuhan Sumber Daya Manusia Bidang TIK Pada Sektor Usaha Besar di Indonesia Tahun 2020.* Kominfo.

Qureshi, M. I., Iftikhar, M., Abbas, S. G., Hassan, U., Khan, K., & Zaman, K. (2013). Relationship between Job Stress, Workload, Environment and Employee Turnover Intention: What we know, what should we know. *World Applied Sciences Journal, 23 (6)*.

The World Bank. (2018). *Preparing ICT Skills for Digital Economy: Indonesia within the ASEAN context.* Jakarta: The World Bank Group Occupies.

Uglanova, E., & Dettmers, J. (2013). Sustained Effects of Flexible Working Time Arrangements on Subjective Well-being. *Journal of Happiness Studies, 14*(4), 1–25.

Varelland, R. J., Blanchard, C. M., Mageau, G. A., Koestner, R., Ratelle, C., & Leonard, M. (2003). Les passions de l'a^me: On obsessive and harmonious passion. *Journal of Personality and Social Psychology, 85*(1), 756–767.

Weiss, D. J., Dawis, R. V., & England, G. W. (1967). Manual for the Minnesota Satisfaction Questionnaire. *Minnesota Studies in Vocational Rehabilitation, 22*(120).

Wang, C., Xu, J., Zhang, T. C., & Li, Q. M. (2020). Effects of professional identitity on turnover intention in China's hotel employees: The mediating role of employee engagement and job satisfaction. *Journal of Hospitality and Tourism Management, 45*(1), 10–22.

Contemporary Research on Business and Management – Noviaristanti (Ed.)

Analysis of the disparity factor of the Net Asset Value (NAV) on the Exchange Trade Fund (ETF) and its market price on the Indonesia Stock Exchange (IDX)

A. Rahmanto & B. Wibowo
Master of Management, University of Indonesia, Jakarta, Indonesia

ABSTRACT: Efficient market hypothesis (EMH) explains that the price of a security will always be traded at its fair price. However, this is not the case for ETFs, where mispricing can occur. This research analyzes the disparity factors that can affect mispricing of 42 ETFs on IDX using secondary data. This research uses the ordinary least squares (OLS) regression method with the research hypothesis that disparity factors including the fund age, volume, difference between high and low prices, and IHSG return influence mispricing of ETFs. The result of this research indicates that only three factors (without IHSG return) have significant effects on ETF mispricing with R squared about 3.13%, which is similar to that reported in previous research studies by Atanasova and Weisskopf (2020) and by Shin and Soydemir (2010). The aim of this research is to help investors, investment managers, and dealers invest in and develop ETFs in Indonesia.

1 INTRODUCTION

In Indonesia, according to Zulfikar (2016), the first ETF was implemented in Indonesia at the end of 2007. Until 2011, only a few investors knew about ETFs in Indonesia. At the time of this research, in mid-2020, the ETFs in Indonesia showed an increase in their number. There were 45 ETFs listed on the Indonesia Stock Exchange (IDX), but only 42 ETFs were with the underlying asset of stock. Theoretically, the net asset value (NAV) of an ETF should be no different from its market price based on the efficient market hypothesis (EMH). However, some ETF studies have found mispricing of ETFs. Looking at the many ETFs that are listed on the IDX but still have limited enthusiasts investing in them, this research will examine the following question: "what are the disparity factors that affect mispricing of ETFs on the IDX?"

2 LITERATURE REVIEW

There are some previous studies discussing topics related to mispricing between ETF and NAV market prices that have been carried out with different data and results.

Engle and Sarkar (2006) showed that the premium (discount) for domestic ETFs is generally small and temporary, and the standard deviation is smaller than the bid-ask spread. They used the Kalman Filter Space model, dyna model, GARCH model, and error-in-variables model. For the data, they used 21 domestic and 16 international ETFs, both daily and intraday, during the period of April 2000 to September 2000. The result for international ETFs shows that the premium (discount) is much bigger and lasts for up to several days. This difference is due to a higher cost of creation and redemption, and the bid-ask spread is wider.

Petajisto (2017) analyzed inefficiencies in the pricing of ETFs using time series and equal value-weighted of volatility, cross-sectional dispersion, & volatility, and t-statistics with daily data at the American ETF market from early 2007 until the end of 2014. The result shows a new approach to detect mispricing in ETFs by measuring it in relation to the market price of a peer

DOI 10.1201/9781003196013-19

group that will eliminate the old NAV problems. This approach reduces the premium effect on both international holders and liquidity, but it still fluctuates between 100 and 200 bps. This test confirms that the creation of an active trading strategy aims for the exploitation of mispricing because the strategy creates significant profits before the transaction costs are highly statistically significant. Atanasova and Weisskopf (2020) analyzed the relationship between ETF premium (discount) and market liquidity using the premium and illiquidity ETF regression model, GMM, PVAR model, VAR Granger causality, Wald test, and P test statistic. For the data, they used 584 daily international ETFs from early 2012 until the end of 2017. The result shows that a lower ETF liquidity leads to a large price difference between the ETF price and the value of the underlying securities. The effect remains significant after controlling for the characteristics of the ETF, holding cost, and overall macroeconomic conditions.

3 METHODOLOGY

3.1 *Sample design*

In analyzing ETF mispricing, this paper uses the Eviews 9 and the ordinary least square (OLS) model. This research uses data stream sourced from Thomson Reuters Eikon, historical data from 42 ETFs listed on the IDX, and IHSG return with a daily retrieval period that started from the beginning of each ETF established until September 2020. Data taken by ETFs with its underlying asset type in the form of shares includes NAV and market price of the ETF at the daily closing on the IDX, some trading data for each ETF such as volume, and the difference between high and low prices. The number of data observations used in this research was 12,894.

The variables used in this research consist of dependent and independent variables. The dependent variable is the ratio of the price–NAV difference over its NAV (ETF mispricing ratio). The independent variables are fund age (age of ETF since it began trading on the IDX), volume (number of trading transactions in an ETF on the IDX), difference between high and low prices, (difference between the highest and lowest prices on a trading day), and IHSG return (daily return of the IHSG index). In this research, some variables, namely, ETF mispricing ratio and fund age, need to be processed first.

3.2 *Model specification*

This research began by calculating and looking for the formation of descriptive data from the mispricing of 42 ETFs in Indonesia on the annual period category from 2014 to 2020 because the growth of ETFs started around these years. This category was divided based on the ETF mispricing ratio when above 0% (premium) and below 0% (discount) to indicate the position of ETF mispricing observation. At this initial stage, it was intended to describe the mispricing conditions for ETFs on the IDX. The first step was calculating the ETF mispricing ratio as follows:

$$\text{ETF mispricing}_{\text{f,t}} = \frac{P_{f,t} - NAV_{f,t}}{NAV_{f,t}} \tag{1}$$

where $ETF\ mispricing_{f,t}$ is the ratio of the price–NAV difference over its NAV; $P_{f,t}$ is the market price of an ETF f at time t; and $NAV_{f,t}$ is the NAV of an ETF f at time t.

The next step was to calculate the fund age, starting from the beginning of the marketed ETF until the last day of trading in September 2020. The formula for the calculation is as follows:

$$Fund\ Age_{f,t} = Date\ Observation_{f,t} - Date\ Established\ ETF_{f} \tag{2}$$

where $Age_{f,t}$ is the age of an ETF f at time t; $Date\ Observation_{f,t}$ is the date of observation of an ETF f at time t; and $Date\ Established\ ETF_{f}$ is the established date of an ETF f.

After obtaining the processed variable data, the variables were tested to determine the factors that resulted in ETF mispricing using the OLS model. In previous research, Atanasova and Weisskopf (2020) and Shin and Soydemir (2010) used the OLS regression model with panel data. Yet, in this research, due to limited data and unbalanced ETF trading time, the OLS regression model was applied using the cross-sectional data type of 42 ETFs. The following is the regression model and the assignment of variables to Eviews.

$$|PD|_{f,t} = \alpha_f + \beta_{f1} FUND\_AGE_{f,t} + \beta_{f2} VOLUME_{f,t} + \beta_{f3} \Delta HL_{f,t} + \beta_{f4} IHSG_{f,t} + \varepsilon_{f,t} \quad (3)$$

where $|PD|_{f,t}$ is the absolute value of *ETF mispricing*$_{f,t}$ or the absolute value of the ETF market price at time t over its NAV minus one; $FUND\_AGE_{f,t}$ is the age of ETF f at time t; $VOLUME_{f,t}$ is the transaction volume of ETF f at time t; $\Delta HL_{f,t}$ is the difference between the high and low price of the ETF f at time t; and $IHSG_{f,t}$ is the return of the IHSG index at time t.

4 RESULTS

Based on the annual period category from 2014 to 2020, the calculation of observation data from 42 ETFs, which had the underlying asset of stock, shows that each ETF has an indication of mispricing each year. This finding is shown below in Table 1.

Table 1. ETF mispricing ratio per year.

	Number of ETF Per Year						
	7 ETF	8 ETF	9 ETF	12 ETF	22 ETF	35 ETF	42 ETF
Mispricing ETF Ratio	2014	2015	2016	2017	2018	2019	2020
Premium Observation	504	747	795	673	1177	1442	2109
Discount Observation	441	344	384	322	589	878	1021
Total of Data Observation	945	1101	1179	995	1766	2320	3120
Average Mispricing (%)	0.04	0.40	0.46	0.53	0.32	0.27	1.67
Average Absolute Mispricing (%)	0.45	0.76	0.92	0.97	0.82	0.87	2.46

Table 1 explains that every year, ETFs in Indonesia experience mispricing. Comparison of the number of observations of premium and discount from year to year shows that most of them experience the number of premium observations more than the number of discount observations. It is found that the mispricing range with the largest number of observations occurred in 2020, with 42 ETFs whose mispricing range of premium was between greater than 0.5% and less than 1%. The average ETF mispricing in all annual periods shows a premium or positive value, and only in 2020, it had the largest total observations. This was because the number of ETFs had increased and reached 42 compared to the previous year.

The results of the regression model estimation from Table 2 show that only the fund age, volume, and difference of high and low prices variables have a significant effect on ETF mispricing. The result for the fund age variable also has a significant effect on the research of Atanasova and Weisskopf (2020). Based on the work by Hubbard et al. (2010), fund age describes the reputation of mutual funds in managing their portfolios. Because the majority of ETFs in Indonesia are under 5 years old, there are still few ETFs that have a reputation in managing their portfolios and allowing for ETF mispricing. The next variable, volume, has a significant effect on ETF mispricing based on research by Caginalp and DeSantis (2017). Using turnover in which there is a trading volume component, their research shows that there is a mispricing effect on ETF if there is a large number of ETF stock transactions. Because the ETF market in Indonesia is considered inefficient, it does not rule out the possibility of a large enough ETF volume transaction. Furthermore, the variable

Table 2. Regression model result.

Variable	42 ETF Coefficient
FUND_AGE	−0.0017***
VOLUME	−2.06E-09*
HL	0.0007***
IHSG	−0.064
Constant	0.0164
R^2	0.0313
F-stat	104.26***
Observations	12,894

*, **, and *** denote significance at 10%, 5%, and 1% levels, respectively.

difference of high and low prices also shows a significant effect on ETF mispricing. In the research study of Caginalp and DeSantis (2017), the variable difference between high and low prices was used as a component in the relative difference between high and low prices, which can affect ETF mispricing in terms of volatility. The last variable, IHSG return, does not show any influence on ETF mispricing. However, this finding is different from the research of Marshall, Nguyen, and Visaltanachoti (2013), where they found that the index return has an influence on ETF mispricing. This probably happens because the ETF market in Indonesia is still relatively new. Moreover, the ETF data in this research consist of active ETFs that do not fully follow or refer to an index. The result of R squared of this research is 3.13%. This finding is similar to the research of Atanasova and Weisskopf (2020) and Shin and Soydemir (2010), where they got an R squared of around 6%–7%. However, several variables and regression models used in this research are different from their research, which does not include the expense ratio, AUM, bid-ask spread, NAV momentum, and the volatility of each stock from the portfolio of an ETF.

5 CONCLUSION

ETFs in Indonesia are still a relatively new investment product, and there are not many volume transactions on the stock exchange. In this research, it is found that ETF investment instruments in the Indonesian ETF capital market still indicate inefficiencies in the market. With this mispricing, the ETF market in Indonesia shows inefficiency. This can be seen from the mispricing of every 42 ETF listed on trading days in the exchange market. Most of the mispricing of ETFs in Indonesia is premium or overvalued for the annual period category from 2014 to 2020. In this research, several mispricing factors, namely, fund age, daily trading volume, and the difference between the highest and lowest prices show significant results on mispricing of ETFs, while the IHSG return variable has no effect on ETF mispricing. However, the influence of this factor does not really describe the occurrence of the mispricing because it only produces an R squared of 3.13%. These results are not much different from the research of Atanasova and Weisskopf (2020) and Shin and Soydemir (2010). The implication of this research is that 2020 was the right year for investors to invest in ETFs because in the future, ETFs will further grow in Indonesia, and profits are still being made from the mispricing condition of ETFs with a premium value.

REFERENCES

Atanasova, C. & Weisskopf, J. P. 2020. The price of international equity ETFs: The role of relative liquidity. *Journal of International Financial Markets, Institutions and Money* 65.

Shin, S. & Soydemir, G. 2010. Exchange-traded funds, persistence in tracking errors and information dissemination. *Journal of Multinational Financial Management* 20: 214–234.
Zulfikar. 2016. *Introduction to the capital market with a statistical approach*. 1st ed. Yogyakarta: Deepublish.
Engle, R. & Sarkar, D. 2006. Premiums-Discounts and Exchange Traded Funds. *The Journal of Derivatives* 13: 27–45.
Petajisto, A. 2017. Inefficiencies in the pricing of exchange-traded funds. *Financial Analysts Journal* 73: 24.
Hubbard, R. G., Koehn, M. F., Ornstein, S. I., Van Audenrode, M. & Royer, J. 2010. The mutual fund industry: Competition and investor welfare. New York: Columbia University Press.
Caginalp, G. & Desantis, M. 2017. Does price efficiency increase with trading volume? Evidence of nonlinearity and power laws in ETFs. *Physica A* 467: 436–452.
Marshall, B. R., Nguyen, N. H. & Visaltanachoti, N. 2013. ETF arbitrage: Intraday evidence.
Journal of Banking & Finance 37: 3486–3498.

Contemporary Research on Business and Management – Noviaristanti (Ed.)

Does psychological empowerment and perceived organizational support influence organizational citizenship behavior in the pharmaceutical industry? Affective organizational commitment as a mediator

A.S. Usada & R. Rachmawati
Faculty of Economic and Business Universitas Indonesia, DKI Jakarta, Indonesia

ABSTRACT: The objective of this paper is to examine the relationship among psychological empowerment (PE), perceived organizational support (POS), and organizational citizenship behavior (OCB), while considering the mediating factor of affective organizational commitment (AOC) in the pharmaceutical industry in Indonesia. By employing a convenience sampling method with the help of a cross-sectional online survey, the data were collected from 270 employees working in pharmaceutical industries. SPSS, structural equation model, and LISREL were used for statistical analysis. The result of the study strongly supported the relationship among PE, POS, and AOC. However, an insignificant relationship was found between POS and OCB. The results also reported a significant relationship between AOC and OCB and between PE and OCB. Moreover, the relationship between POS and OCB was found to be mediated by AOC. These results help pharmaceutical industries to develop strategies to increase OCB in their company.

1 INTRODUCTION

In early 2020, Covid-19 spread throughout the world, including Indonesia (WHO, 2020). Many sectors were affected by the pandemic. One of the affected sectors is the pharmaceutical industry. To meet the needs of public health, the government (the Ministry of Industry) encourages the pharmaceutical industries to improve their performance. (Hakim, 2020). This affects employees who work in the industry. They should give an excellent performance to increase the effectiveness of the organization. To achieve effectiveness, it is necessary to improve the quality of human resources by building the desired character of organizational behavior and reflecting good citizenship. A previous study showed there are several factors affecting organizational citizenship behavior, such as affective organizational commitment (AOC), psychological empowerment (PE), and perceived organizational support (POS). The objective of this study was to examine structural relationships among PE, POS, AOC, and OCB in pharmaceutical industries.

2 LITERATURE REVIEW

Organizational citizenship behavior is an attitude of employees that aims to contribute more to the company or organization outside of their role (Bateman & Organ, 1983). In 1994, Organ stated that OCB is a behavior exhibited by organization members without being given any compensation if they do work sincerely and sanctions if they do not do the work. The work is done without prior direction given and listed in the job description. Chen and Francesco (2003) argued that affective commitment has a positive effect on performance and OCB. The increasing affective commitment

 DOI 10.1201/9781003196013-20

will also increase OCB from employees. Affective commitment plays the role of a mediator in the relationship between POS and OCB (Gaudet and Tremblay, 2017). Employees who possess a high POS level will be more committed to their organization and motivated to contribute to the organization to achieve its goals.

Affective commitment can also be the mediator for the relationship between PE and OCB (Srivastava et al, 2016). Affective commitment not only improves OCB, but is also one of the consequences of psychological empowerment felt by employees in their workplace. Employees with a high PE level will increase their affective commitment and motivation to give their best to achieve the organization's goals.

3 RESEARCH METHODOLOGY

This study is a quantitative research. The respondents were employees who worked in the pharmaceutical industries in all provinces throughout Indonesia. The data were collected using online questionnaires. The questionnaire required respondents to rate the importance of each statement on a 6-point Likert scale. To measure the psychological empowerment, this study used 12 questions that were adopted from Spreitzer (1995). To measure the perceived organizational support measurement, this study used 8 questions that were adopted from the Rhoades and Eisenberger (2001) research study. The AOC was measured by using 8 questions generated from the research done by Allen and Meyer (1990). Lastly, to measure the OCB, this study used 24 questions obtained from the research conducted by Podsakoff et al. (1990). The collected data were then processed using the SEM method with the help of the LISREL application.

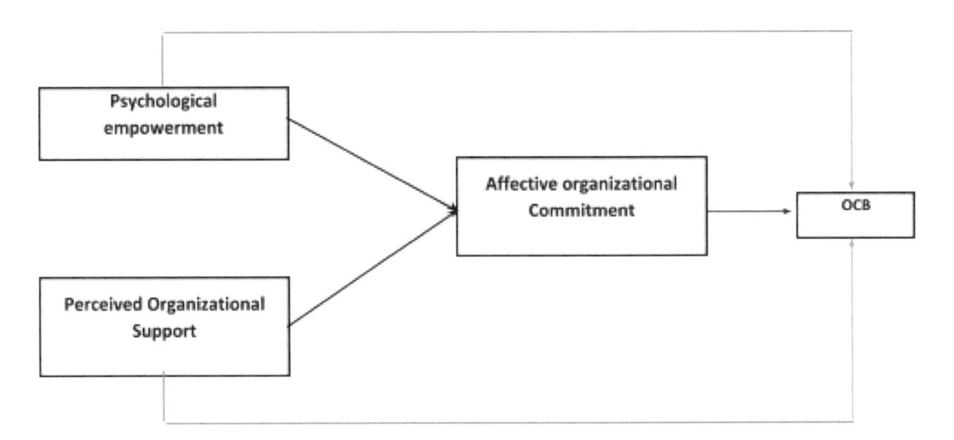

Figure 1. Research model

4 RESULT

4.1 *Descriptive*

The results of the survey showed that there were 109 (40,37%) male employees and 161 (59,63%) female employees from the 270 questionnaires distributed. The majority of the respondents belonged to the 21–30 years age group (77.78%). Almost 56.30% of respondents work in the national private sector; 27,40% work in the foreign private sector; and 16,30% work in state-owned enterprises.

4.2 *Validity and reliability test*

Most of the indicators were valid. However, two indicators of OCB fell below the desirable cutoff of SLF; thus, they were omitted. The reliability test showed that the CR value of all dimensions/variables was above 0.7, and the AVE value of all dimensions/variables was more than 0.5. This shows that all of the dimensions/variables used are reliable.

4.3 *The structural model*

Based on the structural models, this study has seven good fit indexes (RMSEA, ECVI, NNFI, NFI, RFI, IFI, and CFI). The hypothesis test was carried out to determine the influence among the variables used—psychological empowerment (PE), perceived organizational support (POS), affective organizational commitment (AOC), and organizational citizenship behavior (OCB). The estimation results of the overall model based on the SLF and T-values can be seen in the following table:

Table 1. Path diagram results for the structural model.

Path	SLF	T-value		Conclusion
PE → OCB	0.33	4.06		Hypothesis is accepted
POS → OCB	0.06	0.64		Hypothesis is rejected
PE → AOC	0.33	5.70		Hypothesis is accepted
POS → AOC	0.64	9.56		Hypothesis is accepted
AOC → OCB	0.39	3.23		Hypothesis is accepted
Path	**Direct effect**	**Indirect effect**		**Conclusion**
		a	B	
PE → AOC → OCB	PE → OCB t-value: 4.06 (Supported)	PE → AOC t-value: 5.70 (Supported)	AOC → OCB t-value: 3.23 (Supported)	*Complementary Mediation*
POS → AOC → OCB	POS → OCB t-value: 0.64 supported)	POS → AOC t-value: 9.56 (Supported)	AOC → OCB t-value: 3.23 (Supported)	*Indirect only mediation*

Source: Result of statistical data using LISREL

The results of the model testing had several outputs. When viewed from the results of the hypothesis test, there is one hypothesis that is rejected. It shows that perceived organizational support has no direct effect on OCB. However, in hypothesis 6, the results show that the effect of POS on OCB must be mediated by an affective organizational commitment. These results are in line with research conducted by Gaudet and Tremblay (2017), which shows that the POS variable will have an effect on OCB if it is mediated by AOC. Furthermore, the results proved that PE was found to have a positive effect on OCB. The results are in line with previous research conducted by Jha (2014), which states that PE had a positive and significant effect on OCB. His study also showed that AOC mediated the relationship between PE and OCB. These results indicate that someone who has a high psychological empowerment level can increase his/her involvement to perform extra roles in the organization. When employees are confident in doing their own work, they will be happy to do additional roles in their organization. An employee who has a high PE level will also influence their perception of commitment to their organization, which of course, will provide many benefits for the organization.

The idea that POS increases employee involvement in OCB behavior is not supported in this study. Therefore, even though management increases the level of POS in the company, it does not guarantee that these employees will be involved in OCB. The fact is that POS is not proven to directly increase OCB, but the effect of POS can be increased by adding affective commitment to OCB. An employee who is noticed by the company for his/her good behavior will increase his/her emotional attachment, which in turn will ensure that he/she will engage in OCB behavior.

5 CONCLUSION AND SUGGESTION

This study contributes to the understanding of the factors that influence OCB by examining the role of POS and PE on affective commitment. The results of this study suggest that someone who is trusted by the company will get a welfare and strong affective commitment to the company and thereby influence their OCB behavior.

Pharmaceutical industries can increase the OCB of their employees by giving their employees confidence, because of which they will have a strong influence on their work results. Furthermore, pharmaceutical industries can also give them a strong sense of belonging by conducting open and encouraging discussions between employees and management.

To increase the affective commitment of an employee, the company should give attention to its employees. The company must also respect the values and goals of each individual by being fair to all employees. The fair attitude can be demonstrated by not discriminating between each employee in the distribution of resources and opportunities to attend training held by the company.

REFERENCES

Allen,N.J., & Meyer, J.P. (1990). The measurement and antecedents of affective, continuance, and normative commitment to the organization. *Journal of Occupational Psychology.* 63,1–18

Bateman, T.S., & Organ, D.W. (1983). Job satisfaction and the good soldier: the relationship between affect and employee "citizenship". *Academy of Management Journal.* 26(1),587–595

Chen, Z.X and Francesco, A.M. (2003). The relationship between the three components of commitment and employee performance in China. *Journal of Vocational Behavior,* 62(3), 490–510

Gaudet, M.C., & Tremblay, M. (2017). Initiating structure leadership and employee behaviour: The role of perceived organizational support, affective commitment and leader-member exchange. *European Management Journal.* 35(2017), 663–675

Hakim, R.N. (2020). Hadapi Pandemi Covid-19, ketua DPR Ingatkan Pemerintah benahi Sektor Kesehatan Nasional. https://nasional.kompas.com/read/2020/06/30/16180461/hadapi-pandemi-covid-19- ketua-dpr-ingatkan-pemerintah-benahi-sektor

Jha, S. (2014). Transformational leadership and psychological empowerment: Determinants of organizational citizenship behaviour. *South Asian Journal of Global Business Research.* 3(1)

Podsakoff et al. (2000). Organizational citizenship behaviour: A critical review of the theoretical and empirical literature and suggestions for future research. *Journal of Management.* 26(3),513–563

Rhoades, L., Eisenberger, R., & Armeli, S. (2001). Affective commitment to the organization: The contribution of perceived organizational support. *Journal of Applied Psychology. 86(5),825–836*

Spreitzer, G.M. (1995). *Taking Stock: A Review of More Than Twenty Years of Research on Empowerment at Work. The Handbook of Organizational Behaviour.* Cooper and Bariling (Eds.). Sage Publication Srivastava, A.P. and Dhar, R.L. (2016). Impact of leader member exchange, himan resource management practices and psychological empowerment on exta role performance: The mediating role of organizational commitment. *International Journal of Productivity and Performance Management.* 65(3),351–377

WHO (2020). WHO *Corona Virus Disease (Covid-19) Dashboad.* https://covid19.who.int/ (Accessed on 4 august 2020)

Contemporary Research on Business and Management – Noviaristanti (Ed.)

Analysis of the factors affecting the intention of using mobile payments among millennials with the unified theory of acceptance and use of technology II

A.E.S. Sinaga & D.A. Chalid
Master of Management, University of Indonesia, Jakarta, Indonesia

ABSTRACT: Mobile payments are financial services that minimize payments through conventional ways and use technology to meet payment and transaction needs. However, the services provided have not been carried out comprehensively because there are factors that affect the user's intention to adopt technology in terms of transactions. This study aims to understand the determinants of users' intentions in using mobile payments, specifically among millennials. This research was conducted using the user acceptance approach, the unified theory of acceptance and use of technology II (UTAUT2). A survey was conducted with 310 respondents who were given a questionnaire regarding the factors of research findings, namely, the perceived ease of use, perceived usefulness, social influence, trust, promotional offers, and perceived risk, which affect behavior intention to use mobile payments. This research is expected to help both users and mobile payment providers to develop features according to the factors that have been studied.

1 INTRODUCTION

The development of the internet has encouraged many industries, from businesses to financial service providers such as banks, to switch to digitalization with the emergence of financial technology or fintech. Fintech is currently in development and has become a challenge for previous conventional financial institutions; fintech provides services that are fast, direct, and cheaper in real time. Based on data compiled by EY in 2019, there was 75% use of fintech for money transfer and digital payments worldwide. In 2015, the use of fintech as a payment method was 18%, and in 2017, it increased by 50%; in 2019, 75% of fintech users in the world were using fintech services for mobile payments and transfers. However, the intention of users in using the mobile payment service is not clear, and the sustainability of features used in fintech is yet to be found. The use of these services has many determining factors that encourage consumers to use digital payment services and other features provided by fintech companies.

This research provides further information to determine the factors that support technology adoption in the financial service industry. It also helps in determining the user's intention to use financial technology services, especially in mobile payments and transactions, through the user's acceptance approach related to system or technology and consumer behavior. This paper discusses the background, literature review, research methods, results, and conclusions.

2 LITERATURE REVIEW

Financial technology or fintech is a technology that eliminates or reduces the costs of financial intermediaries (Das 2019). The significant change and use of fintech are in mobile payments, where consumers carry out transactions with payment processing. Current fintech companies that

 DOI 10.1201/9781003196013-21

serve mobile payments collaborate with telecommunication providers or are directly related to the telephone network associated with these telephone users. In addition, they also collaborate with certain supermarkets and stores that provide cash replenishment in e-wallets, which can also be served through some convenience stores and some banks that offer online transactions in collaboration with these digital payment providers (Napitupulu et al. 2017). There is a combination of approaches called the unified theory of acceptance and use of technology II or UTAUT2 (Venkatesh et al. 2003). The UTAUT2 theory has four determinants of user behavior intentions: perceived usefulness, perceived ease of use, social influence, and facilitating conditions.

In the social cognitive theory, self-efficacy is the key to behavior initiation and behavior defense (Bandura 1977). Self-efficacy is central to making a goal and determining a target. Personal innovativeness (Kim et al. 2010) refers to a person's tendency to try a new information system. Perceived usefulness is defined as the possibility of user subjectivity to the belief in the usefulness of a system and to improve user performance, for example, in mobile payments. Perceived ease of use refers to the likelihood that the user has expectations on the target system which do not require much effort (Davis 1989). Social influence is the extent to which users believe that something can have a significant impact on other people, which leads to a belief that someone must use a technology. Trust is an element of user behavior intentions, where it is the belief that the mobile payment service provider will provide services meeting the user expectations (Shin 2009). Promotional offers are actions in offering lower prices in a short period of time to increase the effectiveness of product sales to cost-sensitive users or consumers (Sunny & George 2018). The risk in adopting a technology also affects consumer intentions where perceived risk (Li & Huang 2009) is a construct of a situation where there is the possibility of loss and unexpected consequences.

H1 and H2: There is a positive effect of self-efficacy and personal innovativeness on perceived ease of use on user behavior intentions in using mobile payments.

H3, H4, H5, H6, and H7: There is a positive effect of perceived ease of use, perceived usefulness, social influence, trust, and promotional offer on user behavior intentions in using mobile payments.

H8: There is a negative effect of perceived risk on user behavior intentions in using mobile payments.

3 RESEARCH METHODS

3.1 *Data collection and analysis*

The data were collected through a survey of fintech users, who provided answers to questions based on the specified variable indicators. The assessment for each question from the questionnaire used a Likert scale. The sampling technique used was convenience sampling. This study's population was all the fintech users who have used mobile payment services. The limitation is that the respondents, namely, millennials, were born between 1980 and 2000, that is, they were in the age range of 20 to 40. Before conducting the research, research testing was conducted through questionnaire's validity and reliability testing. The number of research samples (Hair et al. 2009) representing research was a minimum of five times the research indicators or a maximum of ten times the research indicators.

In this study, the distribution of questionnaires consists of 31 indicators, making the minimum required research respondents 155 and a maximum of 310. The data processed in descriptive statistical analysis with data processing techniques used the Statistical Package for Social Science (SPSS). The data validity test shows that the investigated data have the same data as the object under study. As for the validity test, some criteria must be suitable, namely, if $r \geq 0.30$, then the questionnaire items are valid. In addition, research also requires reliable data, that is, the data, when used to measure the same research object several times, should provide the same data results. A variable construct is said to be reliable if it gives a Cronbach Alpha value of >0.6 (Ghozali 2013). In this study, data analysis used the partial least squares (PLS) approach. PLS is a component or variant based on the structural equation modeling (SEM) equation.

4 RESULTS AND DISCUSSION

After collecting and processing data using SmartPLS to test the questionnaire's validity and reliability and to test the hypotheses, the research was conducted. Overall, most of the respondents were 20–24 years old (81.3%), and most of respondents were based in Jakarta (133 people or 42.9%).

Table 1. Hypotheses test.

	Original Sample (O)	Sample Mean (M)	Standard Deviation (STDEV)	T Statistics (|O/STDEV|)	P-Values
Perceived Ease of Use → Behavior Intention to Use	0.163	0.164	0.069	2.357	0.019
Perceived Risk → Behavior Intention to Use	0.120	0.125	0.038	3.106	0.002
Perceived Usefulness → Behavior Intention to Use	0.284	0.286	0.058	4.927	0.000
Personal Innovativeness → Perceived Ease of Use	0.416	0.418	0.045	9.249	0.000
Promotional Offers → Behavior Intention to Use	0.172	0.168	0.067	2.568	0.011
Self Efficacy → Perceived Ease of Use	0.347	0.347	0.045	7.760	0.000
Social Influence → Behavior Intention to Use	0.132	0.134	0.044	3.007	0.003
Trust → Behavior Intention to Use	0.210	0.205	0.046	4.568	0.000

Table 1 shows the hypotheses testing, which is a test of each independent variable affecting the dependent variable. This test was also carried out using SmartPLS to find out whether the hypotheses were acceptable. The hypotheses testing, which shows the results of all variables, shows an influence on behavior's intention to use mobile payments. The self-efficacy relationship shows that the higher the belief that users can use technology, the higher the influence on perceived ease of use will be because they have confidence that the technology can be used easily. Personal innovativeness has a positive relationship with a person's decision to do online shopping, a decision on a technological innovation's adoption behavior. Being innovative plays a significant role in increasing confidence in the ease of use and influencing the intention to use mobile payment behavior in fintech.

The mobile payment is used by consumers for shopping transactions and for sending money digitally in real time. The use of mobile payment provides convenience and speed to users. The perceived usefulness variable shows that user confidence in using a specific system will improve its performance or help it become more productive so that someone will intend to use the technology. The effect of social influence shows that someone who gives a positive opinion about technology will influence other users to use it. Trust also has a significant effect; many users feel uncertain when using mobile applications, so fintech companies help users believe that the services provided can be used for transactions according to user expectations. Promotional offers are factors where many companies offer promotions as sales incentives when launching a product to potential users, for example, promotional offers such as cashback and discounts. Furthermore, they also cooperate with other merchants related to mobile payments, such as restaurants, shopping places, etc. The variable perceived risk has an inversely proportional relationship. If the risk of using mobile payments grows higher, it will reduce behavior intention because it does not provide a sense of security in user transactions.

5 CONCLUSIONS

The purpose of this study is to determine factors that influence behavior's intention to use mobile payments. This research focuses on the millennial generation because of the fact that the majority of the users of mobile payments are from the millennial generation. The UTAUT2 model uses several independent variables, namely, self-efficacy, personal innovativeness, perceived ease of use, perceived usefulness, social influence, trust, promotional offers, and perceived risk. Meanwhile, the dependent variable is behavior intention to use. Based on data collection and processing, it was found that each independent variable becomes a factor that influences the user's intention to use mobile payments.

There are limitations to this study. First, the cross-sectional data collection only represents a particular place. This research was conducted in Jabodetabek (Greater Jakarta) only, so other regions may show different results. Second, the study is also limited to the millennial generation because most users are from the millennial generation. It is suggested that further research studies use different variables with different sample categories. Despite these limitations, we are confident that this study has covered the intention factors of using mobile payments among millennials.

REFERENCES

Bandura, A. 1977. Self-efficacy: Toward a unifying theory of behavioral change. *Psychological Review*. https://doi.org/10.1037/0033-295X.84.2.191

Das, S. R. 2019. The future of fintech. *Financial Management*. https://doi.org/10.1111/fima.12297

Davis, F. D. 1989. Perceived usefulness, perceived ease of use, and user acceptance of information technology. *MIS Quarterly: Management Information Systems*. https://doi.org/10.2307/249008

Ghozali, I. 2013. Aplikasi Analisis Multivariate Dengan Program IBM dan SPSS. In *Aplikasi Analisis Multivariate dengan Program IBM SPSS 19*.

Hair, J. F., Black, W. C., Babin, B. J., Anderson, R. E., & Tatham, R. L. 2009. Multivariate Data Analysis. In *New Jersey*.

Kim, C., Mirusmonov, M., & Lee, I. 2010. An empirical examination of factors influencing the intention to use mobile payment. *Computers in Human Behavior*. https://doi.org/10.1016/j.chb.2009.10.013

Li, Y.-H., & Huang, J.-W. 2009. Applying Theory of Perceived Risk and Technology Acceptance Model in the Online Shopping Channel. *World Academy of Science, Engineering and Technology*.

Napitupulu, S., Rubini, A., Khasanah, K., & Rachmawati, A. 2017. Kajian Perlindungan Konsumen Sektor Jasa Keuangan: Perlindungan Konsumen Pada Fintech (Studies on Consumer Protection in the Financial Services Sector: Consumer Protection at Fintech). In *Otoritas Jasa Keuangan*.

Shin, D. H. 2009. Towards an understanding of the consumer acceptance of mobile wallet. *Computers in Human Behavior*. https://doi.org/10.1016/j.chb.2009.06.001

Sunny, P., & George, A. 2018. Determinants of Behavioral Intention To Use Mobile Wallets–a Conceptual Model. *Journal of Management (JOM), 5*(5).

Venkatesh, V., Morris, M. G., Davis, G. B., & Davis, F. D. 2003. User acceptance of information technology: Toward a unified view. *MIS Quarterly: Management Information Systems*. https://doi.org/10.2307/30036540

Contemporary Research on Business and Management – Noviaristanti (Ed.)

Business coaching: Implementation of digital marketing and addition of online channels in Indonesian MSME's F&B business during the 2020 pandemic

A. Adawiyah & H. Suhaimi
Faculty of Business and Economics, University Indonesia, DKI Jakarta, Indonesia

ABSTRACT: This paper aims to assist MSMEs in Indonesia in marketing activity management during the COVID-19 pandemic by implementing certain strategies to increase their promotion. The data were collected using the business coaching method. In this study, qualitative research was used to map MSMEs' conditions and problems. Afterward, corrective actions were given to implement remedial solutions. It was found that the survival solutions that can be applied by the companies include creating online channels, adding new product variations, and managing Instagram accounts to support marketing activities. After the implementation of the solutions, it was shown that their sales increased significantly and it has helped the company to survive during the difficult times caused by the pandemic.

1 INTRODUCTION

Coffee shop business, a popular culinary business, comprises many MSMEs in Indonesia. SiCangkir Coffee, founded in 2017, is one of the MSMEs in the food and beverages sector that were operational during the COVID-19 pandemic. Many MSMEs, especially restaurants and cafés, encountered difficulties in running their business during the pandemic. Hence, understanding MSMEs' condition holistically is necessary to find the best solution for their challenges. In this case, several internal and external analyses are required to ensure the coffee shop businesses' survival. This study focuses on several important and feasible aspects necessary to improve SiCangkir Coffee's performance, namely, digital marketing, new product development, and online channel.

2 LITERATURE REVIEW

A theoretical basis is required to ensure the appropriate direction of the study. The theoretical bases of this study are as follows: Taiminen and Karjaluoto (2015) define digital marketing as a new marketing approach. The approach is not merely a traditional marketing channel supported by digital elements, but all marketing activities that utilize digital channels to promote products (Batra & Keller, 2016). Instagram is a mobile social network for photo and video sharing. Ištvanic et al. (2017) estimated that Instagram has 700 million active users word widely. Furthermore, Ištvanic et al. (2017) suggest that Instagram is a great medium for advertising, as this platform can tell a story about a product or brand visually and interestingly. According to Cannon (2014), new product development consists of several stages: idea generation, idea screening, concept development and evaluation, business analysis, product development, test marketing, and commercialization. Walsh and Godfrey (2000) and Kim and Lennon (2012) define electronic commerce as shopping online via the internet. Unlike offline stores that allow consumers to interact directly with the seller, online stores facilitate communication between consumers and sellers through a technical interface.

 DOI 10.1201/9781003196013-22

3 RESEARCH METHOD

This study is a descriptive qualitative research conducted through a business coaching method. The primary and secondary data were collected from interviews, observations, and SiCangkir Coffee's internal data. The business coaching will be implemented through four steps, namely, (1) internal and external analysis; (2) TOWS, gaps analysis, and Pareto analysis; (3) alternative solutions and decision making; and (4) implementation and monitoring.

4 RESULTS AND DISCUSSIONS

The result shows that the coffee shop challenges should be addressed by several solutions: opening online channels, adding new product variations, and managing online promotion via Instagram. In Instagram online promotion, the most important aspect is the images. The posts should be bright, interactive, personal, and simple, while the layout needs to look professional. The findings show that the attempt to improve SiCangkir Coffee's Instagram profile structure has shown promising results, proven by increased engagement on the business's Instagram profile.

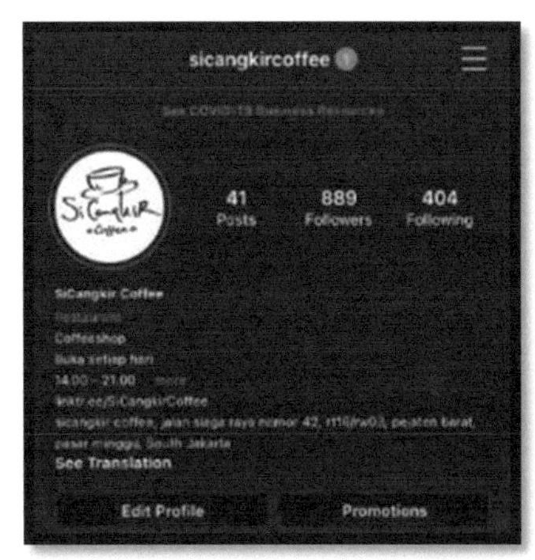

Figure 1. Comparison between before and after instagram improvement from Instagram.com/SiCangkir Coffee.

SiCangkir Coffee will add 'one-liter bottle packaging' as a product variation. According to Crawford and Benedetto (2015), this one-liter bottle packaging product addition can be categorized as new-to-the-firm products or new product lines. To complete the gap from the previous analysis which stated that MSMEs needed to add an online channel in order to keep it operational during a pandemic, the solution offered was to register MSMEs as a GoFood partner.

5 CONCLUSION

MSMEs must implement the right strategy to survive and operate optimally during the COVID-19 pandemic. The implementation results demonstrated that opening online channels, adding product variations, and managing online promotions can increase MSMEs' sales and keep them operational during the COVID-19 pandemic.

REFERENCES

Ištvanic, M., Milic, D. C., & Krpic, Z. (2017). Digital Marketing in the Business Environment. Digital Marketing in the Business Environment. Retrieved June 9, 2018.

HM Taiminen, H Karjaluoto (2015). The usage of digital marketing channels in SMEs. Journal of Small Business and Enterprise Development

J Walsh, S Godfrey (2000). The Internet: a new era in customer service. European Management Journal. Elsevier

Kim, J. H., & Lennon, S. (2012). Music and amount of information: do they matter in an online apparel setting?. The International Review of Retail, Distribution and Consumer Research, 22(1), 55–82.

Contemporary Research on Business and Management – Noviaristanti (Ed.)

The importance of social media marketing for MSMEs in Jakarta to meet the challenges in the new normal era

A.I. Molle
Marketing Management, Universitas Surabaya, Indonesia

ABSTRACT: In Indonesia, the COVID-19 pandemic hit MSMEs, which are key to the national economy. In Jakarta, the capital city of Indonesia, many MSME actors needed to face large-scale social restrictions. During times like this, one of the strategies that could potentially spur its productivity is social media marketing. The increasingly massive and widespread use of social media has made it an ideal marketing tool to reach consumers without direct contact. In this new normal era, consumer behavior has also shifted. It can be seen from the increasingly high purchasing patterns of consumers on social media. It was later found that the implementation of social media marketing could help MSMEs in Jakarta to maintain their productivity, while still complying with the restrictions. Thus, the readiness of MSME actors in Jakarta to develop their social media marketing effectively is required to reduce marketing costs and to increase business profits.

1 INTRODUCTION

COVID-19 pandemic has had a massive impact on all aspects of people's lives. The economy has been greatly affected by the pandemic. The Minister of Finance Sri Mulyani said that there are three major economic impacts due to this pandemic, including the decrease of people's purchasing power and household consumption, the uncertainty of the pandemic that has weakened investment, and Indonesia's exports that have to be stopped due to the weakening of world economy. Due to these impacts, the income was reduced because outside activities decreased, which also lowered buyers' wants. The pandemic's impact was experienced not only by large industries, but also by micro, small, and medium enterprises (MSMEs) in Indonesia (Achmad, Azhari, Esfandiar, & Nur, 2020).

In the globalization era, technological and communication development is increasingly sophisticated. Now, business people choose social media as their main choice to market their product or service. By taking it into account, social media has become the main pillar in delivering information. One of the advantages of social media is that it has a lot of potential for the advancement of a business. Social media can be used to communicate in business, help to market products and services, communicate with customers and suppliers, complement brands, reduce costs, and perform online sales (Achmad, Azhari, Esfandiar, & Nur, 2020).

Social media channels, such as Twitter, Facebook, and YouTube, have grown rapidly over time. In addition, its adoption rate has soared, and it has generated a great number of users in less than 10 years (Al-Deen & Hendricks, 2013). Currently in Indonesia, internet usage is increasing because people can access the internet through various means including handheld devices or smartphones. This convenience has shifted human perspectives and behavior to change from manual to electronic. As of January 2019, the percentage of internet users in Indonesia was 56% of the total population. It shows an increase compared to February 2018 with 143.26 million people or equivalent to 54.7% of the population. On the whole, the number of internet users in Indonesia has increased by 17.3

DOI 10.1201/9781003196013-23

million compared to last year, or a 13% increase. In terms of internet usage, it was also found that Indonesia ranked fourth after India (up 97.8 million), China (up 50.6 million), and the United States (up 25.3 million) (Maulidasari & Damrus, 2020).

Currently, social media is one of the main sources of information in the form of digital media. The current pandemic conditions also affect the increased social media usage. Subsequently, more people rely on social media to stay up-to-date and look for new information. Some of the social media channels that are widely used today include Facebook, Twitter, Instagram, and TikTok. During the pandemic, it is undeniable that businesses around the world are experiencing profit decline. To mitigate this issue, social media marketing can be used as a solution because social distancing and activity limitation have forced many consumers to shop online (Anand, 2020).

Due to the pandemic and the new normal condition, business actors in the business and economic sectors have to find a method to survive. One of them is by changing their sales strategy. For example, actors in the food industry have employed online motorcycle taxi services to promote their foods and to deliver them online. In addition, because people are required to stay at home, business actors also have promoted their business through television and digital platforms such as Tokopedia, Shopee, Bukalapak, Traveloka, and others (Maulidasari & Damrus, 2020).

During the new normal era, MSMEs are finding it difficult to develop their business and market their products. To address this issue, it would be interesting to discuss the advantage of social media marketing as an effective marketing tool in the new normal era that can be employed by business actors, especially MSMEs in Jakarta.

2 LITERATURE REVIEW

In the modern society, social media has significantly influenced MSME product marketing. The use of the internet by MSMEs facilitates exchange of information, instantly stimulates efficiency, and helps solve problems at an early stage (Marjolein, et al., 2015). The use of social media is considered to have a positive impact on a business; particularly, it can reduce substantial marketing costs and improve good relationships with customers (Mokhtar, et al., 2016). Effectively, social media also provides an e-marketing platform that allows potential consumers to easily know the specifications of the products without the need to meet directly with the sellers. Therefore, it can be argued that social media is an effective marketing medium, providing communication facilities for MSMEs and consumers (Sugiarto, 2018).

Marketing is defined as the process of introducing a product to the general public and potential buyers. Currently, marketing can be done through the internet, for it is effective and efficient. Today, marketing can be divided into internet marketing and digital marketing. Internet marketing focuses on marketing activities that target internet users. It takes advantage of the internet to attract customers. On the other hand, digital marketing employs technologies, including the internet, and utilizes them in a more complex process. It not only attracts customers, but also develops several marketing strategies such as using social media for endorsement, webinars, or a free e-book to attract awareness (Juliana, Djakasaputra, Pramezwary, & Hutahaean, 2020).

In Indonesia, MSMEs, as regulated in Law No. 9, require several criteria. According to the law, MSMEs need to have a turnover of less than 200 million rupiahs, excluding land and buildings, annual net sales of 1 billion rupiahs or less, must be owned by Indonesian citizens, must be an independent economic entity, not a branch or subsidiary of a large or medium-sized company, and must not be directly controlled by a large or medium-sized company. MSMEs must be managed by an individual, cooperative organization, or legal entity.

Wiku Adisasmito, the chairman of the Expert Team for the Task Force for the Acceleration of Handling COVID-19, explained that the new normal is a change in behavior, wherein normal activities can continue to be carried out but with the addition of implementing health protocols to prevent Covid-19 transmission. (Kompas.com, 20 Nov 2020). According to him, the most important aspect of the new normal is the capability to adjust to life patterns. Socially, people will experience a new normal method and be required to adapt to the activities and work. In addition, they must reduce physical contact with other people, avoid crowds, and carry out a series of activities from home (Bramasta, 2020).

3 DISCUSSION

This study was conducted on several MSMEs' social media accounts in Jakarta who were able to survive in the new normal period. The MSMEs' are as follows:

1. @bloominplum
   Bloominplum sells their creativity in an aesthetically painted mirror. Its product is interesting because of its uniqueness. Initially, there was only one consumer who bought the product. However, the owner created content and uploaded it on TikTok. Consequently, the product was widely recognized by people. People especially noticed the packaging of the product. In addition, the business owner had also given several small gifts. Now, the account has 38 thousand followers, and more than 200 aesthetic mirrors have been sold.
2. @seasukitomyum
   The second MSME is a food business named Sea Suki Tomyum. The business sells dishes consisting of tom yum sauce, crabstick, beef ball, sweekiaw, and vegetables. This owner used social media to market the business. It can be seen that the owner first promoted the products in March 2020 on Twitter, and it was retweeted by one influencer who had many followers. As a result, sales began to increase with the increase in interest. It can be seen that the MSME was able to take advantage of the situation and survive in the new normal period. At the time when many people missed eating food that is usually available in restaurants, the owner found the opportunity and sold it online so that people can enjoy the food that they have missed. Now, the Instagram account of Sea Suki Tomyum has around 500 followers, and every day, the MSME makes deliveries to Jakarta and other areas that can be reached within a day.
3. @eiffleboutique
   In the new normal era, wearing masks is an obligation and a habit. During this time, people change their masks every day as if they change their clothes and wear clean ones. It can be a great opportunity to sell them masks. However, at the same time, it is also tough because of many competitors. Eiffel Boutique is one of the MSMEs in Jakarta, with an Instagram account @eiffelboutique. They took the opportunity from this new normal period by selling cloth masks that can be used repeatedly with various interesting motifs. People nowadays certainly need masks to use if they have to go out. Therefore, a fashionable mask or a mask with a good and unique motif will certainly attract many customers because people are bored with plain masks. Eiffel Boutique's Instagram account has endorsed several other accounts of people who have many followers to promote their existence. Currently, Eiffel Boutique's Instagram followers have reached more than 250 thousand with more than 10 thousand recorded mask sales.

4 CONCLUSION

From several examples taken from the social media accounts of MSMEs in Jakarta, it can be seen that social media marketing has an enormous effect on their product sales. In the new normal era, business actors are required to be creative and innovative to make a stable sale and to attract buyers. The examples mentioned above are three of the MSMEs that have utilized social media for marketing. Because social media users are increasing, there is a high possibility for marketed products to be known by many people. Moreover, some business actors have collaborated with accounts that have a large number of followers to make their products be recognized more quickly and increase their sales.

REFERENCES

Achmad, Z. A., Azhari, T. Z., Esfandiar, W. N., & Nur, N. (2020). Pemanfaatan Media Sosial dalam Pemasaran Produk UMKM di Kelurahan Sidokumpul, Kabupaten Gresik. *Jurnal Ilmu Komunikasi*, 17–31.

Al-Deen, H. S., & Hendricks, J. A. (2013). *Social Media usage and impact.* United Kingdom: Lexington Books.

Anand, S. (2020, July). *Importance of Social Media Marketing in the Post-COVID-19 Era*. Retrieved from SocialPilot: https://www.socialpilot.co/blog/social- media-marketing-in-post-covid-19-era

Bramasta, D. B. (2020, May 20). *Mengenal Apa Itu New Normal di Tengah Pandemi Corona*. Retrieved from Kompas: https://www.kompas.com/tren/read/2020/05/20/06310 0865/mengenal-apa-itu-new-normal-di-tengah- pandemi-corona-?page=all

Juliana, Djakasaputra, A., Pramezwary, A., & Hutahaean, J. (2020). *Marketing Strategy in Digital Era.* Pekalongan: Nasya Expanding Management.

Maulidasari, C. D., & Damrus. (2020). Dampak Pemasaran Online di Era Covid-19. *Jurnal Bisnis dan Kajian Strategi Manajemen*, 233–245.

Mokhtar, N.F. 2015. Digital marketing Adoption by SmallBusiness Enterprises in Malaysia. IJBSS. 6(1).

Mokhtar, N.F., R.A.H. Zuha., and M. Abu. S. A. H. (2016).

Applying Technology Organization and Environ-ment (TOE) Model in Social Media Marketing Adoption: The Case of Small and Medium Enter-prise in Kelantan, Malaysia. The Social Science.11(21). 5139–5144

Sugiarto, I. (2018). Obstacles and Challenges in the Development of MSMEs: Case Study. *Budapest International Research and Critics Institute-Journal* , 93–98.

Tamtomo, A.B. (2020). INFOGRAFIK: Panduan Protokol Kesehatan Pencegahan Covid-19 untuk Sambut New Normal. www.kompas.com. Diakses pada 20 November 2020.

Contemporary Research on Business and Management – Noviaristanti (Ed.)

Second generation succession in cross-cultural family business alliance: A case study

B.Y. Pratama
Diponegoro University, Semarang, Indonesia

ABSTRACT: This research offers empirical insights on family business succession, a complex and challenging issue seen from cognition, structure, culture, and resource aspects, and identifies patterns related to the succession process structure. An overarching model to evaluate resources such as social capital and successor behavior, sustainability, and development of the cross-cultural family business alliance was developed and used as a reference that influences the succession process's pattern and length. Insights into the succession process contribute to a layered understanding of the cross-cultural family alliances' patterns and business behavior. The connected patterns explain the role of predecessors and successors changing over time during the transition process. The succession process is a formal structured process for defining the business framework, and these practices comprehend various outcomes relevant to the stages of generational ownership and the dynamics of family relationships. In conclusion, this research proposes an agenda for advancing research on cross-cultural family business alliance.

1 INTRODUCTION

The discussion about laws, norms, and values embedded in a cultural context shapes how a company is managed and directed. Certain institutional arrangements offer opportunities to advance the field of management. A family business research is gaining legitimacy gradually as an independent field of study. Over time, some family businesses have begun to lead to strategic alliances involving two or more families of different cultures. In current management research, a research related to the family business and strategic alliances is considered a well-established and prestigious field.

Joint ventures can be a rewarding experience, even though there are many challenges involved, and adding a cross-cultural element is one of the challenges. Collaborative partnerships have long been researched and applied. They range from micro to multinational enterprises and pursue multiple goals. The structure of a family business alliance with an ordinary business is distinctly different. In control of a family business alliance's management and strategic decisions, the family has the dominant position. In practice, the family business is usually governed by two main components: holding a large proportion of equity ownership and involving the dominant managerial control over the business (Luan et al., 2018).

In managerial control, where the founding family has dominance, the business activities are expected to develop with each alternation of generations. Founders can play a critical role in business succession to avoid risks. Business alliances of families with different cultures can reach a vulnerable point at the time of succession. Therefore, proper governance mechanisms are essential in this process. Different attitudes and views toward strategic decisions often arise in succession. The transfer of ownership is seen as an isolated event at the last stage of the transition process and simultaneously acts as an instrument for empowering new leaders.

Family businesses tend to prefer internal candidates over external candidates for succession (Schell et al., 2018). In this case, management control and monitoring mechanisms are deemed necessary by the previous generation to prepare a business transition to the successor (Shen & Su,

DOI 10.1201/9781003196013-24

2017). The family business founder's social capital is a valuable resource and a critical success factor because it has an impact on the performance of the family business in the future (Pearson et al., 2008). Resource management in many family businesses is considered complicated, as family members often participate simultaneously in family and business interactions at home and work (Kidwell et al., 2018).

This phenomenon is a serious concern because every company, especially the one with a cross-cultural alliance business model, has unique challenges and risks. Identifying the difference between management and ownership succession needs to be the focus of more research on the family business. Considering the phenomenon mentioned above, innovative research that can be used as a reference for business succession is needed, especially in a family business context. One of the efforts to identify the problems is a case study. This approach is an in-depth investigation of contemporary phenomena in real contexts, especially when the boundaries of the context are not clear and may involve important contextual conditions (Yin, 2017). With this approach, strategies that can be used as a reference for the next generation's succession can be described correctly.

2 LITERATURE REVIEW

Succession or regeneration in family businesses is one of the most challenging things to do (Merwe et al., 2009). Passing down a company to the next generation is a challenge (Zareie, 2011). It is because succession is a sensitive topic for several companies (Dhewanto & Tirdasari, 2012). Constraints in regeneration and succession planning are quite challenging in family businesses (Milan, 2012). Only about 40% of family businesses can pass the baton of leadership to the next generation, which makes succession a significant concern (Whatley, 2011). Different viewpoints between generations and the existence of a conflict in the family are some of the obstacles to be overcome for successful succession or regeneration (Merwe et al., 2009).

One of the significant factors in determining the continuation of a family business is the success of the planning process and strategy in the succession. Because the easier it is, the more successful will be this transformation in deciding the success and long-term sustainability of these generations (Merwe et al., 2009). The inability to manage the succession process can result in a decrease in the effectiveness of the business and can lead to bankruptcy (Milan, 2012).

3 METHODOLOGY

This research employs a qualitative research method. The qualitative research method was used to gain a better understanding of the experiences of participants with a certain life context (Arcidiacono et al., 2009). In addition, the qualitative research method emphasized the exploration of individual experiences, describes phenomena, and builds theories (Cope, 2014). Therefore, this research investigated and revealed the judgments of a person (participant) and their relationship to their work and experiences (Arcidiacono et al., 2009).

The choice of the qualitative research method depends on the research questions asked and which method will be studied (Yin, 2009). A case study was chosen because there are many 'how' questions posed in a social phenomenon, and an in-depth description of the phenomenon is needed (Yin, 2009). The purpose of this case study is not to generalize the results, but to prove the existence of a theory (Woodside & Wilson, 2003). In conclusion, the case study method was chosen because an in-depth understanding of the life phenomenon of the succession process in a family business is needed.

The data were collected from respondents or participants from different inter-ethnic (Overseas Chinese and Javanese) family business alliances in Central Java. The overview of succession focuses on four main elements: cognition, which includes the strategy and learning process; structure, which includes interactions with employees; culture, which includes norms and values; and resources, which include status and knowledge. The family businesses that become participants in this research were family businesses that have a total of more than two hundred employees. The participants in this research were the first and second generations of family businesses of each ethnicity. They are the successors of the company or are being prepared to become the next generation of the company.

4 RESULTS AND DISCUSSION

Four respondents were examined in this research, the four of whom were successors and prospective successors in their respective companies. The four respondents in the research were referred to as Respondent 1 (R1), Respondent 2 (R2), Respondent 3 (R3), and Respondent 4 (R4). In this research, it was found that R1 and R2 were carrying out a succession process in their family businesses, while R3 and R4's succession was still in progress. R1 and R2 have sat as the top leaders of their family businesses. Currently, R1 and R2 felt the need to carry out the succession process. R2 and R3 are the second generation because the founders are still active in the business as the highest decision makers.

In profile, the four respondents have several similarities and differences. The respondents studied were the second generation of each family of Overseas Chinese and Javanese ethnics. All respondents were male. Another similarity between the two successors was that they have got a bachelor's degree.

The difference between the four respondents lies in ethnicity, experiences within the company, and position in the company. Two respondents (R1 and R3) were of Chinese descent, and the other two respondents (R2 and R4) were of Javanese ethnicity. Based on experiences in the company, R1 and R2 have formed an alliance for more than twenty years. Meanwhile, R3 and R4 have been placed in management only for one year as managers.

The process of succession or regeneration of family companies in Semarang was carried out differently between the companies. Companies owned by R1 and R2 are family companies that have launched a succession process. Meanwhile, the purpose of the succession carried out by R1 and R2 was seen from the elements of cognition, structure, culture, and resources of the successor in maintaining sustainability and developing the business alliance.

5 CONCLUSION

Cognition includes strategies and learning processes of companies, where the highest decision maker in the family company is a member of the owner's family. Decisions related to the company's strategic plan and investment rest with the core family members, namely, the founder, husband/wife of the founder, and the children of the founder. It is also what that differentiates family companies from non-family companies.

The founder assesses the structure of the company and guides the next generation regarding dealing with subordinates and superiors. The founder expects the successor of the company to show a professional attitude. In family companies, the board of directors and top management are limited to family members.

The culture that exists in the family company, including the problem-solving method, will influence the decision making of the company. Acting in a professional manner and not differentiating family and non-family workers are other characteristics of a family company. Values from family cultures with ethnic differences are instilled in the succession process. Family members outside the family have the same position as other employees and have the authority, rights, and obligations in accordance with their respective jobs.

The family company was founded with a variety of resources, especially human resource management (HRM). The status and knowledge of the next generation in running a business cannot be separated from social values. Choosing a business sustainability strategy is important and needs to be mastered by the successor. The company's status, including social values such as empowering the surrounding community and making employees the main assets, is a characteristic of family companies that other companies usually do not have.

From this research, several things were found in connection with the succession process in three companies in Semarang:

a. Succession planning;
b. Succession process time;
c. Motivation driving successor's interest;
d. Selection of successors in the company;
e. Training given in the succession process;
f. The relationship between successors and previous generations (incumbent);
g. The stages in the succession process;
h. The changes brought by the second generation; and
i. The evaluation of the succession process.

REFERENCES

Arcidiacono, C., Procentese, F. & Di Napoli, I. 2009. Qualitative and Quantitative Research: An Ecological Approach. *International Journal of Multiple Research Approaches 2009* Vol.3, 163–176.

Cope, D.G. 2014. *Methods and Meanings: Credibility and Trustworthiness of Qualitative Research.* Oncology Nursing Forum Vol. 41, No. 1 January 2014.

Dhewanto, W. & Tirdasari, N.L. 2012. *Family Business Succession in Indonesia: A Study of Hospitality Industry.* Asia Pacific Business Innovation and Technology Management, Procedia – Social and Behavioral Sciences 57 (2012) 69–74.

Kidwell, R.E., Eddleston, K.A. & Kellermanns, F.W. 2018. *Learning bad habits across generations: How negative imprints affect human resource management in the family firm.* Hum. Resour. Manag. Rev. 28, 5–17.

Luan, C.-J., Chen, Y.-Y., Huang, H.-Y. & Wang, K.-S. 2018. *CEO succession decision in family businesses – A corporate governance perspective.* Asia Pac. Manag. Rev. 23, 130–136.

Merwe, S., Venter E. & Ellis, S.M. 2009. *An Exploratory of Some of Determinants of Management Succession Planning in Family Business.* Management Dynamics Volume 18 No. 4.

Milan, H. 2012. Succession Planning and Generational Transition: The Greatest Challenges for Family-owned Businesses. *Journal of Eastern Europe Research in Business and Economics,* 1–11.

Pearson, A.W., Carr, J.C. & Shaw, J.C. 2008. *Toward a theory of familiness: A social capital perspective.*
Entrep. Theory Pract. 32, 949–969.

Schell, S., Hiepler, M. & Moog, P. 2018. It's all about who you know: The role of social networks in intra-family succession in small and medium-sized firms. *J. Fam. Bus. Strategy 9*, 311–325.

Shen, N. & Su, J. 2017. Religion and succession intention-Evidence from Chinese family firms. *J. Corp. Finance 45*, 150–161.

Woodside, A.G. & Wilson, E.J. 2003. Case Study Research Methods for Theory Building. *The Journal of Business & Industrial Marketing* Vol.18 2003 493.

Whatley, L. 2011. A New Model for Family-Owned Business Succession. *Organization Development Journal* Vol. 29 No. 4, Winter 2011.

Yin, R.K. 2009. *Case Study Research: Design and Methods 4th ed.* Applied Social Research Methods Series Vol.5. Sage Publications: California.

Yin, R.K. 2017. *Case study research and applications: Design and methods.* Sage Publications: California.

Zareie, M. 2011. Analysis of Effective Factors on Family Business Transition to the Next Generations in Iran: Strategic Management Perspective. *International Conference on Financial Management and Economics* vol.11.

Contemporary Research on Business and Management – Noviaristanti (Ed.)

Financial statements and financial performance analysis: Implementation in an Indonesian MSME-based aesthetic beauty clinic

C.P. Wijaya
Master of Management, Universitas Indonesia, Indonesia

ABSTRACT: The aim of this study is to make financial statements and conduct financial performance analysis of one of the micro, small, and medium enterprises (MSMEs) in Indonesia, namely, Luxe Dental and Aesthetic Center, which is engaged in the aesthetic beauty clinic industry. This study was conducted by using gap analysis and Pareto analysis. The results show that the financial statements consisted of income statement, balance sheet, and cash flow statement. Financial performance analysis was performed using common-size financial statements and profitability ratios. Creating financial statements and conducting financial performance analysis are useful to map and describe the condition of an MSME financial performance and to increase its ability to evaluate its performance against competitors.

1 INTRODUCTION

Micro, small, and medium enterprises (MSMEs) around the globe significantly contribute to the economy. They contribute to around 90% of the businesses around the world and are responsible for 60%–70% of employment and up to 50% of the gross domestic product (GDP) of the countries (United Nations Year 2018 about MSMEs). However, in running their business, many MSMEs often face obstacles in terms of financial management. This brings up some constraints associated with the development of MSME financial strategies, such as difficulty in raising capital (Auken & Lema, 2003).

Financial arrangements need to be made so that the company can survive, avoid bankruptcy, compete with other companies, maximize sales, minimize costs, maximize profits, and maintain a steady increase in income. One of the things that need to be done to make it easier to manage finances is to make financial statements (Ross et al. 2013). The ease of making financial statements is important for MSMEs (Halabi et al. 2010).

Moreover, a company needs to analyze its financial performance to assess its current performance and compare it with that of other companies to help it find best practices to improve its performance (Weygandt et al. 2012). The purpose of financial performance analysis is to evaluate the financial performance of the company by studying the relationship between operational decisions and financial performance (Higgins, 2016).

One of the micro, small, medium enterprises (MSMEs) in Indonesia, namely, Luxe Dental and Aesthetic Center, which is engaged in the aesthetic beauty clinic industry, faced problems regarding creating financial statements and conducting financial performance analysis. With the availability of the financial statements and by conducting financial performance analysis, it can clearly understand its financial performance and find best practices to support its growth.

DOI 10.1201/9781003196013-25

2 LITERATURE REVIEW

The purpose of financial statements is to provide information on the financial position and financial performance of an entity, and they help a large number of users make economic decisions, particularly those who are not in a position to request specific financial reports to obtain the information needed. Financial statements also show the accountability of the management for the resources entrusted to it (Weygandt et al. 2012).

The balance sheet is a report that captures the position of the assets, liabilities, and equity of a company for a specific time period (Higgins, 2016). The income statement is a report that contains a summary of the income and expenses of a company (Ross et al. 2013). The cash flow statement is a statement that provides information about the cash position that results from cash inflows and cash outflows, which are usually grouped into operating, investing, and financing activities for a certain period (Ross et al. 2013).

The purpose of financial performance analysis is to evaluate the financial performance of the company by studying the relationship between the company's operational decisions and its financial performance (Higgins, 2016). In this study, there are two tools performed, namely, common-size analysis and financial ratio analysis.

The common-size analysis is used to describe each account in the financial statements using a percentage as the basis value. This standardized financial statement describes the accounts in the statement of financial position as a percentage of assets and describes the accounts in the income statement as a percentage of sales (Higgins, 2016). The advantage of the common-size analysis is that it allows a company to compare its performance with other companies of different sizes. The results of this financial statement analysis are called common-size financial statements, consisting of common-size balance sheet and common-size income statement.

Financial ratio analysis is done by comparing a company's financial ratios, which are calculated based on the financial figures reported by the company. Financial ratios are calculated to obtain comparable results for different companies and over time, regardless of size. The financial ratio analysis used in this study was profitability ratios. Profitability ratios measure the income or operating success of a company for a given period of time (Eklund et al. 2003).

Profitability ratios used in this study are profit margin, return on assets (ROA), and return on equity (ROE). Profit margin is obtained by dividing sales by net income. This ratio is particularly important to calculate because it reflects the company's pricing strategy and its ability to control operating costs. ROA is obtained by dividing net income by total assets. ROA is a basic measure of the efficiency with which a company allocates and manages its resources. ROE is obtained by dividing net income by total equity (Higgins, 2016).

3 RESEARCH METHODS

This study is a descriptive and qualitative study and was conducted through a business coaching method. The data were obtained using both primary and secondary data from MSME business process including marketing, finance, operation and human resources from 2018 to 2019. The data were acquired from interviews, observations, and desk research.

The data were analyzed through gap analysis and Pareto analysis. Gap analysis was conducted to compare the level of mismatch between ideal conditions and actual conditions by using analysis tools such as Business Modal Canvas (BMC), PESTEL, five forces model of competition, service marketing mix, segmenting, targeting, and positioning (STP), SWOT, and TOWS.

Pareto analysis is a priority scale carried out based on the level of urgency and significance of its impact on problems in an MSME.

The threshold for the accumulated problems is 80%. The result of Pareto analysis shows that creating financial statements and conducting financial performance analysis are the most significant and most feasible for an MSME to carry out under current conditions. Upon this result, it is agreed that the implementations to be executed are as follows: (1) creating financial statements and (2) conducting financial performance analysis.

4 RESULTS AND DISCUSSIONS

The MSME's financial statements taken from the MSME financial data from 2018 to 2019 are given. The income statement consisted of sales, operating expenses, and tax expenses. The balance sheet consisted of cash, prepaid rent, equipment, and accumulated depreciation in the assets side and equity and retained earnings in liabilities and equity. The cash flow statement consisted of sales, operating expenses, and tax expenses in operating activities, purchase of equipment and payment of rent in investing activities, and issuance of equity in financing activities.

The following is the result of the comparative common-size financial statements. The result of the common-size income statement shows that the MSME had an increase in financial performance, as indicated by an increase in the net income from 2018 to 2019 by 11.2%. Meanwhile, its competitor had a decrease in financial performance, as indicated by a decrease in the net income from 2018 to 2019 by 0.3%. This shows that the operational efficiency of the MSME had increased compared to its competitor from 2018 to 2019.

The result of the comparative profitability ratio is as follows. The result shows that the MSME had a better financial performance compared to its competitor with an increase in profit margin from 2018 to 2019, greater by 10.9%. This shows that the operational performance of the MSME was increasing from 2018 to 2019, greater than its competitor. Return on assets (ROA) and return on equity (ROE) from 2018 to 2019 were greater by 10.4% compared to its competitor. Increase in the company's ROA shows that there was an increase of 12.7% in the efficiency of using its assets to generate profits, which was greater than its competitor. Increase in ROE shows that the MSME was generating greater return on investment (by 1.7%) from 2018 to 2019 from the equity invested by the owner than its competitor.

The results of this study confirm the benefits of financial statements to reduce information asymmetry of financial information provided by MSMEs. It is believed that creating financial statements will increase MSMEs' access to funding through harmonious and quality financial information (Parera & Chand, 2015).

The results also show that financial ratio analysis is important. This shows whether a company is on the path of improvement. An increase in a company's profitability ratio indicates a better financial performance. A high profit margin means that the company has the ability to absorb cost increases without passing it on to consumers at an increased price. Return on assets (ROA) measures the success of a company in converting its assets into profit. Return on equity (ROE) measures the success of a company in generating returns on investment that has been issued by shareholders (Mahajana & Sarkar, 2017).

Financial performance comparisons are used to map information from economic events to financial statements and thus capture similarities in how two firms explain similar transactions (De Franco et al., 2011). Comparability has the potential to expand company information because companies often rely on information from financial reports when formulating their own strategies. Comparability is believed to enhance the characteristics of financial information that differ from other qualitative characteristics, such as representativeness, relevance, and timeliness.

5 CONCLUSIONS

This study was conducted to create financial statements and to perform financial performance analysis for Luxe Dental & Aesthetic Center. This is used as a basis for the company to assess its financial performance and to compare it with its competitors. The study was conducted through a business coaching method. The data obtained were analyzed with gap analysis and Pareto analysis. The implementation was done by doing common-size analysis for balance sheet, income statement, and profitability ratio analysis, consisting of profit margin, return on assets (ROA), and return on equity (ROE). The results of this study indicate that creating financial statements and conducting financial performance analysis are useful to describe the condition of the MSME's financial performance and can increase its ability to evaluate its performance against its competitor.

The scope of this study, which was carried out through the business coaching method, is limited to the field of financial management of the service company. The results reported herein suggest the need for more analyses and comparable financial information. Finally, this study provides a baseline for additional research. For example, future studies may examine how to evaluate the result of financial performance analysis for future investment in competitive markets.

REFERENCES

Auken, H.E.V. & Lema, D.G.P.D. 2003. Financial strategies of Spanish firms: A comparative analysis by size of firm. Journal of Small Business and Entrepreneurship 17(1): 17–30.

De Franco, G., Kothari, S. P., & Verdi, R. (2011). The benefits of financial statement comparability. Journal of Accounting Research 49: 895–931.

Eklund, T., Back, B., Vanharanta, H., & Visa, A. (2003). Using the self-organizing map as a visualization tool in financial benchmarking. Information Visualization 2(3): 171–181.

Halabi, A. K., Barret, R., & Dyt, R. 2010. Understanding Financial Information Used to Assess Small Firm Performance. Qualitative Research in Accounting & Management 7(2): 163–179.

Higgins, R. C. (2016). Analysis for financial manager. New York: McGraw Hill-Irwin.

Hossain, M.M., Ibrahim, Y., Uddin, M.M. 2020. Finance, financial literacy and small firm financial growth in Bangladesh: The effectiveness of government support. Journal of Small Business and Entrepreneurship: 1–26.

Mahajan, S. & Sarkar, M. 2007. How does financial performance of MNCs in the automobile sector compare with Indian companies? An analysis using financial ratios. Paradigm 11(2): 38–45.

Perera, D. & Chand, P. 2015. Issues in the adoption of international financial reporting standards (IFRS) for small and medium-sized enterprises (SMES). Advances in Accounting 31(1): 165–178.

Ross, S. A., Westerfield, R. W., & Jaffe, J. 2013. Corporate finance. New York: McGraw-Hill-Irwin. Weygandt, J.J., Kimmel, P.D., & Kieso, D.E. 2012. Accounting principle. New Jersey: John Wiley & Sons, Inc.

Contemporary Research on Business and Management – Noviaristanti (Ed.)

The effect of job resources and public service motivation on affective commitment: The mediating role of work engagement

D.H. Rajagukguk & P.M. Desiana
Faculty of Economics and Business, Universitas Indonesia, Jakarta, Indonesia

ABSTRACT: This paper aims to show the antecedents and consequences of work engagement (WE) in the government sector. Specifically, this paper examined whether public service motivation (PSM) and job resources (JR) are significantly related to work engagement in the context of the government sector. Furthermore, this paper examined whether WE mediates PSM and JR's relationship toward affective commitment (AC). This study used a cross-sectional survey research design. Data were collected from 162 employees who work in the government agency located in Jakarta, Indonesia. SEM-Lisrel was used as a tool in data analysis. In line with expectations, JR and PSM have a positive and significant impact on WE. Besides, WE also plays a role in mediating the relationship between JR and PSM to AC. This study shows JR and PSM's role in influencing employees' WE and AC in the government sector. This study is one of the few studies that provide an overview of JR and PSM's impact on WE and AC for employees in the Indonesian government sector.

1 INTRODUCTION

Referring to the three-component model (TCM) of organizational commitment, Allen and Meyer (1990) divided the types of organizational commitment into affective commitment, normative commitment (obligation-based), and continuance commitment (cost-based). The research found that affective commitment had the strongest positive correlation with job performance, organizational citizenship behavior, and attendance (Meyer, 2014; Mayer, Stanley, & Parfyonova, 2012). Increasing the level of affective commitment is important for organizations during this pandemic time to maintain the performance of employees of the organization. The government organizations are faced with budget constraints and have limitations in allocating their expenditure in an effort to improve employee performance. In a meta-analysis study conducted by Fletcher, Bailey, Alfes, and Madden (2020), it was revealed that work engagement had a role in increasing employee commitment to the organization, especially affective commitment. Employees who experienced a high level of engagement are physically healthier, more satisfied with their psychological needs, and also get more benefits, including being more committed to the organization when compared to employees with low level of engagement (Borst, 2018). One concept that can explain the driving factors of work engagement is the job demands–resources (JD-R) framework (Lesener, Gusy, Jochmann & Wolter, 2020). Job resources include support from superiors, autonomy, and performance feedback. They are intended as "good things" at work since they encouraged performance achievement and self-development as well as reduced the impact of psychological burdens that were considered to have the most positive impact on work engagement and affective commitment (Schaufeli, 2017; Dominguez, Chambel, & Carvalho, 2020; Albrecht & Marty, 2020). In the context of the public service sector, public service motivation is seen as a "major psychological resource" (Bakker, 2015), which is expected to encourage a high level of engagement (Lavigna, 2013). A research shows that a high level of public service motivation had a positive effect on public servants' level of engagement (Borst, Kruyen, & Lako, 2019; Cooke, Brant, & Woods, 2019; Gross, Thaler, & Winter, 2019). Besides, based on research, it was also revealed that public

DOI 10.1201/9781003196013-26

service motivation positively and significantly affected affective commitment (Borst, Kruyen, & Lako, 2019; Potipiroon & Ford, 2017). Albrecht and Marty (2020) revealed that work engagement mediated the relationship between job resources and employees' affective commitment in a more general and diverse sector. Meanwhile, in the context of the public sector, Borst, Kruyen, and Lako (2019) concluded that engagement positively mediates the relationship between public service motivation and employees' affective commitment in the public sector.

2 LITERATURE REVIEW

Affective commitment is defined as a desire to remain as a member of an organization because of emotional attachment and involvement with that organization. Affective commitment is considered to have the strongest positive impact on the results achieved by the organization, and there have been a lot of research studies focusing on the relationship of results with the affective commitment (Meyer, 2014). Work engagement is a condition related to a positive psychological state, satisfaction, and motivation. Engaged employees had great energy, were enthusiastically involved in work, and had full concentration on work (Bakker, *et.al.*, 2008). Schaufeli (2017) stated that every job has job demands and job resources. There are many types of "good things" in work that can be categorized as job resources. Several studies suggest that work autonomy, job feedback, and support from superiors have consistently affected work engagement and affective commitment (Schaufeli, 2017; Dominguez, Chambel, & Carvalho, 2020; Albrecht & Marty, 2020). Perry and Wise (1990) defined public service motivation as the tendency of individuals' motivation in public institutions and organizations. Perry (1996) identified four empirical components of public service motivation, such as attraction to public policy making, commitment to the public interest, compassion, and self-sacrifice.

3 METHOD

Based on literature reviews and previous research, the hypotheses developed in this study are as follows:

H1: Job resources have a positive effect on work engagement.
H2: Job resources have a positive effect on affective commitment.
H3: Public service motivation has a positive effect on work engagement.
H4: Public service motivation has a positive effect on affective commitment.
H5: Work engagement has a positive effect on affective commitment.
H6: Work engagement mediates the relationship between job resources and affective commitment.
H7: Work engagement mediates the relationship between public service motivation and affective commitment.

This study used a non-probability sampling design. The type of sampling used was purposive sampling. The data were obtained by distributing self-administered questionnaires. The measurement used a 7-Likert scale from strongly disagree to strongly agree. The population in this study was 761 employees, and the samples were collected from 162 respondents. The primary data that had been collected were then analyzed using SPSS and Lisrel software.

4 RESULTS AND DISCUSSION

Based on this study's results, it was found that job resources had a positive and significant effect on work engagement, and thus, H1 can be accepted. This supports previous research conducted by Reina-Tamayo, Bakker, and Derks (2017), Dominguez, Chambel, and Carvalho (2020), and Albrecht and Marty (2020). Autonomy, supervisors' support, and feedback on the assigned work led to increase in work engagement of the employees. On the other hand, job resources had a positive but insignificant effect on affective commitment; hence, H2 cannot be accepted. This is different from research performed by Dominguez, Chambel, and Carvalho (2020) and Albrecht and Marty (2020). They stated that autonomy, supervisors' support, and feedback did not directly affect the employee's affective commitment, but the employee felt engaged with increase in organizational commitment. This finding is in line with the JD-R concept proposed by Schaufeli (2017), where job resources increased work engagement, and therefore, employees felt engaged. It was then proven that job resources have a positive impact on the organization, especially the commitment to the organization.

The following result reveals that public service motivation had a positive and significant relationship with work engagement; thus, H3 can be accepted. This finding supports previous research conducted by Borst (2018), Borst, Kruyen, and Lako (2019), and Cooke, Brant, and Woods (2019). Employees who had a high level of public service motivation were more engaged than employees who had a low level of public service motivation. Public service motivation is unique to the government sector. The fourth result denotes that public service motivation had a positive and significant relationship to affective commitment; hence, H4 can be accepted. This finding supports previous research carried out by Borst, Kruyen, and Lako (2019) and Potipiroon and Ford (2017). Therefore, during recruitment, government organizations need to measure public service motivation as one of the criteria. Besides, increased socialization within the organization regarding the importance of work for the community can increase public service motivation, employee work engagement, and affective commitment.

The fifth result indicates that work engagement had a positive and significant effect on affective commitment; thus, H5 can be accepted. Work engagement focuses on the relationship between employees and the work they do, while affective commitment emphasizes the relationship between employees and the organization they are a part of. Therefore, it can be concluded that the high level of work engagement felt by employees for their work increases emotional attachment, involvement, and self-identification within the organization (Schaufeli, 2017; Albrecht & Marty, 2020). Finally, the last result signifies that work engagement mediated the relationship between job resources and public service motivation. The results are in line with a research conducted by Albrecht and Marty (2020) and Borst, Kruyen, and Lako (2019); thus, H6 and H7 can be accepted. It can be seen that work engagement has a significant role in driving positive outcomes for the organization. Therefore, organizations must conduct regular surveys of employee work engagement levels to find out how employees perceive job resources in the organization and the strategies needed to improve public service motivation in employees.

5 CONCLUSION

Providing autonomy, supervisors' support, and feedback to employees can create a positive and significant impact on work engagement, but it does not directly affect affective commitment. They only affect organizational commitment when their work engagement is taken into account. Public service motivation has a positive and significant relationship with work engagement and affective commitment. Therefore, during recruitment, government organizations need to measure public service motivation as one of the criteria. Besides, increasing socialization within the organization regarding the importance of work for the community can increase public service motivation, work engagement, and affective commitment. These results indicate that affective commitment is more influenced by personality characteristics rather than job resources. In contrast, work engagement is determined by personality characteristics and job resources. It confirms the idea that work engagement is more pervasive rather than affective commitment. All of the variables studied (i.e., autonomy, feedback, supervisors' support, and socialization) increase PSM and can be carried out during this pandemic without causing significant increase in expenditure for the employees. It can increase work engagement and affective commitment, leading to the desired organizational outcomes.

REFERENCES

Albrecht, S. L., & Marty, A. (2020). Personality, self-efficacy and job resources and their associations with employee engagement, affective commitment and turnover intentions. *The International Journal of Human Resource Management*, 31(5), 657–681.

Allen, N. J., & Meyer, J. P. (1990). The measurement and antecedents of affective, continuance, and normative commitment to the organization. *Journal of Occupational Psychology*, 63, 1–18.

Bakker, A. B. (2015). A job demands-resources approach to public service motivation. *Public Administration Review,* 75, 723-732.

Bakker, A. B., Schaufeli, W. B., Leiter, M. P., & Taris, T. W. (2008). Work engagement: An emerging concept in occupational health psychology. *Work & Stress*, 22, 187–200.

Borst, R. T. (2018). Comparing work engagement in people-changing and people-processing service providers: A mediation model with red tape, autonomy, dimensions of PSM, and performance. *Public Personnel Management*, 47(3), 287–313.

Borst, R. T., Kruyen, P. M., & Lako, C. J. (2019). Exploring the job demands–resources model of work engagement in government: Bringing in a psychological perspective. *Review of Public Personnel Administration*, 39(3), 372–397.

Cooke, D. K., Brant, K. K., & Woods, J. M. (2019) The role of public service motivation in employee work engagement: A test of the job demands-resources model. *International Journal of Public Administration*, 42:9, 765–775.

Dominguez, D., Chambel, M., & Carvalho, V. (2020). Enhancing engagement through job resources: the moderating role of affective commitment. *The Spanish Journal of Psychology,* 23, 1–12.

Fletcher, L., Bailey, C., Alfes, K., & Madden, A. (2020). Mind the context gap: a critical review of engagement within the public sector and an agenda for future research. *The International Journal of Human Resource Management*, 31(1), 6–46.

Gross, H. P., Thaler J., & Winter, V. (2019). Integrating public service motivation in the job- demands-resources model: An empirical analysis to explain employees' performance, absenteeism, and presenteeism. *International Public Management Journal*, 22(1), 176–206.

Lesener, T., Gusy, B., Jochmann, A., & Wolter, C. (2020). The drivers of work engagement: A meta-analytic review of longitudinal evidence. *Work & Stress*, 34(3), 259–278.

Lavigna, R. J. (2013). *Engaging Government Employees: Motivate And Inspire Your People To Achieve Superior Performance*. New York: Amacom.

Meyer, J. P. (2014). Employee commitment, motivation, and engagement: exploring the links. In Gagne, M. (Eds.), *The Oxford Handbook of Work Engagement, Motivation, and Self- Determination Theory* (pp. 33–49). New York: Oxford University Press.

Meyer, J. P., Stanley, L. J., & Parfyonova, N. M. (2012). Employee commitment in context: The nature and implications of commitment profiles. *Journal of Vocational Behavior*, 80, 225–245.

Perry, J. L. (1996). Measuring public service motivation: An assessment of construct reliability and validity. *Journal of Public Administration Research and Theory*, 6(1), 5–22.

Potipiroon, W., & Ford, M. T. (2017). Does public service motivation always lead to organizational commitment? Examining the moderating roles of intrinsic motivation and ethical leadership. *Public Personnel Management*, 46(3), 211–238.

Reina-Tamayo, A. M., Bakker, A. B., & Derks, D. (2017). Episodic demands, resources, and engagement: An experience-sampling study. *Journal of Personnel Psychology*, 16(3), 125–136.

Schaufeli, W. B. (2017). Applying the job demands-resources model: A 'how to' guide to measuring and tackling work engagement and burnout. *Organizational Dynamics*, 46(2), 120–132.

Contemporary Research on Business and Management – Noviaristanti (Ed.)

The effect of teleworking implementation on work engagement: Job resources as a mediator

D. Afdillah & R. Rachmawati
Master of Management, Universitas Indonesia, Indonesia

ABSTRACT: This study seeks to determine the effect of the implementation of teleworking on work engagement. The hypothesis is that job resources' motivational process (i.e., job autonomy and job variety) can mediate that effect. The data were collected from 271 employees of various state-owned companies in Indonesia. This study employed quantitative research using scaling for data collection and the structural equation model (SEM) to test the data. This study found that (1) the implementation of teleworking does not have a significant effect on working engagement and (2) job resources act as a mediator between the implementation of teleworking and work engagement.

1 INTRODUCTION

The COVID-19 pandemic that began in the beginning of 2020 has caused unprecedented changes in the global economy and the professional world (International Labour Organization, 2020). To curb the disease's spread, the government urged to maintain distance, prohibit gatherings and meetings, and limit out-of-house activities. Many companies have issued work from home (WFH) policies and switched to the teleworking system to respond to this situation. Hence, teleworking has become increasingly common in the pandemic situation and is considered an important aspect of ensuring business continuity. Teleworking has some benefits, such as reducing travel time and distractions, increasing workers' focus on their duties, providing a better work–life balance, and making schedules and workspace more flexible. However, some drawbacks, such as the feeling of isolation (especially for employees who live alone) and loss of contact with fellow employees (International Labor Organization, 2020), are indicated in teleworking implementation.

Previous research conducted by Sardeshmukh, Sharma, and Golden (2012) shows that teleworking has a negative relationship with work engagement. Employees who do teleworking tend to be physically and psychologically separated from the workplace, find it difficult to identify with the company, and consider themselves more independent. The research shows that extensive teleworking can reduce work engagement. The research also includes the motivational process from job resources (i.e., job autonomy and job variety) as mediation for the teleworking effect on work engagement. As teleworking changes, the way employees carry out their work physically and cognitively, the job resources, an important aspect for work and individual goals, and individual work engagement are also affected.

This study seeks to determine the effect of teleworking implementation on work engagement. Particularly, the present study hypothesizes that job resources (i.e., job autonomy and job variety) mediate the effect of teleworking on work engagement. Therefore, the hypotheses in this study are as follows:

Hypothesis 1: Teleworking has a negative and significant effect on working engagement.
Hypothesis 2: Job resources mediate the effect of teleworking on work engagement.

DOI 10.1201/9781003196013-27

2 LITERATURE REVIEW

Teleworking generally utilizes information and communication technology (ICT), such as smartphones, tablets, laptops, or computers, to support work activities carried out in places far from traditional office environments (International Labor Organization, 2017). The positive and negative impacts of the teleworking work system can be measured by evaluating aspects such as remote working satisfaction, family and professional conflicts, the use of communication tools and technology, task efficacy, commitment to the company, work–life balance, relationships with colleagues, stress management, and work concentration (Baert, Lippens, Moens, Sterkens, & Weytjens, 2020).

Work engagement is defined as the process where the organizational members express themselves physically, cognitively, and emotionally when doing jobs (Kahn, 1990). Bakker and Demerouti (2008) define work engagement as a positive, full, and affective employee mind state associated with performing work roles characterized by three aspects: vigor, dedication, and absorption.

Job resources are a motivational process generated from a job demands–resources (JD-R) model (Bakker & Demerouti, 2006). Job resources consist of physical, psychological, social, or organizational aspects of work essentials to achieve work goals, reduce physiological or psychological costs, or stimulate growth, learning, and personal development (Schaufeli & Bakker, 2004). Van de Voorde, Veld, and van Veldhoven (2016) suggest that job resources have two dimensions: job variety and job autonomy. Job variety reflects the extent to which a job offers a series of tasks that use plenty of skills and require creativity and personal input (Grant & Ashford, 2008). Job autonomy is defined according to the extent to which the job provides substantial freedom, independence, and discretion to individuals in scheduling work and determining the procedures used in carrying out the work (Hackman and Oldham, 1976).

3 RESEARCH METHODS

3.1 *Types and data sources*

This study is quantitative research. This study employed a census-based survey using a questionnaire distributed to employees using teleworking to work during the COVID-19 pandemic from various state-owned enterprises in Indonesia. Out of this study's 271 participants, 71.2% were men and 28.2% were women. In terms of marital status, 67.5% were married and 22.5% were single. Most of the participants (31.7%) were 26–30 years old.

3.2 *Measures*

The teleworking effect was measured by 14 items adapted from Baert, Lippens, Moens, Sterkens, and Weytjens (2020). The items include "I am globally satisfied that I am working more at home because of the corona crisis" and "I can do my job more efficiently during the extended homework because of the corona crisis." The respondents indicated the extent to which they agreed with each statement on a five-point scale, ranging from 1 (strongly disagree) to 5 (strongly agree). The Cronbach's alpha coefficient of the teleworking is 0.78.

Job resources (i.e., job autonomy and job variety) were measured by eight items adapted from Van de Voorde, Veld, and van Veldhoven (2016). The items include "Does your work require creativity?" and "o you have freedom in carrying out your work activities?". The respondents' answers were indicated on a four-point scale, ranging from 1 (never) to 4 (always). The Cronbach's alpha coefficient of the job resources is 0.60.

Work engagement was measured by 17 items adapted from the UWES (Utrecht Work Engagement Scale) (Schaufeli, Backer, & Salanova, 2006). The items include "At my work, I feel bursting with energy" and "I find the work that I do full of meaning and purpose." The respondents indicated their answers on a seven-point scale, ranging from 0 (never) to 6 (always). The Cronbach's alpha coefficient of the UWES is 0.97.

3.3 *Data analysis*

The analysis was carried out with the Lisrel 8.8 program. This study assessed the properties of the scales by estimating a confirmatory factor analysis (CFA) model. The resulting purified scales were then used for simultaneous estimation of the structural model for testing the hypotheses. The confirmatory factor analysis (CFA) test shows that all indicators of each variable in this study are valid, as they have a standardized factor loading (SFL) value greater than 0.5. The reliability test using the construct reliability (CR) found that in each variable, the CR value was greater than 0.7 and the average variance extracted (AVE) value was greater than 0.5.

Structural equation modeling (SEM), as implemented in Lisrel 8.8, was used to test structural models by using maximum likelihood analysis. The resulting model produced good fit indices (e.g., chi-square = 591.20, d.f. = 222, comparative fit index (CFI) = 0.93, and root mean square error of approximation (RMSEA) = 0.78), indicating a good fit.

4 RESULTS

This study examines the direct effect of teleworking on work engagement and whether job resources mediate the effect. The findings indicate that teleworking's direct effect on work engagement is rejected, but the indirect effect of teleworking on work engagement through job resources is accepted. Teleworking implementation has a negative and insignificant effect on work engagement because the t-values are below 1.64 (SLF = −0.03; t-values = −0.60); thus, H1 is rejected. Then, when the researchers include the motivational process from job resources as a mediator, the results show that job resources mediate the effect of teleworking on work engagement, providing support to H2. Specifically, teleworking implementation has a positive effect on job autonomy and job variety (SLF = 0.31; t-values = 4.22), and changes in those job resources positively impact work engagement on the dimensions such as vigor, dedication, and absorption (SLF = 0.18; t-values = 2.14). Van de Voorde, Veld, and van Veldhoven (2016) suggest that job resource aspects, such as job variety, can positively increase employee work engagement. These results are consistent with those of previous studies, indicating that teleworking's effect on work engagement is mediated by job resources (Sardeshmukh, Sharma, & Golden, 2012). In this study, job resources fully mediate the influence of teleworking on work engagement.

The current teleworking implementation on employees from various state-owned enterprises in Indonesia is a policy enforced abruptly due to the COVID-19 pandemic. The application requires strong employee management and use of technology. The shift from work from the office to a work from home system accommodated by teleworking will change how, when, and where employees work. However, some companies are not familiar with remote employee management and do not have high technology and information systems to implement teleworking; thus, companies have to make significant practical adjustments to maintain work engagement.

5 CONCLUSIONS AND SUGGESTIONS

In summary, this study provides further evidence that teleworking has a negative but insignificant effect on work engagement and that job resources can mediate this effect.

This study has significant managerial implications. The results imply that to implement teleworking successfully, managers should make an effective teleworking work system management strategy that encourages employees to give their best performance when working remotely. Also, managers should increase job resources for employees to do teleworking by providing various jobs that require multiple skills, holding training or skills development, and giving authority to employees doing the tasks.

The limitation of this study is that several questionnaire items were invalid on the teleworking variable. Besides, data retrieval and processing only use quantitative methods, limiting the exploration of the companies' actual specific conditions.

Therefore, future research should use other descriptive measuring tools to measure the teleworking effect on work agreements. The use of teleworking by employees may occur in the long term; thus, the impact that results from employee hard work needs to be investigated.

REFERENCES

Baert, S., Lippens, L., Moens, E., Sterkens, P., & Weytjens, J. 2020. The COVID-19 crisis and telework: a research survey on experiences, expectations, and hopes. *IZA Institute of Labour Economics*.

Bakker, A. B. & Demerouti, E. 2006. The Job Demands-Resources model: state of the art. *Journal of Managerial Psychology*.

Bakker, A. B., & Demerouti, E. 2008. Towards a model of work engagement. *Career Development International*. 13, 209–223.

International Labour Organization. 2017. Working anytime, anywhere: the effect on the world of the work.

International Labour Organization. 2020. Teleworking during the COVID-19 pandemic and beyond: A practical guide.

Khan, W. A. 1990. Psychological conditions of personal engagement and disengagement at work. *Academy of Management Journal*. 33:4. 692–724.

Sardeshmukh, S. R., Sharma, D., & Golden, T. D. 2012. Impact of telework on exhaustion and job engagement: a job demands and job resources model. *New Technology, Work, and Employment*. 27:3.

Schaufeli, W., B., & Bakker, A., B. 2004. Job demands, job resources, and their relationship with burnout and engagement: A multisample study. *Journal of Organizational Behavior*. 25, 293–315.

Schaufeli, W. B., Bakker, A.B., & Salanova, M. 2006. The measurement of work engagement with a short questionnaire: A cross-national study. Educational and Psychological Measurement. 66:4, 701–716.

Van De Voorde, Veld, & Van Veldhoven. 2016. Connecting empowerment-focused HRM and labour productivity to work engagement: the mediating role of job demands and resources. *Human Resource Management Journal*.

Contemporary Research on Business and Management – Noviaristanti (Ed.)

The risks in the implementation of the Write-Off Ceiling Incentive Policy in retail segment: A case study of ABC Bank

D.V. Astriana
Master in Management, Faculty of Economics and Business, University of Indonesia, Indonesia

D. Wibowo
Dirgantara Marsekal Suryadarma, Faculty of Technology Management & Technopreneur, Universiti Teknikal Malaysia Melaka, Malaysia

ABSTRACT: The Write-Off Ceiling Incentive Policy is part of the business policies of ABC Bank regarding debt collection. This policy was set in 2011 to reduce the number of Non-Performing Loans (NPLs). In addition, it also motivates account officers in all ABC Bank branch offices to collect the loans and achieve NPL recovery, including the written-off ones. However, this policy also faces certain risks in the banking sector. This study aims to investigate the risks affecting the effectiveness of the policy implementation in ABC Bank. This study was conducted through a series of interviews and a literature study in September 2020. The results showed that this policy deals with the following risks: (1) operational risk; (2) legal risk; (3) reputation risk; and (4) compliance risk. Therefore, the right risk treatment is required to manage each risk in order to increase the effectiveness of this policy on debt collection.

1 INTRODUCTION

Lending service is one of the essential businesses for banking institutions conducted by distributing funds to the third party. It is based on the contractual agreement in which the debtor agrees to make the monthly principal and interest repayments at the due date. However, this service can put the bank at credit risk due to the inability of the debtors to fulfill their obligation.

Debtors' failure to fulfill their obligation will cause the occurrence of non-performing loans (NPLs). The occurrence of NPLs decreases the technical efficiency of banking institutions. At the same time, it also increases the moral hazard due to the decline in credit quality and instability in the financial system (Zhang et al. 2016). Therefore, NPLs negatively affect the profitability of the bank (Elekdag et al. 2020; Partovi & Matousek 2019)

When bad-debts occur, the bank has to take actions of credit settlement, namely rescheduling, reconditioning, restructuring, and liquidation (Triandi 2018). If those actions fail to handle the existing bad-debts, the bank should do the next action, namely writing-off those bad-debts (Hariyani 2010). Writing-off bad-debts means the bank removes a certain amount of bad-debts, namely the NPLs, from the balance sheet. Nevertheless, this removal does not erase the right of the bank to re-collect the removed amount of NPLs.

As one of the notable banks in Indonesia, ABC Bank has prepared the procedures of NPL settlement. One of them is the Write-Off Ceiling Incentive Policy. According to the Directors Decree of ABC Bank in 2011 regarding the breakdown of write-off ceiling in the retail segment, this policy was set to motivate account officers of ABC Bank in collecting the loans, including the bad-debts or NPLs. If the NPLs can be recovered, the health of the bank, as well as its profit, will get better. Therefore, ABC Bank must anticipate the occurrence of NPLs and encourage its account officers to recover the existing bad-debts. In addition, ABC Bank also has to anticipate the risks

DOI 10.1201/9781003196013-28

that may affect the effectiveness of the policy implementation in order to increase the income of NPL recovery.

2 LITERATURE REVIEW

2.1 *Credit service in banking institutions*

Act of the Republic of Indonesia Number 10 of 1998 concerning Banking stated that the banking institutions are in charge of collecting funds from people and distributing them for social prosperity. Based on this statement, one of the main functions of banking institutions is to distribute funds in form of credit lending. According to Financial Service Authority Regulation Number 40/POJK.03/2019 regarding the Assessment of Commercial Bank Asset Quality, credit is a lending service (in form of money) between the bank and the third party following a contractual loan agreement. This agreement obligates the debtors to fulfill their obligations regarding the lending contract, including the loan principal, the loan interest, etc.

2.2 *Risks in banking institutions*

According to Financial Service Authority Regulation Number 18/POJK.03/2016 concerning the Implementation of Risk Management for Commercial Bank, the commercial banks are dealing with 8 types of risks, covering:

1. Credit risk, which is a risk of losses due to the third parties' failure to meet their obligations to the banks. Credit risks consist of debtors' failure to meet contractual agreements regarding lending—which causes default and the occurrence of nonperforming loans, counterparty credit risk, and settlement risk. According to Bessis (2015), credit risks consists of the deterioration of the debtors' credit standings—which causes a higher likelihood of default; recovery risks—due to the uncertain recovery values under the default; and counterparty credit risks—the potential occurrence of transaction loss when a party involved defaults.
2. Market risk, which is a risk of losses due to the change of market conditions. According to the Indonesian Bankers Association (2015), market risk is defined as the risk of change in market price in portfolio position and administrative accounts. According to Bessis (2015) market aspects that affect the condition consist of interest rates, equity indexes, or foreign exchange rates.
3. Liquidity risk, which is a risk of losses due to the banks' failure to meet the obligations at the due date from funding sources—cash and its equivalent.
4. Operational risk, which is a risk of loss due to the inadequacy and/or dysfunction of internal processes, human error, system failure, and/or external events that affect bank operations.
5. Compliance risk, which is a risk of losses due to the banks' failure to comply with the regulations, acts, and other requirements for banking institutions. According to the Indonesian Bankers Association (2015), this risk is related to capital adequacy, the valuation of productive assets quality, the impairment of losses reserve, and other risks related to certain policies.
6. Legal risk, which is a risk of losses due to the legal lawsuit and/or weakness in juridical aspects. The causes of legal risks are the unavailability of the constitutional regulations—to support the business policy—and the inadequacy of the contractual binding process (Indonesian Bankers Association (2015).
7. Reputation risk, which is a risk of losses due to the decrease of stakeholders' trust due to the negative perceptions regarding the bank. According to Solvency II, the Comite' Europe's des Assurances (CEA), and the Groupe Consultatif Actuariel Europeen (2007) in Gatzert et al. (2016), reputation risk happens due to business events that trigger the occurrence of adverse publicity, and also from other risks inherent in the business activity.
8. Strategic risk, which is a risk of losses due to the inaccuracy in decision making and failure to anticipate the change in the business environment.

It is important for banking institutions to have an adequate risk management system to reduce the loss possibility due to the occurrence of risks arising from all Bank's business activities. Indonesian Bankers Association (2015) stated that while it is obligated by the authority, isk management is also necessary for banking institutions to create their value and reach their targets.

3 RESEARCH METHOD

This study investigated the risks that affect the effectiveness of the Write-Off Ceiling Incentive Policy implementation in ABC Bank. It is carried out by (1) interviewing three expert panels of ABC Bank to learn about the practice of the policy as well as the risks, and (2) conducting a literature study to find out the ideal standard of policy implementation based on regulations.

The interviewees, namely the three expert panels of ABC Bank, consist of 2 (two) Executive Vice Presidents and 1 (one) Senior Manager. The tenures of both Executive Vice Presidents are at least twenty years, while that of the Senior Manager is at least ten years. Therefore, they are expected to show their competencies and experiences with this policy.

4 RESEARCH FINDINGS

Based on the interview with the representatives of ABC Bank, it was found that the risks affecting the effectiveness of the Write-Off Ceiling Incentive Policy implementation in ABC Bank are as follow:

1. Operational risk
   According to one of the Executive Vice Presidents of ABC Bank, the operational risks are related to the limited number of account officers and the miscommunication between two types of account officers.
   "The limited numbers of account officers will result in the work overload, as they are responsible not only for credit collection but also for credit expansion. Therefore, ABC Bank divided its account officers into two types: the marketing account officers and the non-performing loan account officers. However, such division also causes miscommunication."
2. Legal risk
   The legal risks are in connection with the occurrence of legal cases during the collection of non-performing loans. According to the Senior Manager of ABC Bank, the following is an example of legal risk regarding the recollection of NPLs.
   "...Such cases occur when a credit settlement requires the auction of the collateral asset, but the debtors are unwilling to give up their assets. This unwillingness causes the debtors to file a lawsuit against ABC Bank. Therefore, the NPL collection has to be postponed until the lawsuit is over."
3. Reputation risk
   According to the Senior Manager of ABC Bank, the reputation risks are associated with the uncooperative manner of debtors and the spread of false news.
   "...There was a case in which a debtor's house was about to be auctioned. However, s/he refused to leave his/her house. S/he spread false news claiming that s/he was mistreated by ABC Bank, while the truth was that s/he was the one breaching the contracts."
4. Compliance risk
   According to one of the Executive Vice Presidents of ABC Bank, compliance risks occur when the authority set regulations prioritizing credit relaxation over credit collection.
   "...For example, when the Financial Service Authority released Regulation Number 11/POJK.03/2020 concerning the National Economic Stimulus as a Countercyclical Policy on the Impact of the Covid-19 Pandemic Outbreak, the NPL collection will be less effective than it was during the normal period as the convenience of credit is prioritized. Consequently, the NPL recovery after the release of this regulation will be lower than before."

5 CONCLUSION AND SUGGESTION

It is an undeniable fact that NPLs occur and cannot be avoided by banking institutions. To reduce the number of NPLs to a tolerable range, the Write-Off Ceiling Incentive Policy was set and implemented in ABC Bank. However, some risks affect the effectiveness of the policy implementation in ABC Bank. Those risks are (1) operational risk; (2) legal risk; (3) reputation risk; and (4) compliance risk.

Effective risk management is necessary to reduce the negative impact of those risks on the collection of the NPL recovery. Specific actions are required to handle each risk in the implementation of the Write-Off Ceiling Incentive Policy in order to increase the NPL recovery and the effectiveness of the policy implementation. Operational risk can be reduced by recruiting more account officers and bridging the communication between two types of account officers to reduce the miscommunication. Legal risk can be reduced by making sure that the credit contract is bonded properly, including the administrative files completeness and the documentation during the credit contract binding. Compliance risk can be reduced by setting procedures in credit settlement—based on the regulation—that can benefit both debtors and ABC bank. Meanwhile, reputation risk can be reduced through preventive actions. Further research can be conducted by discussing the risk management process to increase the NPL recovery.

REFERENCES

Bessis, J., 2015. Risk management in banking.pdf, Fourth. ed. John Wiley & Sons, West Sussex.

Elekdag, S., Malik, S., Mitra, S., 2020. Breaking the Bank? A Probabilistic Assessment of Euro Area Bank Profitability. J. Bank. Financ. 120, 105949.

Gatzert, N., Schmit, J.T., Kolb, A., 2016. Assessing the Risks of Insuring Reputation Risk. J. Risk Insur. 83, 641–679.

Hariyani, I., 2010. Restrukturisasi dan penghapusan kredit macet. Elex Media Komputindo, Jakarta.

Indonesian Bankers Association (2015). Manajemen Risiko 1. PT Gramedia Pustaka Utama, Jakarta. Otoritas Jasa Keuangan, 2016. Peraturan Otoritas Jasa Keuangan Nomor 18/POJK.03/2016.

Partovi, E., Matousek, R., 2019. Bank efficiency and non-performing loans: Evidence from Turkey. Res. Int. Bus. Financ. 48, 287–309.

Tentang Penerapan Manajemen Risiko Bagi Bank Umum 1–31.Triandi, A., 2018. Analisis Tentang Restrukturisasi Kredit untuk Menghindari Terjadinya Kredit Macet (Studi pada PT). Bank Rakyat. Repos. Institusi USU. Universitas Sumatera Utara.

Zhang, D., Cai, J., Dickinson, D.G., Kutan, A.M., 2016. Non-performing loans, moral hazard and regulation of the Chinese commercial banking system. J. Bank. Financ. 63, 48–60.

Contemporary Research on Business and Management – Noviaristanti (Ed.)

Business in times of crisis: A strategic update during COVID-19

F. Faidal
Diploma School of Entrepreneurship, Faculty of Economics and Business, University of Trunojoyo Madura, Indonesia

ABSTRACT: We report on turning policymakers' attention to saving people's lives. At the same time, the pandemic crisis is threatening the survival of firms on a global scale, with potentially devastating societal and economic outcomes by novel exogenous shocks, it is shocking transcends past experiences and the implications for SMEs. Companies must innovate and adapt because failing to respond fast could result in great losses. Put succinctly, in the face of a pandemic, business model innovation is unavoidable. The reasons behind these shifts can be multifaceted: changing the product line to keep operations going, turning threats into opportunities, diversifying customer segments, scoring competitive advantage, boosting reputation, or cutting costs, to name a few. However, not all businesses have the luxury of pivoting their business models overnight.

1 INTRODUCTION

Data from the Ministry of Cooperative and Small Medium Enterprises (SMEs) show that in the year 2018, there were 64,194,057 SMEs in Indonesia, which employed 116,978,631 workers. Indonesia is dominated by SMEs; they have become the backbone of its economy. They were seriously affected in terms of not only their production and income, but also the amount of energy and work that must be reduced because of this pandemic (Pakpahan, 2020). SMEs have less resilience and flexibility in the face of a pandemic, which is due to several things like low level of digitization, difficulty in accessing the technology, and a lack of understanding about the strategy to survive in the business world (OECD, 2020). SMEs are required to able to adjust themselves because businesses that survive are businesses that are responsive to the development of the times.

During the Covid-19 pandemic, the government issued Government Regulation No. 21 of 2020 on the Limitation of Large-Scale Social Interactions to restrict the movement of people, and they were expected to remain at home if they did not have urgent matters to attend to. It had an impact on those SMEs whose operations were limited because of the reduction of consumers. In the research by Hardilawati (2019) and Setyorini et al. (2019), it was concluded that e-commerce has a positive and significant effect on improving marketing performance and income of SMEs.

SME actors also need to adjust themselves and to condition the sale of goods and products. Improvement of the quality of products and customization of services are required to attract consumers. According to Gary (2013), "the quality of products is their ability to demonstrate overall durability, reliability, accuracy, ease of operation, and repairability". According to Tjiptono (2011), quality of service is the level of excellence that SMEs are expected to show in meeting and conforming to the wishes and expectations of the customers. According to the research by Tripayana and Pramono (2020) and Lestari and R (2019), the quality of products and quality of service significantly impact the purchases of consumers and increase customer satisfaction and customer loyalty.

This research is carried out to answer what SMEs can do and the strategies they can use to be able to maintain their business and be responsive to changes in the business climate during the Covid-19 pandemic.

DOI 10.1201/9781003196013-29

2 METHOD

This research uses a qualitative method. According to Semiawan (2010), the method of research qualitative is a kind of method of research that is most appropriate to capture the perception of man only with the contact directly and mind open, and through the process of inductive and interaction symbolic humans can know and understand something. The research was conducted in Pekanbaru, with SMEs being the object of research. The data collection technique used in this research is a participatory observation with an exploratory step, that is, using one of the recommended qualitative data collection techniques to obtain descriptive data (Gunawan, 2017). Sources of data used are primary data in the form of observations and secondary data in the form of data collected, processed, and presented by other parties in the form of books and previous research results related to SMEs. After doing participatory observation and conducting research, data associated with the theory, the opinion of the experts, and the results of the previous study were analyzed. In the next section, the results of the findings of the study and recommendations based on the results that could be adopted by SMEs are provided.

3 DISCUSSION

Based on the results of observations, the average SMEs, especially those in the culinary field, experienced a decrease in turnover during the Covid-19 pandemic. It was due to the decrease in people's activity outside their home, difficulty in obtaining raw materials due to transport constraints, and decline in people's confidence in products that were prepared outside their homes. SMEs are one of the buttresses the economy because too much supply fieldwork, with their Covid-19's, also began there who do layoffs or laying off employees while for companies/businesses they should be closed while the time.

According to the findings of other researchers based on observation, not all SMEs experienced a drop in turnover of sales or had to shut down their business; there are SMEs that are still stable and that experienced an increase in turnover of sales because they adapt themselves in terms of their products and use marketing strategies to survive. Several things can be done by SMEs, including choosing an open line of products and using a new or renewed system of marketing them, because businesses that survive are the ones that respond well to changes in their environment. Some of the things that can be done by SMEs are as follows:

E-commerce: Research studies by Hanum and Sinarasri (2017) and Ningtyas et al. (2015) show that e-commerce has a positive and significant effect on improving the performance of SMEs. E-commerce yang carried out by SMEs associated with a reduction in the cost of transaction and coordination activities of the economy that is more close between colleagues' business. Technology can be used in business operations to reduce costs and attain company objectives. Hoffman and Fodor (Pradana, 2016) states that SMEs can apply the principles of 4C: connection, creation, consumption, and control. They can improve return on investment (ROI) of the companies, and it can be measured by getting feedback or reviews from the consumer, who can share or recommend the product they used to other people.

The main purpose of using e-commerce by the SMEs is to increase profits, their market share, and the number of areas where sales is conducted; with the use of e-commerce, SMEs can obtain new consumers. SMEs can use e-commerce as a portal not only to sell but also to build relationships; they can also make it a platform for learning. SMEs can also observe how their competitors do sales and adopt their methods.

3.1 *Digital marketing*

Digital marketing is promotional activity and market search through digital media online by utilizing various means such as social networking. (Purwana et al., 2017). Businesspeople use digital marketing to market their products via social media such as Instagram, Facebook, and Twitter.

Besides, digital marketing can also be done on e-commerce and many other media. The rapid development of technology makes digital marketing a tool that must be understood and studied by SMEs.

Research by Hendrawan et al. (2019) shows that digital marketing has a positive and significant effect on increasing the sales performance of SMEs. 70 percent of employers say digital marketing will become a major platform for all communications in terms of marketing and offline stores will be complementary because of the power of digital marketing to reach more consumers. It is also in line with the research that was carried out by Purwana et al. (2017), who stated that businesspeople must use digital marketing to develop their business. SMEs can also create social media accounts and do promotions regularly to increase trust in the company; they can also increase creativity in marketing.

When using digital marketing, SMEs are forced to always learn and be open to technology, which is rapidly growing. They should also use suitable media and the way of communication that is appropriate to the selected market segment, so that marketing will be more effective and not target the wrong audience.

3.2 *Product and service quality improvement*

Research studies by Sustainable and R (2019) and Tripayana and Pramono (2020) stated that the improvement of the quality of product and quality of service has a positive and significant effect on shaping the satisfaction of consumers and creating customer loyalty. In the future, businesspeople, during pandemics, must pay attention to the dimensions of the quality of their products and improve them; it will increase consumers' confidence.

Product quality is defined as the ability of the product to meet consumer needs and desires (Kotler, Philip, and Armstrong, 2012).

It is important for SMEs to improve of quality of products regularly to meet the needs, desires, and expectations of the consumers. According to Garvin (1998), there are eight dimensions in determining the quality of the product: (1) performance of the product, (2) additional features or attributes that complement and enhance product functionality, (3) reliability or the ability of the products to survive changes in the business environment in the period specified, (4) suitability or how good the products are with respect to the standard that exists in the industry, (5) power resistance or durability of products in terms of technical and economic value or how quickly they can be repaired if problems crop up, (7) product aesthetics, that is, how the product is seen, heard, or felt, and (8) perceptions of product quality, which includes brand reputation and other factors that can influence consumer perceptions.

3.3 *Customer Relationship Marketing (CRM)*

In the future, during a pandemic, SMEs should focus not just on obtaining new customers, but also on retaining existing customers, increasing their satisfaction and loyalty. Loyal customers of SMEs will not buy products from their competitors because they already have confidence in their products. One of the ways the SMEs can survive the decline in business operations is to do customer relationship marketing. Customer relationship marketing is a concept of strategic marketing that seeks to establish a long-term relationship with the customer such as maintaining a solid relationship that is mutually beneficial to both; it can increase sales and create loyal customers. According to research by Farida et al. (2017), customer relationship marketing has a positive and significant effect on improving the marketing performance of SMEs through improving the quality of relationships and entrepreneurial orientation. The better the quality of the relationship between SMEs and their consumers, suppliers, and others, the better their ability to improve their marketing performance. Also, business actors who dare to take risks already have experience in business and are flexible toward business, can increase networks, and foster trust from consumers, so that consumers will be retained. The results obtained in the study by Hardilawati (2019) are different, who stated that CRM has a positive but not significant effect on the improvement of the performance of SMEs. It is because the number people SMEs have is limited, and they cannot carry out CRM to the desired extent.

4 CONCLUSION

The Covid-19 pandemic caused instability in the Indonesian economy, especially in SMEs. SMEs felt the impact immediately in the form of decrease in turnover of sales due to the large-scale social restrictions announced by the government; people were urged to stay at home, which led to a temporary halt in SME business operations. In the future, SMEs must have strategies in place to survive a pandemic and adjust to the conditions they find themselves in.

There are several recommended strategies SMEs can use to survive a pandemic and maintain their business operations: (1) conducting sales via e-commerce because people are now familiar with online shopping, (2) using digital marketing to obtain new customers, (3) providing products and services that are of excellent quality, and (4) conducting customer relationship marketing to increase the confidence of consumers and foster their loyalty.

5 SUGGESTION

The suggestions that can be recommended are as follows:

SMEs can adopt strategies that have been mentioned in this study to survive a pandemic; they are expected to adapt themselves to changes in their business environment because businesses that survive a pandemic are the ones that respond well to changes in their environment better in terms of their products, marketing, and sales systems as well as the use of technology that supports the business. The government authorities are expected to continue to provide education in the form of socialization or training to the businesspeople and establish a network of communication for SMEs that can be easily monitored, so that people who are employed in SMEs can keep honing their skills. Also, researchers in the future can expand the study and see the effectiveness of the strategies SMEs can use to survive a pandemic that were formulated in this study.

REFERENCES

Farida, N., Naryoso, A., & Yuniawan, A. (2017). Model of Relationship Marketing and E-Commerce in Improving Marketing Performance of Batik SMEs. Journal of Management Dynamics, 8 (1), 20–29. https://doi.org/10.15294/jdm.v8i1.10408

Garvin, DA (2008). Managing Quality: The Strategic and Competitive Edge. The Free Press. Gary, PK, and. (2013). Marketing Management (14th ed.). Pearson Education Limited.

Gunawan, I. (2017). Qualitative Research Method. In Bumi Aksara (5th ed.).

Hanum, AN, & Sinarasri, A. (2017). Analysis of the factors that influence the adoption of e-commerce and their influence on the performance of SME. Maksimum Journal, Vol. 1 (No. 1), 1–15.

Hardilawati, WL (2019). Customer Relationship Marketing Model, Innovation, and E-Commerce in Improving SME Marketing Performance in Pekanbaru. Journal of Accounting and Economics, 9 (2), 213–222.

Hendrawan, A., Sucahyowati, H., Cahyandi, K., Indriyani, & Rayendra, A. (2019). Influence Digital Marketing Towards Asti Gauri SMES Product Sales Performance in Bantasari Cilacap District. Administration and Secretarial Journal, 4 (1), 53–60.

Kotler, Philip, and Armstrong, G. (2012). Principles of Marketing (15th ed.). London: Pearson. Ningtyas, PK, Sunarko, B., & Jaryono. (2015). Analysis of Factors That Affect Adoption of E-Commerce And Its Effect On SME Performance, UNSOED Journal, 21, 95–107.

Pakpahan, AK (2020). COVID 19th and Implications For Enterprises Micro, Small, and Medium Enterprises. Journal of Entrepreneurship, 24(4), 1–8.

Purwana, D., Rahmi, R., & Aditya, S. (2017). Utilization of Digital Marketing For Businesses Micro, Small, and Medium Enterprises (SMEs) in the Village Malaka Sari, Duren Sawit. Journal of Civil Society Empowerment (JPMM), 1 (1), 1–17. https://doi.org/10.21009/jpmm.001.1.01

Setyorini, D., Nurhayati, E., & Rosmita. (2019). Effect ofTransaction Online (e-Commerce) Against Increased Profit SMEs (Study Case SMEs Processing Iron Ciampea Bogor, Jawa Barat). Journal of Management Partners, 3 (5), 501–509.

Tjiptono, Fandy. (2011). Strategi Pemasaran. Edisi 3. Yogyakarta: ANDI.

Tripayana, S., & Pramono, J. (2020). Quality Product, Service, and Loyalty Customers Where Satisfaction For Variable Intervening in the SME Start-Up Tourism Kombucha Brewing Co., JEMAP Journal, 1(1), 22–45.

Contemporary Research on Business and Management – Noviaristanti (Ed.)

Understanding the influence of a parasocial relationship and beauty influencers' credibility on brand credibility and repurchasing intention on the media platform Instagram

F.F. Fadilah & Y. Alversia
Faculty of Economics and Business, Universitas Indonesia, Jakarta, Indonesia

ABSTRACT: Nowadays, digital influencers are being hired by many companies to market their brands in the form of electronic word of mouth strategy. The companies do this because they consider this as a tool to provide information about their brands to a broader section of people. This study investigates how the perceived attributes of digital influencers on Instagram can influence their parasocial relationship and credibility that could impact brand credibility and repurchasing intention of consumers. This study employed a quantitative approach with a total of 243 samples. The data were collected using purposive sampling and analyzed using the structural equation method (SEM). It was found that attractiveness and attitude homophily were positively related to the parasocial relationship and influencers' credibility. On the other hand, influencers' credibility was positively related to brand credibility. It was shown that brand credibility and parasocial relationship positively influenced consumers' repurchasing intention. However, no significant effect of influencers' credibility on the other variables can be found.

1 INTRODUCTION

The increasing use of technology and the internet, including the use of social media, makes communication and information easier to access. Many businesses take advantage of social media as a marketing strategy to advertise or promote their products to attract and retain users. Constantinides (2014) produced a marketing strategy through passive and active approaches, who explained that an active approach is a strategy that implements social media as a marketing and public relation platform. In this case, Childers, Lemon, and Hoy (2018) explain that it can be done by making digital influencers as endorsers of a company's brand or products. Non-celebrity endorsers, according to Abidin (2016), are known by only a certain group of people and are considered to have more power in the online context because they are considered to be more credible and approachable (Djafarova and Rushworth, 2017). The non-traditional celebrities will then be referred to as a social media influencer in the realm of social media.

Many people who followed a social media influencer feel that there is a strong social relationship between them. This is called a parasocial relationship. A parasocial relationship can be predicted from several determinants, including attractiveness and attitude homophily. According to Colliander and Dahlen (2011), when the followers feel that they have similarities with digital celebrities, along with repeated exposures to the latest content uploaded on their social media, the digital celebrities will be perceived as a reliable source of information and their advertisement will be viewed as a credible purchasing information that will influence their purchasing decision assessment. Influencers who are used as a form of communication tool will construct the credibility they have as a measurement of the credibility of the brand that they are promoting. In other words, the credibility of an endorser will be transferred to the brand being promoted (Wang et al., 2017). In this case, parasocial relationships, influencers' credibility, and brand credibility will lead users to want to purchase the product. In this study, purchasing intention will be intended for those who have previously purchased a product or have a repurchasing intention.

DOI 10.1201/9781003196013-30

The purpose of this study was to see the extent of influence that the beauty influencers have as communicators between the brand and their followers on their Instagram page. It was conducted by investigating the parasocial relationship and the credibility of beauty influencers from their followers, which could influence the brand credibility. In addition to the influence of the parasocial relationship, the credibility of a beauty influencer, and brand credibility on repurchasing intention of the products promoted by the beauty influencer, this study also identifies the establishment of a long-term relationship between the brand or products and their consumers.

2 LITERATURE REVIEW

The theory of the parasocial relationship was first introduced by Horton and Wohl (1956) and was originally defined as an experience of a face-to-face relationship between the audience and performers on television, movie, and radio. Nowadays, a parasocial relationship is inferred as psychological associations built by media users unilaterally with the characters or celebrities of the media being used. A parasocial relationship can be seen from several determinants, including attractiveness. It occurs because social media influencers have an attractive value and have something in common with their followers (Giles, 2002). In this case, Rubin and Step (2000) divided attractiveness into physical and social attractiveness. Another determinant used in a parasocial relationship is the attitude homophily or the similarity of behavior between the beauty influencers and their followers (Giles, 2003; Eyal and Rubin, 2003; Lee and Watkins, 2016). Therefore, the following hypotheses were developed:

H1: Physical attractiveness has a significant positive effect on the parasocial relationship.
H2: Social attractiveness has a significant positive effect on the parasocial relationship.
H3: Attitude homophily has a significant positive effect on the parasocial relationship.

Credibility is defined as a source that is perceived as having knowledge, skill, or experience relevant to a communication topic and can be trusted to give an unbiased opinion or present objective information on the issue (Belch and Belch, 2012). Credibility is one of the important characteristics in which the audience can feel that the communicator has an impact on their decision to purchase products. Physical attractiveness in the beauty industry can be considered as an important selling point. On the other hand, attitude homophily or similarity in behavior can also be considered as a determinant factor of credibility because of the perceived value shared between the followers and the influencers (Sokolova and Kefi, 2019). Therefore, the following hypotheses were developed:

H4: Physical attractiveness has a significant positive effect on beauty influencers' credibility.
H5: Attitude homophily has a significant positive effect on beauty influencers' credibility.

Influencers who are used as a form of communication tool will make their credibility as a measure of the credibility of the brand they are promoting. Brand credibility is the believability of the position of a product owned by a brand. The credibility of a brand depends on the willingness and ability of the company to deliver what has been promised (Erdem et al., 2006). According to Wang et al. (2017), when the endorsement consists of several signals from the brand being promoted, the credibility of the endorser will then be transferred to the brand. Therefore, the following hypothesis was developed:

H6: The credibility of beauty influencers has a significant positive effect on the credibility of the brand being promoted.

Repurchasing intention is a decision made by an individual to repurchase a product or a service from the same company by considering the current and possible circumstances (Hellier et al., 2003). In this study, repurchasing intention was used as a form of anticipation if consumers or respondents have previously made purchases of products or brands that are promoted or recommended by the beauty influencers even before the influencer promotes or recommends it. Sokolova and Kefi (2019) stated that the parasocial relationship and the credibility of beauty influencers can influence their followers to buy the product being promoted if the influencers could be relied on and trusted (Wathen and Burkell, 2002). Meanwhile, brand credibility, according to Jeng (2016), is a signaling tool that can increase consumer's decision confidence and increase their buying interest. Therefore, the following hypotheses were developed:

H7: The parasocial relationship has a significant positive effect on the repurchasing intention of the product being promoted.
H8: The credibility of beauty influencers has a significant positive effect on the repurchasing intention of the product being promoted.
H9: Brand credibility has a significant positive effect on the repurchasing intention of the product being promoted.

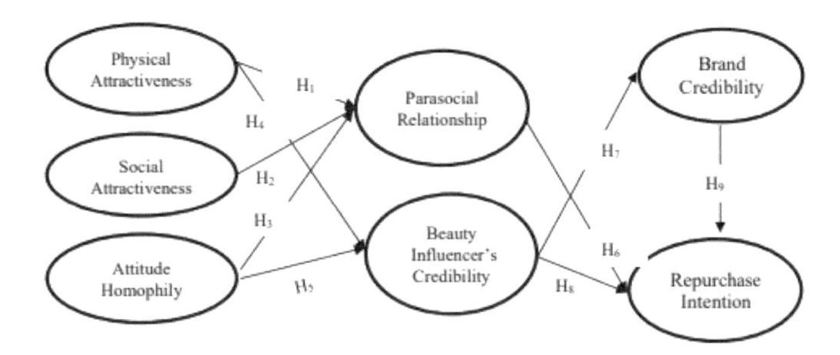

Figure 1. Research model.

3 RESEARCH METHOD

A quantitative approach was used in this study by conducting surveys through online questionnaires, and the results were analyzed using the Likert scale 1–6 (1 = strongly disagree; 6 = strongly agree). This study was conducted in Indonesia, consisting of all active users of Instagram without gender restrictions. Using the purposive sampling technique, sampling will be determined by several characteristics such as the following: at least 18 years old, an active user of Instagram, a follower of at least one beauty influencer on Instagram, and has bought beauty and body care products (makeup, skincare, or sun care) that has been promoted or recommended by the beauty influencer on Instagram in the last six months. Within a week of data collection, 277 respondents had filled out the survey. After the screening process, 243 surveys were selected to be used in this study.

4 RESULTS AND DISCUSSION

To test the hypotheses, this study employed a structural equation model (SEM) using AMOS 24. Physical attractiveness (C.R = 4.628, $p < 0.001$), social attractiveness (C.R = 3.409, $p < 0.001$), and attitude homophily (C.R = 4.361, $p < 0.001$) showed a significant positive effect on the parasocial relationship between beauty influencers and their followers. Physical attractiveness and attitude homophily also showed a significant positive effect on the influencers' credibility, with values of C.R = 6.841, $p < 0.001$ and C.R = 5.601, $p < 0.001$, respectively. Influencers' credibility (C.R = 6.364, $p < 0.001$) showed a significant positive effect on brand credibility, but not on repurchasing intention (C.R = 0.027, $p > 0.500$). However, repurchasing intention was influenced by both the parasocial relationship (C.R = 3.062, $p < 0.005$) and brand credibility (C.R = 8.026, $p < 0.001$). Most of the proposed hypotheses are supported, except for H8.

A parasocial relationship could influence a repurchasing intention from a user or a follower because from this relationship, audiences or followers can feel that they have the same values and behavior as the beauty influencers. Therefore, they will feel the closeness to those influencers and will buy products because of the perceived closeness, which is stated and conceptualized (Sokolova and Kefi, 2019). Influencers here are also endorsers and the endorsement consists of several signals from a brand being promoted. Therefore, the influencers' credibility will be transferred to the promoted brand (Wang et al., 2017). According to Swait and Sweeney (2008), brands can provide additional value to companies, one of which is increasing the bond between the company and consumer, leading to a long-term relationship. In other words, it can be said that high brand credibility can increase consumer's commitment and sustainability.

5 CONCLUSION AND IMPLICATION

The results of this study showed that repurchasing intention was only affected by the parasocial relationship and brand credibility and not by influencer's credibility. It was found that companies that were using Instagram as a platform and beauty influencers as endorsers of their product or brand could be benefited from a parasocial relationship built by the beauty influencer with their followers, which would lead to a long-term relationship between companies and their consumers. The findings are expected to be useful for the beauty influencers to develop their ability to promote and recommend the beauty products or brands they use to influence their followers on Instagram.

REFERENCES

Abidin, C. (2016). Visibility labour: Engaging with In?uencers' fashion brands and #OOTD advertorial campaigns on Instagram. *Media International Australia*, vol. 161, issue 1.

Belch, G. E., & Belch, M. A. (2012). *Advertising and Promotion. An Integrated Marketing Communication Perspective. 10th ed.*, McGraw-Hill, Boston.

Childers, C. C., Lemon, L. L., & Hoy, M. G. (2018). #Sponsored #Ad: Agency perspective on in?uencer marketing campaigns. *Journal of Current Issues & Research in Advertising*, vol.1, issue 17.

Colliander, J., & Dahlen, M. (2011). Following the fashionable friend: The power of social media – Weighing Publicity effectiveness of blogs versus online magazines. *Journal of Advertising Research*, vol. 51.

Constantinides, E. (2014). Foundation of Social Media Marketing. *Procedia-Social and Behavioral Sciences*, vol. 148.

Djafarova, E., & Rushworth, C. (2017). Exploring the credibility of online celebrities' Instagram pro?les in in?uencing the purchase decisions of young female users. *Computers in Human Behavior*, vol. 68.

Erdem, T., Swait, J., & Valenzuela, A. (2006). Brand as Signals: A Cross-Country Validation Study. *Journal of Marketing,* vol. 70, issue 1.

Eyal, K., & Rubin, A. M. (2003). Viewer aggression and homophily, identification, and parasocial relationships with television characters. *Journal of Broadcasting & Electronic Media*, vol. 47, issue 1.

Giles, D. C. (2002). Parasocial interaction: a review of the literature and a model for future research, *Media Psychology*, vol. 4.

Hellier, P. K., et al. (2003). Customer Repurchase Intention: A General Structural Equation Model. *European Journal of Marketing,* vol 37, issue 11/12.

Jeng, S. (2016). The influences of airline brand credibility on consumer purchase intentions. *Journal of Air Transport Management,* vol. 55.

Lee, J. E., & Watkins, B. (2016). YouTube vloggers' influence on consumer luxury brand perceptions and intentions. *Journal of Business Research*, vol. 69, issue 12.

Rubin, A. M., & Step, M. M. (2000). Impact of motivation, attraction, and parasocial interaction on talk radio listening. *Journal of Broadcasting & Electronic Media*, vol. 44, issue 4.

Sokolova, K., & Kefi, H. (2019). Instagram and YouTube bloggers promote it, why should I buy? How credibility and parasocial interaction influence purchase intention, *Journal of Retailing and Consumer Services*.

Wang, S. W., Kao, G. H., & Ngamsiriudom, W. (2017). Consumers' attitude of endorser credibility, brand and intention with respect to celebrity endorsement of the airline sector. *Journal of Air Transport Management,* vol. 60.

Wathen, C. N., & Burkel, J. (2002). Believe it or not: Factors Influencing Credibility on the Web. *Journal of American Society of Information Science and Technology,* vol. 53, issue 2.

Contemporary Research on Business and Management – Noviaristanti (Ed.)

Business coaching: Implementation of the push strategy and digital marketing improvement in an Indonesian MSME restaurant during the COVID-19 pandemic

F. Amelita & H. Suhaimi
Faculty of Business and Economics, University of Indonesia, Indonesia

ABSTRACT: The increasing number of micro, small, and medium enterprises (MSMEs) in Indonesia has contributed positively to Indonesia's economy. However, many MSMEs are unable to successfully manage their business. To address this issue, this study was conducted focusing on an Indonesian MSME restaurant that was experiencing difficulties in managing its business during the pandemic. It aims to identify and analyze the MSME's push strategy and digital marketing promotional activities and to find solutions to improve its strategy and marketing promotion. One of the obstacles was in managing its marketing strategy and activities by implementing a strategy to increase its promotion. The data were collected using business coaching methods and qualitative research to investigate the actual condition and problems. Later, the actions that were approved by the MSME were taken to overcome the problems. After conducting several external and internal analyses, a push strategy was applied to increase consumers' awareness of the MSME's products. In addition, MSME's digital marketing promotion through Instagram and a website was optimized to promote its products. It was proven that the promotion successfully improved their exposure and customer engagement.

1 INTRODUCTION

According to the Indonesian Young Entrepreneurs Association, the COVID-19 pandemic has affected the culinary industry and has decreased its turnover by 30% (Hamdani, 2020). Bunda Bread, which was established in 2017, is an example of the MSMEs that were competing in the food and beverage sector during COVID-19. Entrepreneurs want their business to grow so that the business could continue to run. Attracting new customers is important for small and medium enterprises to sustain future business growth (Klaassen, 2016). However, MSMEs are challenged to find a method for building awareness to accelerate their business growth during the pandemic. According to Satell (2015), to build awareness, a brand could focus on a push strategy. The push strategy is used to promote products to get consumer attention. To increase customer awareness that is affected by the pandemic, Bunda Bread could add promotional activities by giving product information and services to customers. MSMEs are generally not widely known because their primary marketing relies on word of mouth without maximizing digital marketing through social media, including Instagram and websites. Increasingly developing technology has greatly affected marketing practices in various parts of the world. As a result, digital marketing and traditional marketing began to be integrated. In this transitional era, for the integration of digital and traditional marketing, a new marketing approach is needed, known as Marketing 4.0 (Kotler et al., 2016).

Although there are many problems faced by MSMEs, this study only focuses on the improvement of several aspects that are important and feasible for culinary MSMEs to improve, namely, push strategy and digital marketing. The push strategy could increase customer awareness, which can significantly affect product improvement. Meanwhile, digital marketing promotion through social

DOI 10.1201/9781003196013-31

media is considered one of the most cost-effective methods to generate exposure, increase upselling, build a partnership, and improve customer engagement. Based on the mentioned background, this study aims to (1) identify and analyze the actual condition of the MSME's push strategy and digital marketing promotion activities and (2) seek and implement the feasible solution to improve the MSME's push strategy and digital marketing promotion.

2 LITERATURE REVIEW

2.1 *Push strategy*

Belch and Belch (2003) defined a push strategy as a way to push products made by a company through distribution channels that they use aggressively to sell and promote to the consumers. One example of a push strategy is to use personal selling by salesmen. According to the Corporate Finance Institute (2020), the push strategy refers to a strategy in which a company tries to bring its products to consumers and 'push' these products to them. It is usually used to receive and increase product exposure, product demand, and consumer awareness. An example of using a push strategy is to bring the products to the consumers. The consumers are introduced to, or reminded of, the products through one of several available marketing methods, such as point of sale (POS) displays and direct selling to customers.

2.2 *Digital marketing*

Digital marketing is a term used to describe all marketing activities that utilize digital channels to promote products (Batra & Keller, 2016). Channels that are commonly used are websites, search engine marketing, social media marketing, content marketing, email marketing, mobile marketing, and banner advertising. Digital media are used as a tool to promote products to increase brand awareness.

2.2.1 *Instagram*

Instagram is a great medium for advertising because it can tell a story about a product or brand in a visually interesting way (Ištvanic et al., 2017). According to Macharty (2018), an easy method to attract Instagram followers is by maximizing the 150 word limit on the Instagram bio. Instagram users should put a website or point-of-sale place on their profile page that could be visited with just one click. In addition, the profile picture on the Instagram page is also important; it should not show only a blank image. It is also suggested to use a caption on each photo update to describe the brand personality and its unique tone of voice. In addition, placing a hashtag (#) in the description would place the uploaded photo with other photos containing the same hashtag and turn the hashtag into a clickable link that displays the associated set of photos. Then, the most important thing about content on an Instagram profile is to post a bright and sharp photo. On Instagram, all contents have a single personality that coordinates to all content. Therefore, to coordinate all the content, it is necessary to have a planogram that could tell a story about the brand value.

2.2.2 *Website*

A website is a form of online marketing. If the company website is prepared properly, it would be easier to be found on search engines, such as Google, and would be accessible to consumers who have purchasing intention (McPheat, 2011). Companies can use the 7C framework as a guide for designing a customer interface (CI) when creating a website, namely, context, content, community, customization, communication, connection, and commerce (Rayport & Jaworski, 2001).

3 RESEARCH METHODOLOGY

This study employed a descriptive and qualitative method and also used a business coaching method. Both the primary and secondary data were acquired from interviews, observations, and Bunda Bread's internal data. The business coaching was implemented for one year through four steps, namely, (1) internal and external analysis; (2) TOWS, gaps analysis, and Pareto analysis; (3) alternative solutions and decision making; and (4) implementation and monitoring.

4 RESULTS AND DISCUSSION

Several tools and frameworks were utilized to analyze internal and external factors to get a clearer insight into Bunda Bread's actual business condition. In addition, given the owner's expectations, it was crucial to gain customer awareness during COVID-19. Interviews were conducted to understand not only the customer perspective but also customer needs. Delivering information on the products and services to customers has not been carried out properly. Based on observation, Bunda Bread's marketing only relied on word of mouth without maximizing the use of existing brochures and digital marketing; websites and the Instagram account were not maintained. During the COVID-19 pandemic, business opportunities could still be exploited by utilizing technology, so that MSMEs could still serve customers without them having to leave the house. According to the results of several analyses that were conducted, there are several strategies developed using the TOWS matrix, Gap analysis, and Pareto analysis: (1) implementation of the push strategy and (2) carrying out their promotional activities through online channels, especially Instagram and website, that is, digital marketing improvement.

4.1 *Implementation of the push strategy*

The implemented push strategy makes it easier to communicate with customers because the customer could independently find out the details of the product on the products for sale display which is located at the bread display. Employees also expressed the convenience of having a products for sale display related to conditions in Jakarta that were not conducive due to COVID-19, so that the sellers could also maintain a safe distance from the customers while selling. Similar to the implementation of direct selling, provision of brochures and also improvements to the GoFood display encouraged informed customers to order bread in the future and place orders using GoFood. Based on these conditions, a push strategy can be continued to gain more benefits. However, due to time constraints, direct selling as part of a push strategy requires supervision to maintain the communication between the employee and the customers considering what information that must be provided and obtained.

4.2 *Digital marketing improvement*

Changing Bunda Bread's personal Instagram account to a business account enabled Instagram Insights. It was found that Bunda Bread's Instagram has created 735 impressions, 131 profile visits, and 18 website visits in a week. Based on the improvement results of Bunda Bread's Instagram profile structure, the increase in digital marketing through Instagram improvements had been achieved from their engagement on Instagram. Additionally, its followers have increased from 102 to 875. Besides Instagram, the results of the Google Business report showed that the increase in digital marketing through website improvements had been achieved from the engagement on their website. After making improvements to the website, it was found that the viewers increased from 263 to 2,500. Of the 2,500 views, 780 views were using Google, and 1,700 were using Google Maps. This finding proves that Bunda Bread's Instagram and website are worthy of customers' participation because they provided useful information for the customers.

5 CONCLUSION

Through business coaching, it was revealed that Bunda Bread had several problems that were typically faced by an MSME during the COVID-19 pandemic. Studies found that the push strategy developed a method that can respond to the market by providing approaches to acquire customers, which begins with building product awareness. Further, Bunda Bread's digital marketing, including its Instagram and website exposure and engagement, has improved. Based on the results, this study is expected to provide additional insights into the application of the push strategy and digital marketing strategy in MSMEs, especially in the restaurant sector during COVID-19.

Based on external and internal analyses, Bunda Bread, an MSME, was facing several issues in its financial reporting system, human resources management, and operational facilities. However, this study only focuses on marketing aspects. Thus, other issues can be improved by other future studies to leverage the MSMEs' position among their competitors.

REFERENCES

Batra, R., & Keller, K. L. (2016). Integrating marketing communications: New findings, new lessons, and new ideas. *Journal of Marketing, 80*(6), 122–145.

Belch, G. E., & Belch, M. A. (2003). *Advertising and promotion: An integrated marketing communications perspective*. The McGraw- Hill.

Corporatefinanceinstitute.com. (2020). Push Marketing Strategy. *Corporatefinanceinstitute.com*. Retrieved from https://corporatefinanceinstitute.com/resources/knowledge/strategy/push-marketing-strategy/ (accessed 20 August 2020).

Hamdani, Trio. (2020). Restoran hingga Kedai Kopi Terancam Gulung Tikar Imbas Corona. *Finance.detik.com*. Retrieved from https://finance.detik.com/berita-ekonomi-bisnis/d- 4943285/restoran-hingga-kedai-kopi-terancam-gulung-tikar-imbas-corona (accessed 20 June 2020).

Ištvanic, M., Crnjac Milic, D., & Krpic, Z. (2017). Digital marketing in the business environment. *International journal of electrical and computer engineering systems, 8*(2.), 67–75.

Klaassen, T. J. H. (2016). *How do SMEs attract new Customers to Sustain Future Business growth?: An exploratory study into acquisition practices of small companies* (Master's thesis, University of Twente).

Kotler, P., Kartajaya, H., & Setiawan, I. (2016). *Marketing 4.0: Moving from traditional to digital*. John Wiley & Sons.

McPheat, Sean. (2011). Developing Internet Marketing Strategy. USA: Ventus Publishing

Rayport, J. F., & Jaworski, B. J. (2001). Introduction to E-commerce. New York: McGra- Hill

Satell, Greg (2015). Balancing Push and Pull Marketing. *Raconteur.com*. Retrieved from https://www.raconteur.net/business-innovation/balancing-push-and-pullmarketing (accessed 12 June 2020).

Contemporary Research on Business and Management – Noviaristanti (Ed.)

Predicting a company's performance

G. Kurniawan & R. Rokhim
Faculty of Economics and Business, Universitas Indonesia, Indonesia

ABSTRACT: This study analyzed whether the elbow method and K-means clustering can help investors diversify their investment in different companies. Using clustering in machine learning and big data, a method was generated to categorize a list of companies with similar performances based on their fundamental analysis. Companies' fundamental analysis variables were used to classify stock data using K-means clustering application. In the end, companies with similar stock performances will be grouped, and it will be easier for investors to diversify their investments.

1 INTRODUCTION

There are various considerations investors will have when they are investing in company's stock. With investors' consideration and the risk each of investments has, investing will undoubtedly be more complex and varied. The risk in the stock portfolio itself will vary according to the proportions held by individuals or institutions (Statman, 1987).

According to Statman (1987), the risk of investing in a company's stock can be minimized by choosing different types or categories of stocks. The act of selecting different stocks in a stock portfolio is often known as stock diversification. The risks arising from investments can be minimized by carrying out portfolio selection activities. Portfolio selection activity is an activity to select stocks for investment or diversification. Clustering is an efficient method of selecting stocks for portfolio management. This method is also considered to be able to minimize the risks involved in determining stocks when carrying out portfolio management. One of the clustering methods commonly used is K-means clustering (Ren et al., 2017).

This study focuses on how K-means clustering can be used to diversify stocks to minimize the risk that is inherent in each investment. Five fundamental analysis variables were used to classify stock data from Indonesia (Kompas100). This study also seeks to form a K-means clustering algorithm that can reduce the risk of investing in stocks using the diversification method.

2 LITERATURE REVIEW

Clustering is commonly used to group similar stocks in one group. The clustering method for the stock market has been done in Australia (Aitken et al., 1996), China (Liao & Chou, 2013), and India (Nanda et al., 2010). It is expected to assist in determining investment in the stock market. One of the clustering methods commonly used in machine learning is K-means clustering.

K-means clustering aims at looking for similarities between data and grouping the similar data into a cluster or a group. According to Bishop (2014), K-means clustering divides a data set composed of N observations into K clusters. The distribution of data to each cluster in this method is done by considering the smallest distance between the data and the average of each cluster. The determination of cluster values to be carried out using the K-means clustering method is usually determined subjectively.

DOI 10.1201/9781003196013-32

However, to be more certain, one method is used to ensure that the selection of cluster values is correct. This method is the elbow method. The elbow method has been used by Bholowalia and Kumar (2014) in determining the right number of clusters in the K-means clustering algorithm.

This method is used to observe a graph depicting a cluster value whose sum square of errors drop to a sudden lowest point.

3 METHODOLOGY

3.1 *Research data*

The data used in this study were retrieved from stock data for public companies traded in the Indonesia Stock Exchange (IDX) and from public companies listed in the Kompas100 index. Five companies' financial ratios will be used as a variable to perform the clustering analysis.

3.2 *Research model*

The procedures of this research are as follows:

1. Retrieving stock data from websites commonly used for stocks news;
2. Applying elbow method to determine the correct number of clusters;
3. Inputting the stock data into the K-means clustering algorithm;
4. Analyzing the results of the grouping of stock data into each cluster; and
5. Determining the possible investment strategies that can be chosen by investors for each cluster.

3.3 *Elbow method*

The elbow method was commonly used by analysts to determine the right number of clusters to be included in the clustering algorithm. It can also be applied to the data by determining the sum square of errors in each variable. The sum square of errors for each variable can be calculated by entering the cluster values with iteration (Bholowalia & Kumar, 2014). The following model is the elbow method objective function:

$$SSE = \sum_{K=1}^{k} \sum_{xi \in S_k} \|X_i - C_k\|_2^2 \quad (1)$$

3.4 *K-means clustering*

The clustering method is used to group similar stocks into one cluster. K-means clustering used the smallest distance analysis with its average in classifying similar data. This smallest distance is calculated using the Euclidean distance that is widely used in the K-means clustering algorithm (Bishop, 2014). The following model is the K-means clustering objective function:

$$J = \sum_{n=1}^{N} \sum_{k=1}^{K} r_{nk} \|x_n - \mu_k\|^2 \quad (2)$$

With, $r_{nk} = \{1,\ if\ k = argmin_j \|x_n - \mu_j\|^2, x_n = n\ Data, \mu_k = Mean\ k\ 0,\ others$

The results of this cluster can be used as a reference for providing recommendations to investors in making an investment in certain stocks. It can also be used by investors and other institutions in finding the right investment and diversifying stocks to get the expected profit (Nanda et al., 2010).

4 RESULTS

4.1 *Elbow method results*

The elbow method was used to determine the right number of clusters, which will be used for K-means clustering. There are six clusters that were determined by the elbow method.

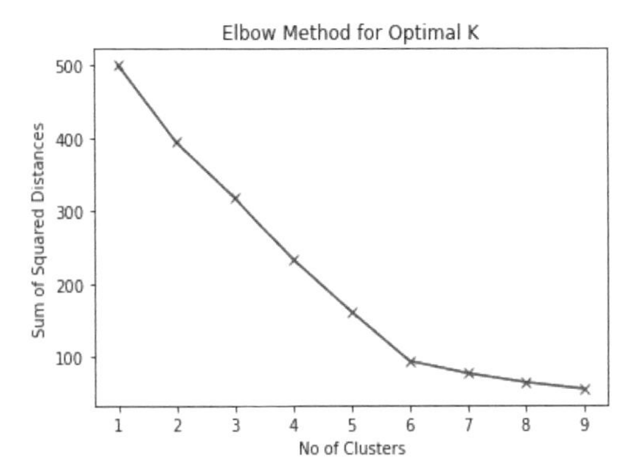

Figure 1. Elbow method.

According to the elbow method, the best cluster to be processed by the K-means clustering method is the sixth cluster.

4.2 *K-means clustering results*

From the six clusters, there are several members of clusters that were achieved using K-means clustering.

Table 1. Number of cases in each cluster

Number of Cases in Each Cluster	
Cluster 1	1
Cluster 2	20
Cluster 3	2
Cluster 4	3
Cluster 5	1
Cluster 6	73

Table 2. Final cluster centers.

	Clusters					
Final Cluster Centers	1	2	3	4	5	6
PBV	8.25287	.52504	−.19753	.64530	1.72116	−.30158
EPS	−.02046	−.12925	6.04085	−.32254	−.98991	−.10300
ROA	4.69248	1.28890	.99357	−.48792	−2.06914	−.39623
ROE	3.83807	.53509	.42988	−.24060	−8.27951	−.08765
P/E Ratio	.18255	−.00418	−.14280	5.03817	−.23212	−.20131

Each cluster from K-means clustering has different characteristics. This characteristic can be seen in the results of K-means clustering with the descriptive analysis on the data.

Cluster 1: High price to book value stock
Cluster 2: Moderate price to book value, ROA, and ROE stock
Cluster 3: High earnings per share stock
Cluster 4: High price to earnings ratio stock
Cluster 5: Low ROE and ROA stock
Cluster 6: Moderately low in all variables stock

According to the analysis, investors can get six clusters with various characteristics. They can do some investment diversification within the six clusters, in which the investment diversification should be based on investors' behavior. In this matter, investors could do investment in some ways that they believe is the strongest or better way to gain profit.

5 CONCLUSIONS AND RECOMMENDATIONS

5.1 *Conclusions*

K-means clustering can be used to cluster the same category of stocks based on company's performance that can be seen through company's fundamental analysis. According to the analysis, K-means clustering grouped company stocks' data into six clusters based on five fundamental ratios. Clusters that were built by K-means clustering can be used by investors to diversify their stock portfolio. Diversification within the stock portfolio can be a way to reduce stock investment risk.

5.2 *Recommendations*

The analysis that was done in this study was based on data from only hundred stocks, which means that the development of this study could be done in another field such as in different index or different stocks' data. Variables used in this study could also be enlarged by other variables suitable to the study. A lot of progress and development could be made with this study. Another conclusion and clustering could also be found if there are different variables and data.

REFERENCES

Aitken, M., Brown, P., Buckland, C., Izan, H. Y., & Walter, T. (1996). Price clustering on the Australian stock exchange. *Pacific Basin Finance Journal*. https://doi.org/10.1016/0927-538x(96)00016-9

Bholowalia, P., & Kumar, A. (2014). EBK-Means: A Clustering Technique based on Elbow Method and K-Means in WSN. In *International Journal of Computer Applications*.

Bishop, C. M. (2014). Bishop - Pattern Recognition And Machine Learning - Springer 2006. *Antimicrobial Agents and Chemotherapy*. https://doi.org/10.1128/AAC.03728-14

Liao, S. H., & Chou, S. Y. (2013). Data mining investigation of co-movements on the Taiwan and China stock markets for future investment portfolio. *Expert Systems with Applications*. https://doi.org/10.1016/j.eswa.2012.08.075

Nanda, S. R., Mahanty, B., & Tiwari, M. K. (2010). Clustering indian stock market data for portfolio management. *Expert Systems with Applications*. https://doi.org/10.1016/j.eswa.2010.06.026

Ren, F., Lu, Y. N., Li, S. P., Jiang, X. F., Zhong, L. X., & Qiu, T. (2017). Dynamic portfolio strategy using clustering approach. *PLoS ONE*. https://doi.org/10.1371/journal.pone.0169299

Statman, M. (1987). How Many Stocks Make a Diversified Portfolio? *The Journal of Financial and Quantitative Analysis*. https://doi.org/10.2307/2330969

Contemporary Research on Business and Management – Noviaristanti (Ed.)

Effect of entrepreneurial self-efficacy and proactive personality on the entrepreneurial intention of post-migrant workers through attitude toward entrepreneurship

G. Rosalina & A. Satrya
Faculty of Economy and Business, University Indonesia, Jakarta, Indonesia

ABSTRACT: The purpose of this research is to examine the effects of entrepreneurial self-efficacy and proactive personalities on entrepreneurial intentions among migrant returnees through attitude toward entrepreneurship. This research employs a survey-based methodology and uses a 27-item questionnaire with five-point Likert for a total sample of 195 migrant returnees. The research was conducted in the East Nusa Tenggara province, Indonesia. The hypotheses derived from the research model were assessed through structural equation modeling using Lisrel. The results show the indirect effect on the relationship between proactive personality and entrepreneurial intention through attitude toward entrepreneurship, while entrepreneurial self-efficacy was found to have a significant negative effect on entrepreneurial intention. By doing so, this research significantly contributes to the literature of entrepreneurship. The results also provide policy makers with important insights on developing entrepreneurial intentions on post-migrant workers and giving attention to key elements of the entrepreneurial process.

1 INTRODUCTION

Entrepreneurship is widely recognized as a determining factor that can improve social and economic development, both in developing and developed countries. Entrepreneurship plays an important role in creating new jobs, thus helping the government in dealing with unemployment and poverty problems (Asad, Ali, & Islam, 2014). According to the data retrieved from the Ministry of Cooperatives and SMEs of The Republic of Indonesia in 2019, it is known that MSMEs contribute 60.34% to Indonesia's GDP. One of the groups in the society that have the potential to be developed into entrepreneurs is migrant workers, who have returned to their country after going abroad to have a job for a certain period of time. A study conducted by Demurger and Xu (2011) found that the migration experience allows them to bring back accumulated human, social, and financial capital that will enable them to set up their own businesses upon returning.

Given the increasing recognition of the role of entrepreneurship as an important driver for economic development and innovation, many studies have focused on developing models for understanding and predicting entrepreneurial intention as an important factor in shaping entrepreneurial behavior. From many studies conducted to identify factors that affect entrepreneurship, personal characteristics and basic psychological conditions are considered to be the most interesting factors that affect entrepreneurial intention (Khuong & An, 2016). Entrepreneurial self-efficacy and proactive personality have emerged as the main psychological constructs in research on entrepreneurship because they can affect entrepreneurial motivation, intention, behavior, and performance (Miao, Qian, & Ma, 2017). Therefore, the purpose of this research is to examine the influence of entrepreneurial self-efficacy and proactive personality on entrepreneurial intention on returned migrant workers mediated by attitude toward entrepreneurship.

DOI 10.1201/9781003196013-33

2 LITERATURE REVIEW

Entrepreneurship intention (EI) is a person's desire to consider entrepreneurship as a career choice, serving as the initial stage in the process of forming entrepreneurial behavior. Ajzen (2011) describes intention as an indication of a person's readiness to perform certain actions. Krueger et al. (2000) defines entrepreneurial intentions as individual beliefs in preparing and realizing entrepreneurial behavior or planned behavior to start a new business (Liu, Lin, Zhao, & Zhao, 2019).

The term entrepreneurial self-efficacy (ESE) comes from the social learning theory developed by Bandura (1977). In this theory, self-efficacy is defined as an individual's belief in their capacity to take actions that can provide certain results, which reflects their belief in their ability to exercise control over motivation, behavior, and the social environment (Yueh, Wu, & Chen, 2020). The definition of self-efficacy in this theory also explains the role of individual beliefs in their ability to create the environment and the expected results from their personal actions.

Some researchers agree that proactive personality (PP) is one of the strong predictors of entrepreneurial intentions. Bateman and Crant (1999) defines proactive behavior as taking initiative action to improve the current state and refuse to passively adapt to current conditions. A proactive attitude is also an important attribute of flexibility and adaptability to face an uncertain future, which is very much needed in the entrepreneurial process (Prabhu, McGuire, Drost & Kwong, 2012).

Attitude toward entrepreneurship (ATE) refers to the degree to which a person gives an evaluation, either positively or negatively, about perception of developing an entrepreneurial career. Someone who has a positive assessment of entrepreneurial behavior tends to try to find and start a new business because they think it will bring benefits for them. Some researchers also support the role of attitude as a mediator between the personal characteristic and self-efficacy with behavioral intentions (Rosique-Blasco, Madrid-Guijarro, & Lema, 2018). Therefore, the proposed hypotheses are as follows:

*H*1. PP has a significant positive effect on ATE.
*H*2. ESE has a significant positive effect on ATE.
*H*3. ATE has a significant positive effect on EI.
*H*4. PP has a significant positive effect on EI.
*H*5. ESE has a significant positive effect on EI.
*H*6. PP has a significant positive effect on EI through mediation of ATE.
*H*7. ESE has a significant positive effect on EI through mediation of ATE.

3 RESEARCH METHOD

3.1 *Types and data source*

This study used quantitative research. The data were collected from the returned migrant workers in Kupang City, East Nusa Tenggara province, Indonesia, with convenient sampling techniques. Participants were asked to complete a questionnaire consisting of 4 sections and 27 questions, covering general and variable questions including indicator questions for entrepreneurial intention, entrepreneurial self-efficacy, proactive personality, and attitude toward entrepreneurship.

3.2 *Measurement scales and data analysis*

This study used the established measures of constructs based on the earlier studies. In this study, the entrepreneurial self-efficacy was measured using questionnaire items from the research of Wilson, Kickul, and Marlino (2007), proactive personality variables were measured using questionnaire items from the research of Seibert, Krant, & Kraimer (2001), attitude toward entrepreneurship was measured using questionnaire items from the research of Solesvik *et al.* (2012), and measurement of entrepreneurial intention variables were measured using questionnaire items from the research of Linan and Chen (2009). Demographic variables including age, gender, education, and marital status will be controlled in this study. The participants were asked whether they agreed or disagreed with a set of sentences across a 5-point Likert scale, where 1 = strongly disagree and 5 = strongly agree. The data were analyzed using SPSS version 25 and Lisrel 8.54 by performing structural equation modeling (SEM).

4 RESULTS AND DISCUSSION

A total of 195 responses were obtained. Most of the respondents were female (74.4%), married (61.54%), had graduated from elementary schools (30.7%), and in the age groups of 31–40 (49.2%). Validity and reliability of each indicator used in the study were tested. Specifically, this study used standardized loading factor, composite reliability (CR), and average variance extracted (AVE) to assess the adequacy of outer-measurement models. The SEM results indicate that the questionnaire items used are valid and reliable because CR values are above 0.7 and SLF and AVE values are also above 0.5 (Hair *et al.*, 2017).

Figure 1 shows the hypothesis testing results indicating that proactive personality (PP) was significantly and positively related to attitude toward entrepreneurship (ATE) (t-values = 10.57, $\beta = .82$). Furthermore, ATE and entrepreneurial intention (EI) showed a positive and significant relationship (7.34, $\beta = 0.83$), but there was no significant relationship between PP and EI (.04, $\beta = .01$). From the analysis, ATE was known to have a full mediated effect in the relationship between PP and EI. From the result, it was known that entrepreneurial self-efficacy (ESE) and ATE did not have a significant and positive relationship (t-value = −.42, $\beta = .02$). ESE and EI were known to have a significant effect but in the opposite direction because the t value obtained was negative (t-value = −3.00, $\beta = -.01$). This result is contrary to the initial hypothesis and several previous studies indicating that ESE has a significant positive effect on entrepreneurial intention. Attitudes toward entrepreneurship has no mediation effect in the relationship between entrepreneurial self-efficacy and entrepreneurial intention.

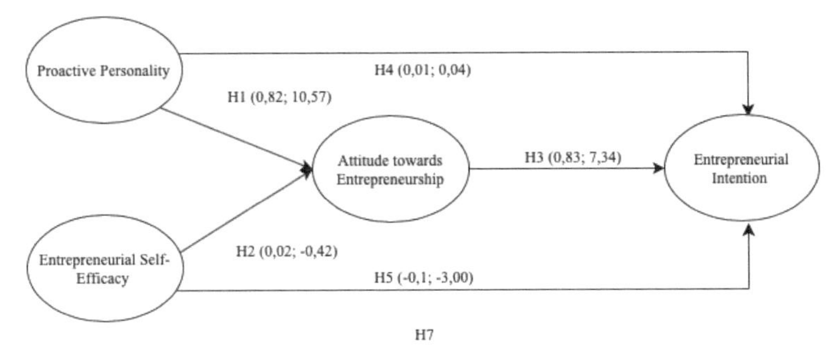

Figure 1. Test result.

This research provides valuable insights into the importance of personality traits in the determination of EIs in post-migrant workers. First, this study clarifies the mediating role of attitude toward entrepreneurship in the relationship between proactive personality and entrepreneurial intentions. The results of the analysis show that attitudes toward entrepreneurship change proactive personality into an intention to start a business. Someone who has a proactive personality will have entrepreneurial intentions if s/he has a positive evaluation toward entrepreneurship and believe that being an entrepreneur is the best career option and will bring benefit. This result also supports previous research by Hu, Wang, Zhang, and Bin (2018) and Rosique-Blasco, Madrid-Guijarro, and Lema (2018). Individuals who have proactivity will be better in identifying opportunities and showing entrepreneurial behavior. Therefore, a proactive personality can be a sign to see someone's tendency to start a business.

There is a significant negative result on the relationship between ESE and EI, which means that an increase in ESE will decrease someone's entrepreneurial intention. This result is contrary to the initial hypothesis and several previous studies showing that ESE has a significant positive effect on entrepreneurial intention. Powers (1991) stated that too high self-efficacy will indeed increase optimism to be successful, but this will make a person reduce their motivation and performance toward certain behaviors. Other research studies found that although ESE is thought to have a positive effect on venture creation, very high self-efficacy might increase risk-taking behavior and decrease their effort to achieve something due to overconfidence. Overconfidence and overoptimism have been shown to have mixed effects on entrepreneurial and organizational outcomes (Sun, Vancouver, & Weinhardt, 2014).

5 CONCLUSION AND SUGGESTION

The study shows that the indirect effect of proactive personality and entrepreneurial intention is fully mediated by the attitude toward entrepreneurship, while entrepreneurial self-efficacy is found to have a significant negative effect on entrepreneurial intention of post-migrant workers. Future research might examine whether individuals with high self-efficacy display cognitive biases that lead to negative effects that may occur as a result and the possibility of undiscovered mediators in the relationship between ESE and EI.

REFERENCES

Asad, A., Ali, H. M & Islam, U. (2014). The Relationship between entrepreneurship development and unemployment reduction in Pakistan. *Global Journal of Management and Business Research*, 14(10), 75–79.

Demurger, S., & Xu, H. (2011). Return migrants: The rise of new entrepreneurs in rural China. *World Development*, 39 (10), 1847–1861

Hair, J.F., Hult, G.T.M., Ringle, C.M., & Sardstedt, M. (2017). *A Primer on Partial Least Squares Structural Equation Modeling* (PLS-SEM). Los Angeles : SAGE

Hu R., Wang L., Zhang W., & Bin P. (2018). Creativity, proactive personality, and entrepreneurial intention: the role of entrepreneurial alertness. *Frontiers in psychology*, 9 (951), 1–10.

Khuong, M.N, & An, N.H. (2016). The factors affecting entrepreneurial intention of the students of Vietnam National University — A mediation analysis of perception toward entrepreneurship. *Journal of Economics, Business and Management*, 4 (2), 104–111.

Liñán, F., & Chen, Y.-W. (2009). Development and cross-cultural application of a specific instrument to measure entrepreneurial intentions. *Entrepreneurship Theory and Practice*, 33(3), 593–617.

Liu, X., Lin, C., Zhao, G. & Zhao, D. (2019). Research on the effects of entrepreneurial education and entrepreneurial self-efficacy on college students' entrepreneurial intention. *Frontiers in psychology*, 10 (869),1–16.

Miao, C., Qian, S., & Ma, D. (2017). The relationship between entrepreneurial self-efficacy and firm performance: A meta-analysis of main and moderator effects. *Journal of Small Business Management,* 55(1), 87–107.

Prabhu, V. P., McGuire, S. J., Drost, E. A., & Kwong, K. K. (2012). Proactive personality and entrepreneurial intention. *International Journal of Entrepreneurial Behavior & Research*, 18(5), 559–586.

Rosique-Blasco, M., Madrid-Guijarro, A., & García-Pérez-de-Lema, D. (2017). The effects of personal abilities and self-efficacy on entrepreneurial intentions. *International Entrepreneurship and Management Journal*, 14(4), 1025–1052.

Seibert, S. E., Kraimer, M. L., & Crant, J. M. (2001). What do proactive people do? a longitudinal model linking proactive personality and career success. *Personnel Psychology*, 54(4), 845–874.

Solesvik, M. Z., Westhead, P., Kolvereid, L., & Matlay, H. (2012). Student intentions to become selfemployed: the Ukrainian context. *Journal of Small Business and Enterprise Development*, 19(3), 441–460.

Sun, S., Vancouver, J., & Weinhardt, J. (2014). Goal choices and planning: Distinct expectancy and value effects in two goal processes. *Organizational Behavior and Human Decision Processes*, 125(2014), 220–233.

Wilson, F., Kickul, J., & Marlino, D. (2007). Gender, entrepreneurial self-efficacy, and entrepreneurial career intentions: implications for entrepreneurship education. *Entrepreneurship Theory and Practice*, 31(3), 387–406.

Yueh, H.P., Wu, Y.J., & Chen, W. (2020). *The Psychology and Education of Entrepreneurial Development.* Frontiers Media SA: Lausanne

Contemporary Research on Business and Management – Noviaristanti (Ed.)

The role of psychological capital and perceived organizational support on task performance: The mediating effects of individual readiness for change

G.F. Ramdhani & P.M. Desiana
Faculty of Economics and Business, Universitas Indonesia, Indonesia

ABSTRACT: This paper examines the role of psychological capital and perceived organizational support in task performance with individual readiness for change as a mediator. The data for this quantitative study were collected using questionnaires given to 223 employees who work in five telecommunication service provider companies in Indonesia. Structural equation modeling (SEM) using Lisrel software was used to present and analyze the obtained data. The result showed that individual readiness for change acted as a mediator in the causal relationship between psychological capital and perceived organizational support on task performance. However, perceived organizational support had no direct effect on task performance. This study discloses the importance of psychological capital and perceived organizational support in boosting the employees' task performance.

1 INTRODUCTION

The COVID-19 pandemic has brought changes to various aspects of life. As a result of the increasing number of COVID-19 cases in Indonesia, the government had implemented large-scale social restrictions that have resulted in the implementation of work from home to maintain business continuity (Haryono, 2020). The telecommunication service provider is one of the industrial sectors that are considered important to remain fully operational to meet the needs of the community amid the pandemic situation (Fachriansyah, 2020). This sector needs to significantly increase its role to support bandwidth demand due to the increasing data traffic (Haryono, 2020) and also to present additional programs (Evandio, 2020). The government through the Ministry of Education and Culture issued a policy of internet data quota assistance to help the process of distance learning (Jamaludin, 2020).

This condition illustrates that environmental dynamics make organizations always ready to face challenges in embracing change; therefore, individuals and organizations take roles in dealing with the changes (Shah, 2011). However, employees may feel insecure about the changes, which may, inadvertently, affect their performance (Cartwright & Holmes, 2006). Organizations need high-performing individuals to achieve goals, increase productivity, and increase competitiveness (Sonnentag & Frese, 2002). The level of employee performance can depend on the level of their readiness to face changes (Bellou & Chatzinikou, 2015).

Individual readiness to change is shown through support, openness, and commitment to change. One of the factors that can influence these attitudes and behavior is personal characteristics (Palmer, Dunford, & Buchanan, 2017). Organizations can increase their competitive advantage by managing the psychological capital of their employees (Luthans, Youssef, & Avolio, 2007) and become more successful by creating positivity among employees, which can also improve employee performance (Tosten & Toprak, 2017).

DOI 10.1201/9781003196013-34

In addition, factors that play a role in increasing individual readiness for change and facilitate effective performance are the need for organizations to support their employees. Organizational support is an effort to reward, pay attention to, and increase the welfare of each employee (Rhoades & Eisenberger, 2002). The development of a positive impression regarding the support that employees receive will lead to positive outcomes for both the employee and the organization.

2 LITERATURE REVIEW

Koopmans et al. (2011) stated that task performance is an important dimension of individual performance. Task performance is defined as proficiency (that is, competence) or the ability to perform a core task or job, which is also often referred to as job-specific task proficiency, technical proficiency, or in-role performance.

Individual readiness for change is defined as the extent to which individuals feel ready or primed, either mentally, psychologically, or physically marked by participation, promotion, and resistance to participating in organizational development activities (Hanpachern, 1997). Individual readiness for change consists of three dimensions: (1) participation, (2) promotion, and (3) resistance.

Psychological capital is the state of positive psychological development of individuals which consists of four dimensions, namely, (1) self-efficacy, (2) hope, (3) optimism, and (4) resilience (Luthans, Youssef, & Avolio, 2007).

The concept of perceived organizational support comes from organizational support theory, which states that perceived organizational support is a general belief of employees about the extent to which the organization can appreciate their contribution and also care about their welfare (Eisenberger, Huntington, Hutchison, & Sowa, 1986).

3 METHODOLOGY

Data analyzed in this research were collected from five telecommunication service provider companies in Indonesia. One of the criteria for this research were employees who were actively working in the telecommunication service provider companies. The distribution of online questionnaires was carried out in three stages, including the wording test to 10 respondents, the pilot test to 30 respondents, and the main test to 223 respondents.

Fifty questions were adopted from previous relevant studies. Respondents were required to answer the questions on a 6-point Likert scale. The questionnaire used to measure task performance was the IWPQ developed by Koopmans (2015), which consists of five questions. Individual readiness for change was measured using 14 questions adopted from Hanpachern (1997). Psychological capital was measured using 24 questions adopted from PCQ-24 developed by Luthans, Youssef, and Avolio (2007). Perceived organizational support was measured using seven questions by Eisenberger, Cummings, Armeli, and Lynch (1997). The data were analyzed by using the structural equation technique or structural equation modeling (SEM)-Lisrel.

The hypotheses in this study are stated as follows:

H1: Psychological capital has positive and significant effects on individual readiness for Change.
H2: Perceived organizational support has positive and significant effects on individual readiness for change.
H3: Psychological capital has positive and significant effects on task performance.
H4: Perceived organizational support has positive and significant effects on task performance.
H5: Individual readiness for change has positive and significant effects on task performance.
H6: Individual readiness for change mediates psychological capital on task performance.
H7: Individual readiness for change mediates perceived organizational support on task performance.

4 RESULTS

The questionnaires were filled out by 169 men (75.78%) and 54 women (24.22%). The majority of respondents belonged to Generation Y or the 26–40 years age group (75.78%). Most of them were bachelors and graduates (83.4%). Around 77% had been working for the telecommunication service provider companies for 1–5 years. Just over a half of them were working as staffs (51.12%), and more than 75% of them were permanent employees.

The measurement model was assessed using validity, reliability, and goodness-of-fit indices. The validity was evaluated by the value of standardized loading factors (SLFs), while reliability was evaluated by the value of construct reliability (CR) and average variance extracted (AVE). The desirable value of SLF was 0.50; CR was 0.70; and AVE was 0.5. All indicators in this study were above the criteria.

This study used nine goodness-of-fit indicators to test the overall model fit, which were RMSEA, NFI, NNFI, CFI, IFI, RFI, Std. RMR, GFI, and AGFI. There were five GoF indicators that resulted in a good fit. Meanwhile, the rest showed marginal fit and poor fit. Despite the four unsatisfactory GoF indicators found, the overall structural model had a good fit for the data. After ensuring the model, the hypotheses developed for this study were tested.

In this model, perceived organizational support had no direct effect on task performance (SLF = 0.091; t-value = 1.40), denoting that H4 is not supported. However, perceived organizational support was found to have a positive and significant effect on individual readiness for change (SLF = 0.21; t-value = 3.31), signifying that H2 is supported. Psychological capital was found to have a positive and significant effect on task performance (SLF = 0.59; t-value = 7.48) and individual readiness for change (SLF = 0.48; t-value = 7.21), indicating that H1 and H3 are supported. Individual readiness for change had a significant and positive effect on task performance (SLF = 0.12; t-value = 1.83), denoting that H5 is supported. Individual readiness for change was found to fully mediate the effect of psychological capital on task performance; therefore, H6 is supported. Furthermore, individual readiness for change was found to mediate the effect of perceived organizational support on task performance; thus, H7 is also supported.

5 DISCUSSION

The findings of this study showed that psychological capital had a positive and significant effect on individual readiness for change and task performance. This study supported the previous study done by Kirrane, Lennon, O'Connor, and Fu (2016). From the study, it can be said that psychological capital has an important role in organizational change because employees who feel more positive about their own ability to cope with challenges will be more prepared to face changes. Employees with high positive psychological capital will also do work simultaneously, so that they will strive to be successful by completing tasks well.

Perceived organizational support had a positive and significant effect on individual readiness for change. This finding is in line with another study done by Gigliotti, Vardaman, Marshall, and Gonzalez (2018), who showed that perceived organizational support was a key role during organizational change and had an impact on the reaction of the change recipients.

The results are also supported the previous research, which showed that individual readiness for change had a significant positive relationship with performance (Iqbal & Asrar-ul-Haq, 2018). Employees will not compromise on their performance and are ready to accept organizational changes. However, organizational support had no direct effect on task performance. This result did not support the previous study done by Eisenberger, Huntington, Hutchison, and Sowa, (1986), who showed that perceived organizational support would improve employee performance as a function of socio-emotional needs.

Even so, individual readiness for change was found to mediate the effect of psychological capital and perceived organizational support on task performance. Change readiness acts as a mediator among personal characteristics, internal context enablers, and external pressures on performance based on the multilevel framework of the antecedents and consequences of readiness for change (Rafferty, Jimmieson, & Armenakis, 2013).

This study also provides factors that affect task performance by investigating the role of psychological capital and perceived organizational support in individual readiness for change.

However, perceived organizational support had no direct effect on task performance. The findings also imply that the psychological capital and employees' perception of organizational support are important to increase their readiness to face changes and in turn influence their task performance.

REFERENCES

Bellou, V., & Chatzinikou, I. 2015. Preventing employee burnout during episodic organizational changes. *Journal of Organizational Change Management, 28(5),* 673–688.

Cartwright, S., & Holmes, N. 2006. The meaning of work: The challenge of regaining employee engagement and reducing cynicism. *Human Resource Management Review, 16(2)*, 199–208.

Eisenberger, R., Huntington, R., Hutchison, S., & Sowa, D. 1986. Perceived organizational support. *Journal of Applied Psychology, 71*, 500–507.

Evandio, A. 2020, March 18. 'Perang' Layanan ala Operator Telekomunikasi di Tengah Social Distancing. Retrieved September 20, 2020, from Bisnis.com: https://teknologi.bisnis.com/read/20200318/101/1215150/perang-layanan-ala-operatortelekomunikasi-di-tengah-social-distancing

Fachriansyah, R. 2020, April 8. Jakarta will impose stronger mobility restrictions on Friday. Here's what you need to know. Retrieved August 12, 2020, from The Jakarta Post: https://www.thejakartapost.com/news/2020/04/08/jakarta-will-impose-stronger-mobilityrestrictions-on-friday-heres-what-you-need-to-know.html

Gigliotti, R., Vardaman, J., Marshall, D., & Gonzalez, K. 2018. The role of perceived organizational support in individual readiness. *Journal of Change Management, 19(2),* 86-100.

Hanpachern, C. 1997. *The extension of the theory of margin: A framework for assessing readiness for change.* Unpublished Doctoral Dissertation, Colorado State University, Fort Collins.

Haryono, H. 2020, June 3. Pandemi Covid-19 dan industri telekomunikasi. Retrieved September 20, 2020, from indotelko.com: https://www.indotelko.com/read/1591163164/pandemi-covid-19telekomunikasi

Iqbal, A., & Asrar-ul-Haq, M. 2018. Establishing relationship between TQM practices and employee performance: The mediating role of change readiness. *International Journal of Production Economics, 203,*, 62–68.

Jamaludin, F. 2020, September 22. Ramai-ramai Operator Seluler Beri Bantuan Kuota Internet untuk PJJ. Retrieved September 27, 2020, from merdeka.com: https://www.merdeka.com/teknologi/ramairamai-operator-seluler-beri-bantuan-kuota-internet-untuk-pjj.html?page=1

Kirrane, M., Lennon, M., O'Connor, C., & Fu, N. 2016. Linking perceived management support with employees' readiness for change: the mediating role of psychological capital. *Journal of Change Management,* 47–66.

Koopmans, L., Bernaards, C., Hildebrandt, V., Schaufeli, W., de Vet Henrica, C., & van der Beek, A. (2011). Conceptual frameworks of individual work performance: a systematic review. *Journal of occupational and environmental medicine*, 53(8), 856–866.

Luthans, F., Youssef, C. M., & Avolio, B. J. 2007. *Psychological capital: Developing the human competitive edge.* Oxford: Oxford University Press.

Palmer, I., Dunford, R., & Buchanan, D. 2017. *Managing organizational change: A multiple perspective approach, 3rd edition.* New York: McGraw Hill Education.

Rafferty, A., Jimmieson, N., & Armenakis, A. 2013. Change readiness: A multilevel review. *Journal of Management, 39(1)*, 110–135.

Rhoades, L., & Eisenberger, R. 2002. Perceived organizational support: a review of the literature. *Journal of Applied Psychology, 87,* 698–714. doi: 10.1037/0021-9010.87.4.698.

Shah, N. 2011. A study of the relationship between organisational justice and employee readiness for change. *Journal of Enterprise Information Management, 24(3),* 224–236.

Sonnetag, S., & Frese, M. 2002. *Performance Concept and Performance Theory in Psychological Management of Individual Performance.* Sussex: John Wiley & Sons.

Tosten, R., & Toprak, M. 2017. Positive psychological capital and emotional labor: A study in educational organizations.*Cogent Education, 4(1),* 1301012.

Contemporary Research on Business and Management – Noviaristanti (Ed.)

Analysis of Panorama Tour and Bayu Buana Travel's strategies

H. Fransisca & A.I. Molle
Universitas Surabaya, Indonesia

ABSTRACT: Indonesia tourism has a resource potential that can be developed to generate dividends. Supporting factors in tourism include transportation and accommodation. A travel agency needs to be able to search for information on various tourist attractions. This study aims to analyze the performance, strategy, and development of two travel agencies, namely, Bayu Buana Travel and Panorama Tour. The market penetration strategy was applied to analyze Panorama, while the market development strategy was applied to analyze Bayu Buana. The strategy development was applied by Bayu Buana and Panorama with an online tour and travel company, such as holding public events or activities, utilizing email marketing, and placing advertisements on the Internet and to deal with the tight prices of the "services". It was found that each consumer had his method of buying the preferred product. The opportunity is still wide open because there are still many people who still prefer face-to-face transactions.

1 INTRODUCTION

Tourism in Indonesia has the potential to generate dividends which also needs to be optimized from a service provider in tourism places. Each city in Indonesia has the characteristics of a tourist destination. For example, Bali has beautiful beaches, Yogyakarta has beautiful temples, and Bandung has street foods and a shopping district. These aspects are attractions for both domestic and international tourists. In addition, transportation and accommodation in tourist destinations are needed to facilitate the tourists. In this case, a travel agency is one of the facilitators to disseminate tourist areas. Travel and tour agencies can recommend tourist destinations with specific conditions, such as tourist attractions, criteria, or facilities, to tourists. Indonesia has many travel and tour agencies (BPW, *Biro Perjalanan dan Wisata*) that compete against each other. According to ASITA (2011), Bandung had 170 BPWs. Some of the famous BPWs in Bandung are Anta Express Tour & Travel, Java Star Tours & Travel, Jackal Holidays, Panorama Tours, Primajasa Tours, Bayu Buana, and others. The companies use promotion strategies in the form of printing and electronic media to attract customers in order to survive and gain revenue. This study used two BPWs, namely, Panorama Tour and Bayu Buana Travel, to analyze their performance, strategy, and growth in their business development.

2 LITERATURE REVIEW

Strategy: According to David (2011), a strategy is a tool used to achieve long-term objectives. Business strategy includes geographical expansion, diversification, acquisitions, product development, market penetration, divestment, liquidity, and joint ventures.

Types of strategies: According to David (2011), based on its basic conception to apply strategic management, there are four types of strategies: integration strategy, intensive strategy, diversification strategy, and survival strategy

Strength, weakness, opportunity, and threats (SWOT) analysis: SWOT analysis is divided into two types, namely, internal analysis that identifies a company's strength and weakness and

DOI 10.1201/9781003196013-35

external analysis that identifies a company's opportunities and threats. The internal analysis comprises marketing, finance, production and operation, and human resource, while external analysis comprises competition with similar companies, potential new rivals, development of replacement products, supplier bargaining, and customer bargaining power (Kotler, 1997). The SWOT matrix is a technique used to help the manager develop four types of strategy (David, 2011): 1. SO (strength–opportunities) uses internal power to make use of the external opportunity. 2. WO (weakness–opportunities) uses internal weakness to make use of the external opportunity. 3. ST (strengths–threats) uses company power to avoid or decrease external threat. 4. WT (weakness–threats) is the defensive strategy to decrease internal weakness and avoid external threats.

3 METHOD

This study employed a descriptive method. Bogdan and Taylor in Moleon (2004:3) state that a descriptive study is a procedure to produce descriptive data in the form of written or spoken words. The descriptive analysis does not require experimentation because the variable cannot be manipulated. Data collection is not only limited to observing objects because categorizing, ordering, manipulating, and summarizing of data are also carried out to obtain answers to the research questions (Kerlinger in Kasiram, 2010: 120). This study used the observation method. The data were analyzed using SWOT and SWOT matrix to know the strategy of the company. Based on the results of the analysis, the strategy used by Bayu Buana Travel and Panorama Tour was determined and whether or not it was the appropriate strategy.

4 RESULTS AND DISCUSSION

4.1 *Internal analysis of Panorama Tour and Bayu Buana Travel*

Marketing: Both Panorama Tours and Bayu Buana Travel have a website to market their product. On their websites, the companies provide information related to their promotions and activities, for example, their involvement in the Travel Fair.

Finance: Bayu Buana was found to be financially superior to Panorama. Based on the source of funding, Bayu Buana mostly used their own funds, with little funding from outside parties. Therefore, Bayu Buana was more independent than Panorama who relied highly on funds from outside parties.

Production/operation: Panorama had successfully transformed its business from a traditional family company into an integrated travel company. During the study, it was found that Panorama has four pillars, with each playing a role in the company, namely, outbound management, inbound management, transportation management, and MICE business management. In other words, Panorama has continued to transform by owning companies with certain segments to extend their business.

Human resource: Human resource management is one of Panorama's focuses. Panorama has four approaches, namely, attract, develop, deploy, and service. Panorama also has employed media sources to get human resources. At Panorama, human resource works passionately as a Panoramanian, a term used to refer to their human resource people. They have roots in The SPIRIT of Panorama. Human resource planning is one of Bayu Buana's focuses. Bayu Buana intends to have human resources that can make the company an industry leader in the world of tourism. Bayu Buana has grown to become the leader of travel agencies in Indonesia that provide solutions to tourists while traveling.

4.2 *External analysis of Panorama Tour and Bayu Buana Travel*

The rivalry between companies: Bayu Buana and Panorama have competitors that are in similar businesses and that has affected their business progress and sustainability. The competitors of Bayu Buana and Panorama are PT Smailing Tour & Travel and PT Golden Rama Express Tours & Travels. In addition, there are several factors that the company needs to survive, namely, the packaging of tourism product services with a choice of prices and facilities according to the available budget size; quality of service to consumers in terms of CRM, which includes all customer withdrawals quickly, consistently, and efficiently; prices that are competitive with fellow actors; and extensive network by partnering with various countries.

Potential of newcomers: Bayu Buana's effort is to maintain a customer-oriented performance that prioritizes customer satisfaction and promotes a long experience as a travel agency. They also have employed technological advances that have been beneficial for customers. This utilization of technology granted them "The Best Performance" award from ABACUS. Panorama's effort is to commit through The SPIRIT of Panorama, work smart, and make innovation and creation in providing the best service for customers.

Substitute product development: With the increasing need for tourism, more places can be used as tourist destinations. In addition, the advancement of the internet has led to the emergence of new tourist agencies that sell services, such as new tourist routes, passport, and visa processing services, online ticket sales, and hotel vouchers. Nowadays, the development of the internet has slowly changed people's shopping behavior from the conventional method to the modern method. Several online travel sites continue to emerge, and these sites have been widely used by tourists who want to take a vacation. The following are several sites often used by tourists to facilitate their holidays: Traveloka, Tiket.com, Agoda, and Mister Aladin.

Customer bargaining power: Bayu Buana and Panorama are well-known customer preferences. Bayu Buana provides services to consumers, including an online system for airlines, hotel arrangements, travel document management services, car rental, and assistance services at the airport and tour packages. Panorama provides services to three main sectors, namely, the tourism sector, transportation sector, and hospitality sector.

4.3 *Strength, Weakness, Opportunity, Threats (SWOT) Analysis*

Strength: Bayu Buana's strength is their flagship programs such as money-saving tour or holiday savings. In the low season, they apply for early group ticket reservations. Besides, they have cooperation in the form of partnerships both locally and abroad, IT development that is in line with human resources, and a satellite network program to follow community movements that lead to e-commerce. Panorama's strengths are their tour packages on offer, which had cooperative relationships with several overseas business relations and an extensive hotel chain.

Weakness: Bayu Buana's weakness was their MICE program that did not bring revenue, middle-class market segments, and unoptimized website in terms of online inquiry services. On the other hand, Panorama's weakness was their focus on the main business of tourism and hospitality, relatively expensive package, and the website that could not maximally serve product ordering.

Opportunity: Bayu Buana and Panorama had the opportunities in the Asian free market, the distribution channels throughout Indonesia, and exhibitions in malls. Indonesia's resistance to the economic crisis has been quite strong, and the support from the government has encouraged the tourism sector to be an economic driver. Support has been received from the Ministry of Tourism for the Indonesian tourism industry with the brands of Wonderful Indonesia and Pesona Indonesia.

Threats: Bayu Buana and Panorama had the threats in the form of internet technology, which provides consumers with more information so that they know other tour products, the value of the Rupiah currency that is vulnerable to foreign currency values, the influence of unstable global economic conditions, the slow world economic growth, and unfavorable road and weather conditions in Indonesia's tourism sector.

SO (strength–opportunities) strategy: Bayu Buana maximized their service to the customer by introducing their flagship programs, improving their technology and information system of the MICE program, and suppressing cost production as much as possible in the low season to get the maximum margin of income. Panorama maximized customer service by introducing a tour package, creating an attractive tour package that satisfies customers, and carrying out an aggressive market strategy through cooperation with business relations abroad and a wide hotel network. The collaboration can be done by obtaining vouchers from these business relations so that customers would be interested to join the tour packages.

WO (weakness–opportunities) strategy: Bayu Buana conducted training on existing human resources to deal with large markets and competitors. To reach a wider target market, regular promotions in malls were not enough. In this case, online promotion must also be done. Online promotion can be done remotely at a lower cost. In addition, Bayu Buana has also collaborated with foreign travel agents in organizing tours to various countries so that their partners could provide local guides for those countries. Panorama conducted training on their human resources to face the big market and create online promotions to reach the target market. The online promotion is remote and with a relatively lower cost.

ST (strength–threats) strategy: Bayu Buana and Panorama have maintained and improved the company's reputation to strengthen the brand image in the community, maintained competitive prices so that old customers still use Bayu Buana/Panorama as the choice of tour and travel agencies, and provided member service for customers by earning points that could be exchanged for attractive prizes and CRM services to accommodate complaints and suggestions from customers to provide better service.

WT (weakness–threats) strategy: Bayu Buana and Panorama have adapted the latest technology, created an integrated platform for all travel products such as transportation tickets, hotel rooms, and vehicle rentals at the destination, provided after-sales services for customers by providing up-to-date information such as tourism packages during the holiday season and birthday or religious day celebrations card, and collaborated with financial institutions that could facilitate their customers in payment.

5 CONCLUSION AND SUGGESTIONS

Based on the company performance and strategy analysis, Panorama has applied a market penetration strategy to improve the company's internal environment to survive.

Meanwhile, Bayu Buana has applied a market development strategy for the development of the company. Several strategies that can be applied by Bayu Buana are as follows:

a. Developing an existing product such as ticketing. Nevertheless, Bayu Buana has to find another way in marketing its product because there are many substitute products with many promotions that attract customers. Therefore, Bayu Buana could make attractive and popular routes such as the tourism route to Raja Ampat, so that the customers would be attracted to buy it.
b. Removing products with high operational costs and lower income, for example, a hotel that costs a lot. However, if it is maintained, they can make attractive promotions or specific products for certain groups, such as the low-cost budget room for staying.
c. Modernizing the company by introducing new destinations using VR.

Several strategies that can be applied by Panorama are as follows:

a. Analyzing the company supply chain for each division to minimize waste time and improve the human resource from its division.
b. Removing products with high operational costs and lower income and then analyzing the strength and weakness of those products for re-development.

It is suggested that Bayu Buana and Panorama hold public activities or events, use email marketing, and apply advertising on the Internet to face this fierce competition in travel and tour services.

REFERENCES

Coulter, Mary K. 2002. Strategic Management in Action, 2nd eds. New Jersey: Prentice Hall

David, Fred R., 2011. Strategic Manajemen Strategis Konsep, 12th eds. Salemba Jakarta: Salemba Empat.

Kasiram, Mohammad. 2010. Metodologi Penelitian. Yogyakarta: UIN MALIKI-PRESS

Contemporary Research on Business and Management – Noviaristanti (Ed.)

Business continuity and Covid-19 pandemic responses on micro-businesses in Surabaya

M.R. Amalia, I.A. Abdillah & G.C. Premananto
Airlangga University, Surabaya, East Java, Indonesia

ABSTRACT: Business continuity and COVID-19 pandemic response are needed for micro-businesses to deal with these unexpected conditions. The purpose of this study is to determine the COVID-19 pandemic response and suggest a suitable BCP design for micro-businesses in Dolly Ex-localization for anticipating the impact of the pandemic. This research was conducted using a qualitative research approach based on a case study. The results showed that there was no structured planning in dealing with business crises for micro-businesses at Dolly Ex-localization. There are four main risks that are being faced; two of them are at a high level and the other two at a moderate level. Thus, a BCP design that is suitable for them is expected to be useful to minimize all the impacts and risks, to maintain the continuity of critical business processes, and to increase the ability of micro-businesses to recover after crisis and disruption as soon as possible.

1 INTRODUCTION

The rapid spread of the COVID-19 outbreak had an impact on the economy in Indonesia. Appeals for social distancing, physical distancing, working, studying, and worshiping at home, and prohibiting activities that cause crowds certainly hampered the economy. Unlike the previous crisis, micro-business became the foundation for the survival of Indonesian economy. The micro-business is the most affected sector during the COVID-19 pandemic. Based on the results of a survey conducted by a number of institutions and the Ministry of MSMEs, the COVID-19 outbreak had major impacts on the sustainability of MSMEs. This is because the micro-business sector targets the economic cycle of middle to lower class people, whose activities were limited when COVID-19 spread. In fact, the number of micro-business and their contribution to the Indonesian economy is enormous.

This phenomenon is also experienced by micro-business at the Dolly Ex-localization area. Micro-business are the main sector that the Surabaya City Government and GMH Foundation, who act as government partners, to revive the economy in the region after Dolly's closure. Some micro-businesses have even become icons of changes to Dolly Ex-localization, including Samijali, Orumy, and Tempe Bang Jarwo (www.detik.com). These three micro-businesses are also assisted by GMH, which are also the objects of this research.

The purpose of this research is to find out the response of micro-business at Dolly Ex-localization in facing the COVID-19 pandemic and to provide a design of suitable business continuity plan (BCP) for micro-business at Dolly Ex-localization. The objective of BCP is to reduce financial risk and improve the organization's ability to recover as soon as possible from a disruptive event (Broto, 2010). In the next section, we will discuss about their anticipations and risks that may potentially occur, as references for designing the BCP.

2 METHODS

The research method used in this research is qualitative based on a case study. The types and sources of data used are as follows: 1) organizational archive records related with informants;

2) documentation and the results of interview reports conducted by researchers with informants at Dolly Ex-localization. Then, the election of the informant in this study is done using purposive sampling. The specific criteria used in this study are micro-business located at the Dolly Ex-localization area under GMH's development. The focus of this research is to explore all information related to business continuity and pandemic responses that were exhibited by micro-businesses in Dolly Ex-Localization. Then, it can be the basis for designing a business continuity plan and pandemic responses that suitable for them.

According to Qintarah, in the research of Safi'i et al. (2020), the results of the respondents' answers were processed using a risk assessment matrix. This is to determine the limit between acceptable risk and unacceptable risk. This step is carried out by assessing the probability of an event occurring (probability) and the impact of the event (impact) on each identified risk. The levels of probability and impact are divided into high, medium, and low, as explained by Marriet (1999). The answers from respondents for the risk items of probability and impact are summed, and the median value is calculated with the following equation (Safi'i, et al., 2020):

$$Me = x\frac{(n+1)}{2} \tag{1}$$

where Me = Median; n = sum; x = data value.

After the assessment was carried out, the calculation process was then carried out into a matrix by multiplying the results of the impact and the probability for each risk (Safi'i, et al., 2020):

$$\text{Risk} = \text{Impact} \times \text{Probability} \tag{2}$$

As for the classification of the risk level criteria are divided into 3 zones: (1) high in the red zone if the risk score is $4 \leq X \leq 9$, then action is needed to control the risk; (2) medium in the yellow zone if the risk score is $2 \leq X \leq 3$, it is advisable to take action if resources are available; (3) low in the green zone if the risk score is 1, then in this zone, there is no need to take action because it is still acceptable.

After the data were collected, business impact assessment will be carried out to create the business impact analysis matrix. According to Santoso and Gitarini (2017), the business impact analysis matrix includes information about business processes and the impact that they caused. In this case, Santoso and Gitarini (2017) use the level severity of impact and maximum acceptable outage time (MAOT), which is the maximum duration of time to stop operations before affecting operations or causing an insurmountable impact for each business process.

According to Santoso and Gitarini (2017), the classification of MAOT is as follows: a score of 1.0 (MAOT $\leq$ 2 days); 0.8 (3 $\leq$ MAOT $<$ 6 days); 0.6 (6 $\leq$ MAOT $<$ 11 days); 0.3 (11 $\leq$ MAOT $<$ 14 days); and 0.1 (MAOT $\geq$ 14 days). Furthermore, the calculation process is carried out into a matrix by multiplying the results of the severity impact and MAOT on each business process and obtaining an overall critical rating.

$$\text{OCR} = \text{SOI} \times \text{MAOT} \tag{3}$$

where OCR = overall critical rating; SOI = value of effect level severity of impact; MAOT = maximum acceptable outage time.

The critical categorization is divided into 3 zones: (1) high if the score is $70 \leq X \leq 100$; (2) medium if the score is $40 \leq X < 70$; and (3) low if the score is <40.

3 RESULT AND DISCUSSION

Interviews were conducted with five people with predefined respondent criteria through an online forum group discussion with a semi-structured interview guide. The purpose of interview questions in this study was to explore the views, experiences, and expectations of research subjects related to the design of a business continuity plan that is suitable to the needs of the micro-business at dolly ex-localization.

The results of interviews and observations with 3 informants of the micro-business and the representative of GMH foundation can identify several risks due to the COVID-19 pandemic, including the following:

1. Risk of cost: This risk is associated with increasing costs and decreasing income of the micro-business during the COVID-19 pandemic. The risk of decreasing income (costs) on average is IDR 4–6 million, which is at the medium level (2), with its impact on the high level (3). Hence, the value of the risk matrix is Impact x Probability = $3 \times 2 = 6$.
2. Risk of delay: This risk is associated with a delay in production time from before the COVID-19 pandemic. The risk of slowing down the production time (schedule) leads to 1–2 days delay on average, which is at the low level (1), with its impact on the high level (3). Hence, the value of the risk matrix is Impact × Probability = $3 \times 1 = 3$.
3. Risk of production: The risk is associated with the decreasing amount of production in micro-business. The risk of the reducing amount of production is 25–50%, which is at the medium level (2), with its impact at the high level (3). Hence, the value of the risk matrix is Impact × Probability = $3 \times 2 = 6$.
4. Risk of raw material: This risk is regarding the increase in prices of main raw materials for the micro-business. The risk of raw materials results in a price increase of 10–40% from the normal price, which is at the low level (1), with its impact on the medium level (2). Hence, the value of the risk matrix is Impact × Probability = $2 \times 1 = 2$.

Further analysis is focused only on the risk value of more than or equal to 2, which are those in the red and yellow zones that occur at the micro-business. These risks include the risk of decreased income and decreased production amount with a matrix value of 6, the risk of delay in production time (schedule) with a matrix value of 3, and the risk of raw material with a matrix value of 2. Thus, the four risks mentioned above need to be taken into further consideration because they already have an impact on the business.

In this study, the level effects were 1 (minor aspect of organization, severity 10); 2 (moderate aspect in one unit or division, severity 25); 3 (moderate aspect of the organization or many divisions, severity 60); 4 (catastrophic aspect in one unit or division, severity 80); and 5 (catastrophic aspects of the entire organization, severity 100). Furthermore, identification of the maximum acceptable outage time (MAOT), the maximum duration of operational shutdown is carried out before affecting operations or causing irreversible impacts for each business process. The MAOT classification used is 1.0 ($\text{MAOT} \leq 2$ days); 0.8 ($3 \leq \text{MAOT} < 6$ days); 0.6 ($6 \leq \text{MAOT} < 11$ days); 0.3 ($11 \leq \text{MAOT} < 14$ days); and 0.1 ($\text{MAOT} \geq 14$ days).

The business impact analysis matrix table aims to determine the critical level of each business process in micro-business at Dolly Ex-localization. From the table, it appears that most of the business processes are in the medium and high categories, including the process of receiving orders, the process of purchasing raw materials, the production process, the packaging process, and the shipping process. This shows that every business process of the micro-business has a fairly strong dependency relationship. Only one business process is in the low category, which is the weekly evaluation process.

Furthermore, the recovery target time scale describes the time scale for the business recovery target of micro-business at Dolly's Ex-localization returning to its original condition. Some business processes, even though they are classified as a high critical level (order acceptance process, raw material purchasing process, production process, packaging process, and shipping process), can be placed in the category of recovery time of less than 7 days. Meanwhile, the weekly evaluation business process is classified as a low critical level process.

To reduce the risk of deeper losses for the micro-businesses due to the COVID-19 pandemic, their COVID-19 pandemic response has been studied. From the results of the COVID-19 pandemic response checklist, it was understood that out of the total of 19 points, only 2 points have been done by them, 7 points are still in progress, and there are 10 points that have not been done in response to the COVID-19 pandemic. So, it can be concluded that micro-businesses at Dolly Ex-localization, Surabaya, do not yet have a good response and follow-up plan in dealing with the COVID-19 pandemic.

In the initial stage of the recovery plan, use of the BIA that has been made previously is required. At this stage, BIA is used to identify and record the minimum resource requirements that are needed for business operations. By implementing this plan, businesses can determine the important resources that are needed for the company's business activities to run normally. In this study, the identification of short-term and long-term resources needs to be carried out because this pandemic has been going on for 9 months since March 2020.

After identifying minimum short-term and long-term resource requirements, the next step is to explore recovery strategy options to reduce risks and speed up recovery, that is, rebudgeting strategy, safety cash strategy, monitoring and stock opname, backup documentation strategy, new market penetration strategy, media and communication, and then resize and repackaging strategy.

4 CONCLUSION

From the results of this study, it can be concluded that the design of a business continuity plan is needed for micro-businesses at Dolly Ex-localization to face unexpected conditions that can affect their business activities. From the results of the COVID-19 pandemic response checklist, it can be concluded that they do not yet have a good response and follow-up plan in dealing with the COVID-19 pandemic. The purpose of designing the BCP in this study is to provide views to micro-businesses in Dolly Ex-localization in anticipating and responding to the impact of COVID-19 on their activities, which may need to be done at the earliest in the next 6 months, considering that the COVID-19 cases in Indonesia have not decreased until now. The stages of BCP design for micro-businesses at Dolly Ex-localization are risk analysis, business impact analysis, incident response plan, recovery plan, review, and BCP approval.

REFERENCES

Broto, T. W. (2010). *Pentingnya Menyusun Business Continuity Plan bagi Organisasi. Retrieved from BDI Surabaya*: http://bdisurabaya.kemenperin.go.id/2010/10/10/pentingnya-menyusun-business-continuity-plan-bagi-organisasi/.

Yin, R. K. (2009). *Case study research: Design and methods (4th Ed.)*. Thousand Oaks, CA: Sage.

Contemporary Research on Business and Management – Noviaristanti (Ed.)

Green supply chain management in the food industry: A case study during COVID-19

J. Laksmanawati
Diponegoro University, Semarang, Indonesia

ABSTRACT: Green supply chain management (GSCM) plays an important role in and contributes to environmental issues. The conceptual pattern of GSCM focuses on multidisciplinary economics and environmental issues arising from green practice management in the field of the supply chain. Therefore, food industries are supposed to plan their operational procedure from the product packaging design process, selection of the source of raw materials, production process, delivery of the products, and waste management when they plan and arrange the supply chain. By employing a qualitative approach, this study investigated how food industries maintain their operations without neglecting the sustainability aspect and health protocol. The findings contribute to the latest perception of green sustainability during COVID-19, and its implication can be used as references by managers.

1 INTRODUCTION

Environmental and social issues have been receiving a lot of attention. One of these issues is global warming, which has become the spotlight of the World Economic Forum on January 21–24, 2020, in Davos, Switzerland. In this regard, the issue of global warming has become world's interest and struggle due to its increasingly critical impacts felt every year (Daily, 2020). In the last few months, the world has faced an outbreak of an infectious disease caused by 2019 coronavirus infection (COVID-19). The World Health Organization (WHO) stated that the disease (Covid-19) firstly appeared in Wuhan, China, at the end of December 2019 (Rume & Islam, 2020). The outbreak has also affected people's lives globally, and policies must be made to control the spread of viruses, which indirectly contribute to environmental changes.

Plastics have become an important component of products and packaging because they are durable, lightweight, and inexpensive. Plastic packaging is considered practical because it is single-use and faster to dispose. However, plastic waste disposal in the landfill is considered unsustainable from an environmental point of view. This causes the capacity of the final disposal site to decrease, and the garbage accumulating in landfills has an impact because it is difficult to harvest in a short time (Febriansyah, 2019).

During the pandemic, the food industry changed rapidly due to consumers' awareness of traceability in the food supply chain, the origin of raw materials and food safety, the environmental impact of products and processes, and other social issues increase. This is because the food industry can be considered as a link between humans and their natural surroundings. This issue is related to processed food products that are friendly to the environment and safe against viral and bacterial contamination. (Topleva & Prokopov, 2019).

In this regard, the food industry must provide food products that are nutritious and safe, of high quality, and has a longer shelf life. As a result, they use various kinds of plastic packaging combined with aluminum, glass, and paperboard. The use of these various materials has resulted in

DOI 10.1201/9781003196013-37

a multi-component packaging concept that has a major part in the sustainability and social aspects of the society today (Boziaris, 2014). The conceptual pattern of GSCM focuses on multidisciplinary economic and environmental issues arising from environmental management practices in the field of the supply chain. In this case, Lu and Taylor (2016) stated that industries have to consider environmental issues (energy, gas emission, raw material, and waste) when designing and managing their supply chain. This includes the operational stages of the company itself such as product design, selection of raw material sources, production processes, product delivery, and final management of a product. (Cankaya & Sezen, 2019).

2 THEORETICAL BASIS

2.1 *Green Supply Chain Management (GSCM)*

The GSCM concept was first introduced by Michigan State University in 1996 when solving the "environmentally responsible manufacturing" issue. Rooted in supply chain management ideas and development theory, GSCM is considered to be a new subject that has not been studied thoroughly. However, recently, several studies have been carried out to investigate GSCM. According to Rha (2010), green supply refers to how the industries have to consider the environmental context in the field of innovation and purchase. Thus, the system in GSCM is complicated because it is related to suppliers, manufacturers, distributors, retailers, customers, and logistics. GSCM starts with the design of the product and ends in recycling of the final product. Consequently, GSCM covers five aspects, namely, green procurement, green design, green manufacturing, green distribution, green consumption, and recycling.

Integrated green supply chain planning requires the management of the company or organization to determine the inputs, drivers, and initial planning that must be processed for the production, transportation and distribution, packaging, and recycling of green products (Elbaz & Iddik, 2019). Over the last few decades, several innovative practices and technologies have emerged to achieve automation, simplification, optimization, and redesign of the GSCM process. Particularly, the following initiatives have been promoted: (1) raw material procurement, production, re-production, warehouse, supply chain design, and waste management; (2) improving communication and integrating supply chain partners; and (3) supporting the decision-making process at three levels of business (operational, tactical, and strategic) and addressing the needs to identify outputs and impacts obtained socially, financially, and environmentally.

3 METHODOLOGY

In nature, this is a qualitative study that was conducted to explore and obtain empirical evidence about the motivation and implementation of GSCM in the food industry during a pandemic. To investigate the issue, this study employed a case study approach to make comparisons between similarities and differences in cases easier. A case study is a qualitative approach that investigates one or more cases from time to time through detailed data collection involving many sources of information such as observations, interviews, audiovisual materials, documents, and many others. (Creswell, 2007). This approach was chosen because it is an empirical method based on the current phenomena in society, and an in-depth investigation can be conducted when the boundaries between the phenomenon and the environment are less clear. (Yin, 2018). In connection with this statement, Linton et al. (2007) emphasized that a case study is highly recommended for studying sustainable development because it has several advantages including being able to explain the linkages between causal factors and identifying and exploring concepts, core variables, and their meanings. In addition, by implementing this approach, the validity of a study can be improved because the combined data collected can create stronger empirical evidence (Neuman, 2014).

4 RESULTS AND DISCUSSION

4.1 *Green procurement*

Green procurement is the purchase process of raw materials involving suppliers and logistics. Various aspects of both the supplier and logistics must be considered by the company to ensure that the whole process of the supply chain can be implemented successfully and cleanly. Based on the interview results, the food industry players conducted a strict food ingredients selection from suppliers who had obtained business licenses during the pandemic and had implemented health protocols before sending raw materials to be processed.

4.2 *Green design*

Several studies have revealed that 70% to 80% of product performance is determined at the design stage. Meanwhile, the design process generally costs only about 10% of production. Therefore, companies have to consider the ecological and environmental impacts of the product design stage. The product design is expected to minimize energy consumption and environmental pollution. Benson in 1992 stated that the green design includes studies on standard, modular, removable, and retractable designs, as follows:

1. Design standards make structural members relatively fixed so that production difficulties and energy consumption are reduced. In addition, standard designs can reduce the complexity of technological equipment.
2. The modular design is intended to meet the requirements for the rapid development of environmentally friendly products, so that they are easy to install, dismantle, and maintain.
3. The detachable design means that all parts of the structure design are reasonable, accessible as a separate structure, and can be easily removed without damaging the components, thereby reducing environmental pollution.
4. Recyclable designs are designs that can be used repeatedly to its maximum usage. The findings show that several business actors have implemented reusable packaging designs, even though the packaging material is still made from plastic. This method is conducted to reduce the spread of viruses that can be transmitted through the use of non-plastic packaging.

4.3 *Green distribution*

Green distribution is based on logistical processes. With a dramatic increase in the number of logistics and increased traffic flow, the atmospheric environment is highly contaminated. This mainly consists of an evaluation of centralized distribution and consumption of resources and planning of reasonable transportation routes. During the pandemic, several cities have implemented large-scale social restrictions that impact the distribution of food to consumers. As a result, to keep business operations running, industries have used online transportation services that have complied with the requirements of the health protocol (using masks, checking temperatures, etc.). It was done so that the consumers will feel safe without having to leave their home.

4.4 *Recycling*

Since the increase of technological development, products have become complex and their life cycle has become shorter. As a result, they produce consumer waste that is a significant source of environmental pollution. Therefore, a green recycling system is needed to solve this issue. After the "stay at home" campaign was issued by the government several months ago, industry players have been competing to create unique food packaging creations and participating in reducing solid waste. For example, one of the study objects stated that they are using 1-liter bottles as packaging to sell their beverage product.

5 CONCLUSION

Based on the results of the interview with several informants, it can be concluded that the key to successfully implementing GSCM amid the Covid-19 outbreak is the willingness of industry players to be strict in selecting and sorting raw materials from suppliers to comply with the health protocol during the pandemic. In addition, business actors are also required to be creative in creating environmentally friendly packaging (which can be used repeatedly) and to make consumers feel safe with the quality of the products offered.

REFERENCES

Arvanitoyannis, I. S. (2008). *Waste Management for the Food Industries.* London: Elsvier Inc.

Bauman, B. (2019, August 20). *Yale Climat Connection.* Retrieved from The Yale Center for Enviromental Communaction Website: https://yaleclimateconnections.org/2019/08/how-plastics-contribute-to-climate-change/

Boziaris, I. S. (2014). *Seafood Processing Technology, Quality, and Safety.* West Sussex: John Wiley&Sons, Ltd.

Cankaya, S. Y., & Sezen, B. (2019). Effects of Green Supply Chain Management Practices on Sustainability Perfomance. *Journal of Manufacturing Technology Management, 30*(1), 98–121.

Creswell, J. W. (2007). *Qualitative Inquiry and Research Design: Choosing Among Five Approaches* (2nd ed.). United States: Sage Publications, Inc.

Daily, I. (2020, January 23). *Investor.id.* Retrieved from Investor Daily Indonesia Web site: https://investor.id/editorial/pemanasan-global

Elbaz, J., & Iddik, S. (2019). Culture and Green Supply Chain Management: A Systematic Literature Review and A Proposal of a Model. *Management of Environmental Quality: An International Journal*, 483–504.

Febriansyah. (2019, December 6). *Tirto.id Sosial Budaya.* Retrieved from Tirto.id: https://tirto.id/penyebab-perubahan-iklim-fakta-dan-solusinya-emYU

Kosseva, M. R., & Webb, C. (2013). *Food Industry Wastes: Assesment and Recuperation Commodities.* San Dieso, USA: Elsviere Inc.

Linton, J., Klassen, R., & Jayaraman, V. (2007). Sustainable Supply Chains: An Introduction. *Journal of Operations Management*, 1075–1082.

Lu, W. B., & Taylor, M. E. (2016). Which Factors Moderate the Relationship between Sustainability Perfomance and Financial Perfomance? A Meta-analysis Study. *Journal of International Accounting Research*, 1–15.

Neuman, W. L. (2014). *Social Research Methods: Qualitative and Quantitative Approaches.* Essex: Pearson Education Limited.

Rha, J. S. (2010, July 20). The Impact of Green Supply Chain Practices on Supply Chain Performance. Lincoln, Nebraska: Digital Commons University of Nebraska – Lincoln.

Rume, T., & Islam, S. D.-U. (2020). Environmental effects of COVID-19 pandemic and potential strategies of sustainability. *US National Library of Medicine.*

Shafy, H. I., & Mansour, M. S. (2018). Solid Waste Issue: Sources, Composition, Disposal, Recycling, and Valorization. *Egyptian Journal of Petroleum*, 1275–1290.

Topleva, S. A., & Prokopov, T. V. (2019). Integrated Business Model for Sustainability of Small and Medium-sized Entreprises in the Food Industry. *British Food Journal.*

Yin, R. K. (2018). *Case Study Research and Applications: Design and Methods.* California: Sage Publications, Inc.

Contemporary Research on Business and Management – Noviaristanti (Ed.)

The effect of financial inclusion on micro, small, and medium enterprise performance using fintech as a moderating variable

K. Efan, S.R. Basana, & R.S.D. Ottemoesoe
Faculty of Business and Economics, Petra Christian University, Surabaya, Indonesia

ABSTRACT: Financial inclusion can reduce social disparities and poverty and increase economic growth and national financial efficiency. Access to financial services is provided to all segments of the society. Currently, MSMEs are still experiencing difficulties in terms of capital. Access to financial services in banking is also still difficult due to the lack of financial information on the MSMEs when compared to large, more transparent companies; thus, banks have more difficulty in assessing the creditworthiness of MSMEs. This study examines the influence of financial technology (fintech), which is developing rapidly in Indonesia. The variables used in this study are financial inclusion, MSME performance, and the use of fintech e-payment as moderating variables. The results revealed that financial inclusion has a relationship with the performance of MSMEs. Financial inclusion encourages people to access financial products, and people can more freely use their money to make transactions. The fintech variable does not have a moderating effect on the relationship between financial inclusion and MSME performance.

1 INTRODUCTION

Financial inclusion is public access to financial services by increasing the reach, quality, and availability of the existing financial services. Indonesian Bank (BI) explains that this financial service is provided to all segments of society, especially the poor and low-income people, the productive poor workers, migrant workers, and people living in remote areas. According to data from the *Otoritas Jasa Keuangan* (OJK/Financial Service Authority), as many as 54 million Indonesians are still not reached by conventional financial services, such as banks, which means that many people do not have a bank account (Ramadhani, 2020). Currently, MSMEs are still experiencing difficulties in terms of capital. Enny Sri Hartati, who is the Director of the Institute for Development of Economics and Finance, explained that capital is one of the main obstacles faced by MSMEs with a percentage of 60% (Firdaus, 2020). Access to financial services in banking is still difficult due to the lack of financial information on MSMEs when compared to large, more transparent companies, so banks have more difficulty in assessing the creditworthiness of MSMEs and providing loans to MSMEs. Sometimes, the owners of MSMEs have difficulty in making good financial records/bookkeeping, and thus, it is not easy for banks to carry out risk assessments and creditworthiness. Thus, MSMEs can fail to fulfill the number of orders or consumer requests due to a lack of production capital (Setyadi, 2017).

The financial sector in Indonesia is quite developed based on the emergence of many financial technology-based startup companies, or commonly referred to as fintech in recent years (Ardela, 2017). According to BI, the state benefits from the existence of fintech, for example, by encouraging economic policies, increasing the velocity of money, improving the people's economy, and promoting a national financial inclusion strategy. Fintech increases financial inclusion by providing responsible and sustainable access for individuals and businesses to financial products and services such as transactions, payments, savings, credit, and insurance (Batunanggar, 2019). This study examines the effect of financial inclusion on the performance of MSMEs in Surabaya City. It focuses on MSMEs engaged in the food and beverage sector; this sector absorbs 50% of the

DOI 10.1201/9781003196013-38

household income (Setiawan, Emvalomatis, & Lansik, 2012). This study examines the impact of using fintech e-payment on the relationship between financial inclusion and MSME performance. The rapid development of fintech, especially in Indonesia, is worth researching because fintech can increase the development of new companies, facilitate financial services by streamlining energy and time, and support national financial inclusion (Ramadhani, 2020).

2 RELATIONSHIP BETWEEN RESEARCH CONCEPTS

Chauvet and Jaolin (2017) explain that financial inclusion or public access to financial services is beneficial for all types of businesses, and it affects the growth of a business. According to Agnello, Mallic, and Sousa (2012), financial inclusion contributes to economic growth. In addition, Kodan and Chhikara (2013) also found a relationship between financial inclusion and literacy and per capita income (economic growth). These findings show that with a good level of financial inclusion, there will be growth in per capita income, which will be reflected in MSMEs' income. The research conducted by Yanti (2019) explains that the MSME performance will increase if they continue to increase financial inclusion. The increase in financial inclusion or access to banking products will help MSMEs to obtain funds that can be used for business capital, daily needs, expansion, and reserve funds for the future of these MSMEs. Therefore, the first hypothesis is as follows:

H1: Financial inclusion affects the performance of MSMEs.

The research conducted by Rahmi (2018) explains that financial inclusion in Indonesia requires fintech to accelerate its progress in the digital era. By embracing fintech, it will provide great assistance for the Indonesian government to achieve the Go Digital Vision 2020 goals (Rahmi, 2018). The research conducted by Kunt et al. (2013) explains that digital technology can increase financial inclusion; thus, people who are not eligible to take loans from banks will enjoy greater access to financial facilities with fintech. This is supported by the research on the case of microloan startup companies conducted by Leong et al. (2017), where fintech can increase financial inclusion of people who take loans from fintech companies. Fintech reaches out to parties that have been neglected by banks and provides better service to the banks' customers (Batunanggar, 2019). Changing cellular technology has opened opportunities and enabled nearly three billion people without bank accounts to access financial services (Iman, 2018). The research conducted by Soriano (2017) on start-up fintech companies explains that company founders, who have experience using financial services, are customer-oriented (customer-centricity), and have strategic partnerships with financial institutions and e-commerce companies, have a positive correlation with financial inclusion (measured by active customers) and financial performance (measured by annual revenue). Fintech will encourage people to have a bank account before using its services. Afterward, they can use or get services such as payment/credit easily. In addition, by using fintech, there will be data that can be used by banks or other institutions to evaluate the feasibility of providing credit. This financial service will be used to improve its performance in the present and as a reserve fund for the future.

H2: The use of fintech services moderates the relationship between financial inclusion and MSME performance.

3 METHODS

The population used in this study was the MSMEs in Surabaya City. The research sample was selected using the non-probability sampling technique employing purposive sampling. The sample criteria included in this study were MSME owners/managers engaged in the food and beverage sector and using fintech e-payment services. This study used a questionnaire as a way of collecting the data. The questionnaire contains optional questions to obtain information from the respondents. Each question is given a score from 1 (strongly disagree) to 5 (strongly agree). The analysis process carried out in this study was to collect research data by distributing questionnaires to the respondents in order to meet the target sample size required and process the data from the questionnaire that had been obtained by testing the outer model, inner model, and hypothesis testing using the SEM with SmartPLS version 3.0 software.

4 FINDINGS AND DISCUSSION

The respondents in this study were MSMEs in Surabaya City engaged in the food and beverage sector. The research data were obtained using a questionnaire that was distributed online to the respondents via the Google Form link, namely, MSME actors who live in Surabaya City who use fintech e-payment services. However, 32 questionnaires did not match the specified criteria; thus, the questionnaires that met the processing requirements were only 84 questionnaires.

Overall answers regarding financial inclusion statements obtained a mean value of 4.57. The respondents made financial transactions more than twice in a month by looking at the mean of the statement of depositing or withdrawing money at least twice in a month, namely, 4.02 and 4.11. In addition, banks and their facilities are already available in various places, making them easily accessible to the public. Furthermore, the respondents' overall answers regarding the use of fintech e-payment obtained a mean value of 4.36. The respondents felt that the fintech e-payment service was easy to use, indicated by the mean of the statement of the ease of learning, operating, understanding, and mastering the e-payment facility, namely, 4.48, 4.69, 4.37, and 4.46, respectively. Moreover, the respondents often made transactions using e-payment services, as indicated by the mean of 4.18. The respondents' answers as a whole regarding the MSME performance statement obtained a mean value of 4.33. This is indicated by the statement of the increase in income, sales, and business production with a mean of 4.12, 4.17, and 4.11, respectively. In addition, the respondents' payments using e-payment services were already very fast and easy, indicated by the mean of each statement, namely, 4.65 and 4.62.

The next analysis is to test the research hypothesis using the *t*-test, namely, by looking at the value of the *t*-statistic, which is compared to the t-table value of 1.64.

Table 1. Research hypothesis testing.

Research Hypothesis	*t-statistic*	*Original Sample*	*p Values*	Information
Financial inclusion → MSME performance (H1)	2.074	0.279	0.039	Hypothesis is accepted
Moderating effect of the use of fintech e-payment on the relationship of financial inclusion → MSMEs performance (H2)	0.305	0.029	0.761	Hypothesis is rejected

Table 2 depicts that the influence of FI (financial inclusion) and KI (MSME performance) resulted in a *t*-statistic of 2.074, which is greater than the *t*-table of 1.64. In other words, H0 was accepted. Therefore, it can be concluded that financial inclusion affected MSME performance. The value of the original sample was 0.279, which indicated that financial inclusion had a positive effect on MSME performance. This shows that public access to financial products such as banking products is one of the factors that can improve MSME performance. The respondents can access their bank accounts through banks/ATMs and their cell phones. Their access to bank accounts is very important; in this case, the access made is like financial transactions. The results of this study are consistent with research of Chauvet and Jaolin (2017), who stated that financial inclusion has an impact on company growth. In line with Indriyati (2020), financial inclusion also affects MSME performance. This indicates that the better the financial inclusion, the higher will be the level of MSME performance and vice versa. The research by Yanti (2019) also reveals that financial inclusion influences MSME performance.

In addition, Table 2 shows that the moderating effect of using fintech e-payment services and the relationship between financial inclusion and MSME performance had a *t*-statistic of 0.305, and this value was below 1.64, which means that H0 was rejected. Therefore, the use of fintech did not moderate the effect of financial inclusion on the MSME performance. The category of the moderating variable can be seen in the results of the *p* value or alpha. The *p* value obtained was below 0.005 or 5%. If the *p*-value is above 0.005, the relationship is considered insignificant. In Table 2, it can be seen that the p-value of the moderating effect is 0.761, which means it was not significant because it was above 0.005. In addition, the original sample value of the moderation relationship, which was 0.029, indicated that the moderating variable had a positive effect on the relationship between financial inclusion and performance.

5 CONCLUSION

The data analysis results show that financial inclusion has a relationship with MSME performance. Financial inclusion encourages people to access financial products, and people can use their money more freely to make transactions wherever they are, in this case, making transactions in food and beverage businesses in Surabaya City. From the results of hypothesis testing, it was found that the variable use of fintech services did not moderate the relationship between financial inclusion and MSME performance, but it still had a positive, insignificant effect on this relationship. The use of fintech services is not visible because the majority of respondents already have bank accounts and use fintech e-payment services, so that the respondents are familiar with the existing system. Therefore, the effect of using this system does not have a significant impact on its performance.

REFERENCES

Agnello, L., Mallick, S. K., & Sousa, R. M. (2012). Financial reforms and income inequality. *Economics Letter*, 583–587.

Ardela, F. (2017, October 24). *Teknologi finansial: Tengok dulu perkembangan fintech di Indonesia*. From Finansialku: https://www.finansialku.com/perkembangan-fintech-di-indonesia/

Batunanggar, S. (2019). Fintech development and regulatory frameworks in Indonesia. *Asian Development Bank Institute*, 1–12.

Chauvet, L., & Jaolin, L. (2017). Financial inclusion, bank concentration, and firm performance. *World Development*, 1–13.

Iman, N. (2018). Is mobile payment still relevant in the fintech era? *Elsevier*, 72–82.

Indriyati, N. (2020). *Pengaruh inklusi keuangan dan literasi keuangan terhadap kinerja UMKM batik di kabupaten Tegal.* Tegal: Universitas Pancasakti Tegal.

Kodan, A. S., & Chhikara, K. S. (2013). A theoretical and quantitative analysis of financial inclusion and economic growth. *sagepub*, 103–133.

Kunt, A. D., & Klapper, L. (2013). Measuring financial inclusion: Explaining variation in use of financial services across and within countries. *Brookings Papers on Economic Activity*, 279–340.

Leong, C., Tan, B., Xiao, X., Tan, F. T., & Sun, Y. (2017). Nurturing a fintech ecosystem: The case of a youth microloan startup in China. *International Journal of Information Management*, 92–97.

Rahmi, M. (2018). Fintech for financial inclusion: Indonesia case. *1st International Conference on Economics, Business, Entrepreneurship, and Finance (ICEBEF 2018)* (pp. 1–3). Indonesia: Atlantis Press.

Ramadhani, N. (2020, March 03). *6 manfaat fintech yang dapat dirasakan oleh banyak orang*. From Akseleran: https://www.akseleran.co.id/blog/manfaat-fintech/

Ramadhani, P. I. (2020, August 24). *Kurang literasi, ini tantangan besar kembangkan fintech di Indonesia*. From Liputan6: https://www.liputan6.com/bisnis/read/4338030/kurang-literasi-ini-tantangan-besar-kembangkanfintech-di-indonesia

Setiawan, M., Emvalomatis, G., & Lansik, A. O. (2012). The relationship between technical efficiency and industrial concentration: Evidence from the Indonesian food and beverages industry. *Journal of Asian Economics*, 466–475.

Setyadi, A. (2017, May 16). *UMKM sulit dapat modal usaha, ini solusi BI*. From detikfinance: https://finance.detik.com/berita-ekonomi-bisnis/d-3502490/umkm-sulit-dapat-modal-usaha-ini-solusi-bi

Soriano, M. A. (2017). *Factors driving financial inclusion and financial performance in fintech new ventures: An empirical study.* Singapore: Singapore Management University.

Yanti, W. I. (2019). Pengaruh inklusi keuangan dan literasi keuangan terhadap kinerja UMKM di kecamatan moyo utara. *JURNAL MANAJEMEN DAN BISNIS*, 1–10.

Contemporary Research on Business and Management – Noviaristanti (Ed.)

The importance of simple financial statements and Standard Operating Procedures (SOPs) in Indonesian MSMEs during the COVID-19 pandemic

L.I. Zulkarnain & R. Lupiyoadi
Faculty of Economics and Business, Universitas Indonesia, DKI Jakarta, Indonesia

ABSTRACT: COVID-19 pandemic has caused a domino effect on almost all lines. Indonesian MSMEs have been among the worst affected. This study aims at helping MSMEs, especially during the pandemic. The business coaching method with a qualitative approach was used to investigate the actual conditions and problems. Then, the agreed-upon action was taken as a solution to overcome the problem. After conducting several external and internal analyses, interviews, and observations, some of the obstacles faced by MSMEs were revealed. In this study, gap and Pareto analyses were used to emphasize the most important problem, which was the absence of financial statements and standard operating procedures (SOPs). It is important for MSMEs to make simple financial statements and SOPs, so they can forecast their performance and financial health. The results showed that MSMEs could successfully calculate their profitability and can take the right business decisions and make their production processes more efficient.

1 INTRODUCTION

According to the Central Bureau of Statistics data, micro, small, and medium enterprises (MSMEs) play an important role in determining the direction of general policies for economic growth in Indonesia. Based on the Ministry of Cooperatives and SMEs of the Republic of Indonesia's data, both in traditional and modern sectors, MSMEs were able to provide 96.87 percent of employment and employ up to 89.2 percent of the workforce in Indonesia. This is what makes the government to rely on MSMEs as one of the sectors that drive the national economy.

NA Production House is one of the MSMEs that produce birthday supplies and other party supplies in Jakarta, Indonesia. The products produced by NA Production House are party supplies that are mostly made from paper, such as bunting flags, birthday hats, and birthday invitation cards. NA Production House's business has been running for about four years, starting from July 27, 2015. In running this business, the owners acknowledge that their B2B consumers who resell their party supplies have been loyal to them because NA Production House not only guarantees good quality, but they are also capable of maintaining product differentiation as their competitive advantage. However, the high consumer demand has gradually made the owners aware that their current production capacity and capabilities are not sufficient to meet all the demands of their consumers.

In developing this business, the owners wish to solve the problems. To solve the issues, gaps analysis and Pareto analysis were used to emphasize the two of the most important problems. Through business coaching, NA Production House's goals to develop their business through increasing production capacity is expected to be achieved with a focus on problems prioritized on the results of the Pareto analysis: (1) making simple financial statements and (2) making standard operating procedures (SOPs).

DOI 10.1201/9781003196013-39

2 LITERATURE REVIEW

Weygandt et al. (2013) defined accounting as a process of recording all financial events and producing a report that can be used by various parties interested in the economic activity and the conditions of a business. In addition, Kieso (2013) added that accounting provides reliable, relevant, and timely information regarding the efficient measurement of company profitability and an overview of the company's financial health to interested parties.

According to Kasmir (2013), financial statements are reports that show the company's current financial condition or the company's financial condition within a certain period. Financial statements are made so that a company can survive, avoid bankruptcy, compete with other companies, maximize sales, minimize costs, maximize profits, and maintain a steady increase in income (Ross et al., 2013). Meanwhile, according to the Indonesian Accounting Association (2015), the purpose of financial reports is to provide information regarding the financial position, performance, and changes in the financial position of a company that is useful for a large number of users in making decisions. The procedures that can be taken to make simple financial statements according to Anthony (2011) are (1) collecting proof of transactions that occur within the company, (2) recording proof of transactions in the journal, (3) adjusting the journal, and (4) preparing the financial statements.

In addition, according to Griffin (2004), SOPs are a series of systematically structured guidelines for the processes, tasks, and roles of individuals or groups that are carried out daily in an organization. The benefits of SOPs include clarifying the functions and roles of each position in an organization so that there is no misuse of work activities, clarifying the flow of tasks or activities done by each employee, and avoiding failure, doubt, and duplication of work activities in the organization (Rachmi, 2018).

3 METHODOLOGY

This study employed a qualitative method through business coaching. It was conducted for 10 months, starting from February 2020 to November 2020. The primary data were obtained through direct interaction with the object, which was NA Production House, in the form of in-depth interviews, observations, and surveys that were conducted several times to examine the main problems. In addition, other data were collected from a survey of the NA Production House's B2B consumers and the owner's documents, such as daily production capital, daily sales records, and the list of orders made by NA Production House's consumers.

The data were analyzed using business model canvas (BMC) analysis, PESTEL analysis, Five Forces Model of Competition analysis, STP analysis, financial analysis, SWOT analysis, TOWS matrix, and marketing mix analysis. Then, the results from these analytical tools were compiled into a gap analysis table. In addition, this study also employed a Pareto analysis to provide solutions for MSME's problems.

4 RESULTS AND DISCUSSION

After mapping the conditions of the MSME, the business coaching method was conducted to implement the proposed solutions. From the Pareto analysis, two proposed solutions had been approved: first, creating simple financial statements, hence, it can be used by the MSMEs to analyze financial performance and to make business decisions; and second, developing SOPs as a guide for their production employees, so that the production process can be done effectively.

The owners admitted that recording financial statements was something they never routinely did. Therefore, during the four years of running their companies, they had not been able to separate their finances from business finances. The owners also had difficulty in increasing overtime hours or salaries for their employees in times of high demand because they did not know the amount of budget that has to be allocated for the additional work hours.

Through business coaching, sales report data from NA Production House's daily production delivery were collected. The sales report data collection was done by observing the owner's WhatsApp chat history with their B2B consumers. However, not all data can be collected through scrolling the chat history. Thus, several items of data were missing and could not be recorded in the journal entry. From the data, a simple financial statements template in Microsoft Excel, which contains the recording of daily production capital and daily sales, was made, and it can be presented in the form of simple financial statements. The template also has an automatic calculation formula, so the owner no longer needs to calculate them manually.

No.	Tanggal	Jenis Barang	Jumlah Barang	Modal Bahan Baku Produksi	Ongkos Cetak Produksi	Upah Karyawan Per Produksi	Biaya Lain-lain	Total Modal Produksi
1	3 Januari 2020	Bunting Flag HBD Princess	Min 100 lusin	1,842,900	720,000	150,000	115,000	2,827,900
2	4 Januari 2020	Bunting Flag HBD Tayo	Min 100 lusin	1,842,900	720,000	150,000	115,000	2,827,900
3	4 Januari 2020	Bunting Flag Happy Engagement Hitam	Min 100 lusin	1,842,900	720,000	150,000	106,000	2,818,900
4	7 Januari 2020	Bunting Flag HBD Doraemon Pelangi	Min 100 lusin	1,842,900	720,000	150,000	90,000	2,802,900
5	8 Januari 2020	Bunting Flag HBD Spiderman	Min 100 lusin	1,842,900	720,000	150,000	140,000	2,852,900
6	8 Januari 2020	Bunting Flag HBD Sofia	Min 100 lusin	1,842,900	720,000	150,000	90,000	2,802,900
7	9 Januari 2020	Bunting Flag Happy Wedding Bunga	Min 100 lusin	1,842,900	720,000	150,000	115,000	2,827,900
8	10 Januari 2020	Bunting Flag HBD Unicorn Putih	Min 100 lusin	1,842,900	720,000	150,000	106,000	2,818,900
9	11 Januari 2020	Bunting Flag HBD Polcadot Pink	Min 100 lusin	1,842,900	720,000	150,000	122,000	2,834,900
10	13 Januari 2020	Topi Cars, Spiderman, Tayo, Hello Kitty	Min 1.200 pack	2,544,000	920,000	200,000	105,000	3,769,000
11	14 Januari 2020	Bunting Flag HBD Captain America	Min 100 lusin	1,842,900	720,000	150,000	122,000	2,834,900
12	14 Januari 2020	Bunting Flag HBD Pony Pink	Min 100 lusin	1,842,900	720,000	150,000	90,000	2,802,900
13	15 Januari 2020	Bunting Flag HBD Polcadot Hitam Gold	Min 100 lusin	1,842,900	720,000	150,000	90,000	2,802,900
14	16 Januari 2020	Topi Unicorn, Doraemon, Pony, Tsum2	Min 1.200 pack	2,544,000	920,000	200,000	140,000	3,804,000
15	17 Januari 2020	Bunting Flag Happy Engagement Putih	Min 100 lusin	1,842,900	720,000	150,000	115,000	2,827,900
16	20 Januari 2020	Bunting Flag Happy Wedding Orang	Min 100 lusin	1,842,900	720,000	150,000	106,000	2,818,900
17	21 Januari 2020	Bunting Flag Happy Engagement Sakura	Min 100 lusin	1,842,900	720,000	150,000	90,000	2,802,900
18	22 Januari 2020	Topi Cars, Spiderman, Tayo, Hello Kitty	Min 1.200 pack	2,544,000	920,000	200,000	122,000	3,786,000
19	22 Januari 2020	Bunting Flag HBD Unicorn Pelangi	Min 100 lusin	1,842,900	720,000	150,000	122,000	2,834,900
20	23 Januari 2020	Bunting Flag HBD Polcadot Biru	Min 100 lusin	1,842,900	720,000	150,000	140,000	2,852,900
21	24 Januari 2020	Bunting Flag Happy Engagement Glitter	Min 100 lusin	1,842,900	720,000	150,000	90,000	2,802,900
22	25 Januari 2020	Topi Thomas, C. America, Pony, Minion	Min 1.200 pack	2,544,000	920,000	200,000	115,000	3,779,000
23	25 Januari 2020	Bunting Flag HBD Tsum2	Min 100 lusin	1,842,900	720,000	150,000	106,000	2,818,900
24	27 Januari 2020	Bunting Flag HBD Batman	Min 100 lusin	1,842,900	720,000	150,000	140,000	2,852,900
25	28 Januari 2020	Topi Thomas, C. America, Pony, Minion	Min 1.200 pack	2,544,000	920,000	200,000	115,000	3,779,000
26	30 Januari 2020	Undangan All Characters	Min 4.200 pack	4,556,000	1,120,000	200,000	115,000	5,991,000
27								
28								
29								
30								
							GRAND TOTAL	81,376,000

No.	Tanggal	Jenis Barang	Jumlah Barang (dalam lusin/pack)	Harga Jual (lusin/pack)	Total Pendapatan
1	3 Januari 2020	Bunting Flag HBD Princess	102	43,000	4,386,000
2	4 Januari 2020	Bunting Flag HBD Tayo	105	43,000	4,515,000
3	4 Januari 2020	Bunting Flag Happy Engagement Hitam	104	43,000	4,472,000
4	7 Januari 2020	Bunting Flag HBD Doraemon Pelangi	105	43,000	4,515,000
5	8 Januari 2020	Bunting Flag HBD Spiderman	102	43,000	4,386,000
6	8 Januari 2020	Bunting Flag HBD Sofia	102	43,000	4,386,000
7	9 Januari 2020	Bunting Flag Happy Wedding Bunga	104	43,000	4,472,000
8	10 Januari 2020	Bunting Flag HBD Unicorn Putih	101	43,000	4,343,000
9	11 Januari 2020	Bunting Flag HBD Polcadot Pink	105	43,000	4,515,000
10	13 Januari 2020	Topi Cars, Spiderman, Tayo, Hello Kitty	1240	7,000	8,680,000
11	14 Januari 2020	Bunting Flag HBD Captain America	104	43,000	4,472,000
12	14 Januari 2020	Bunting Flag HBD Pony Pink	102	43,000	4,386,000
13	15 Januari 2020	Bunting Flag HBD Polcadot Hitam Gold	104	43,000	4,472,000
14	16 Januari 2020	Topi Unicorn, Doraemon, Pony, Tsum2	1266	7,000	8,862,000
15	17 Januari 2020	Bunting Flag Happy Engagement Putih	102	43,000	4,386,000
16	20 Januari 2020	Bunting Flag Happy Wedding Orang	104	43,000	4,472,000
17	21 Januari 2020	Bunting Flag Happy Engagement Sakura	104	43,000	4,472,000
18	22 Januari 2020	Topi Cars, Spiderman, Tayo, Hello Kitty	1212	7,000	8,484,000
19	22 Januari 2020	Bunting Flag HBD Unicorn Pelangi	102	43,000	4,386,000
20	23 Januari 2020	Bunting Flag HBD Polcadot Biru	103	43,000	4,429,000
21	24 Januari 2020	Bunting Flag Happy Engagement Glitter	104	43,000	4,472,000
22	25 Januari 2020	Topi Thomas, C. America, Pony, Minion	1246	7,000	8,722,000
23	25 Januari 2020	Bunting Flag HBD Tsum2	105	43,000	4,515,000
24	27 Januari 2020	Bunting Flag HBD Batman	102	43,000	4,386,000
25	28 Januari 2020	Topi Thomas, C. America, Pony, Minion	1224	7,000	8,568,000
26	30 Januari 2020	Undangan All Characters	4310	1,700	7,327,000
27					
28					
29					
30					
				GRAND TOTAL	139,481,000

Figure 1. NA Production House's daily production capital and daily sales.

At the end of the month, after recording the daily production capital and daily sales, the reports were recapitulated into the simple financial statement per month. The monthly simple financial statements were limited to the income statement because the income statement was expected to present the benefits obtained by the MSME significantly. In addition, the income statement can also present information on the amount of loss that must be borne by the MSME if the production process experiences errors or failures.

No.	Perkiraan	Kredit	Debit
	Penjualan		139,481,000
	Beban Gaji Karyawan	4,200,000	
	Beban Listrik	800,000	
	Beban Air	120,000	
	Harga Pokok Penjualan (HPP)	70,254,000	
	Beban Lain-lain	2,922,000	
	Total Beban	78,296,000	
	Laba Rugi Bulan Berjalan		61,185,000

Figure 2. NA Production House's simple financial statements (in a month).

From the monthly simple financial statements, the owners can see that the profit generated by NA Production House's production sales was quite significant. Thus, for the recorded profits, the owner can allocate around 30% of the budget for adding overtime hours or salaries for their employees in times of high demand.

After making the financial statements, the absence of SOPs became a problem for the owner. With the high consumer demands, NA Production House was quite overwhelmed in fulfilling the orders on time. Therefore, in this business coaching process, SOPs were created for the production process, so that it could focus on the efficiency of processing time. Therefore, the production capacity of NA Production House could increase, and all the consumer demands could be fulfilled.

With the new SOPs, the target of completing the production process can be achieved more effectively. The process of working on 100 dozen bunting flags, which previously took five hours if it was done by four employees, can now be completed in just three and a half hours with the same number of workers. In addition, the process of working on 1,200 packs of birthday hats, which previously took six hours if it was done by four employees, can now be completed in just four hours with the same number of workers. Meanwhile, the process of working on 4,200 packs of birthday invitation cards, which previously took 8–10 hours if it was done by four employees, can now be completed in just six hours with the same number of workers. The efficiency of this production process has significantly improved because the work system that was previously carried out jointly has been broken down. Therefore, employees already know their respective duties and responsibilities so that they can focus on completing their work without having to wait and rely on one another. Based on these results, the focus of planning for MSMEs to increase their production capacity to meet high market demand has been achieved.

5 CONCLUSION

The results showed that having a simple financial statement and SOPs is important for MSMEs, in this case, for NA Production House. With a simple financial statement, the MSME could assess their performance and financial health of their business, which could support their business decision-making. Furthermore, with the SOPs, the employees understand their respective duties and responsibilities, and so, they can do their work according to the target time. Therefore, the production process efficiency has been improved and the MSME can now focus on increasing their production capacity.

REFERENCES

Cadle, J. (2014). Business Analysis Techniques: 99 Essential Tools for Success, 2nd ed.

Cooper, D. R., & Schindler, P. (2014). Business Research Methods. New York: McGraw-Hill Education.

Gamble, J., Peteraf, M., Thompson, A. (2015). Essentials of Strategic Management, 4th ed., New York: Mc.Graw-Hill.

Greener, S. (2008). Business Research Methods. BookBon.

Griffin, R. W. (2004). Manajemen. Jakarta: Erlangga.

Jones, F. & Rama, D. (2002). Accounting Informations System: A Business Process Approach. South-Western College Publishing.

Kasmir. (2013). Analisis Laporan Keuangan. Jakarta: Rajawali Pers.

Kieso, Jerry, & Weygandt. (2013). Intermediate Accounting. Jakarta: Erlangga.

Kotler, P., & Armstrong, G. (2010). Principle of Marketing. New Jersey: Pearson.

Osterwalder, A. & Pigneur, Y. (2010). Business Model Generation: A Handbook for Visionaries, Game Changers, and Challengers. New Jersey: John Willey & Sons.

Rastogi, Nitank & Trivedi, M.K. (2016). PESTLE Technique-A Tool to Identify External Risks in Construction Projects. India: International Research Journal of Engineering and Technology (IRJET).

Ross, Westerfield, & Jaffe. (2013). Corporate Finance. New York: Mc.Graw-Hill.

Sutojo, S. (2009). Manajemen Pemasaran. Jakarta: PT. Damar Mulia Pustaka.

Tambunan, T. (2012). Usaha Mikro Kecil dan Menengah di Indonesia. Jakarta: LP3ES.

Contemporary Research on Business and Management – Noviaristanti (Ed.)

Factors influencing consumers' attitude and repurchase intention towards Online Food Delivery (OFD) services in Indonesia

M.R. Hakim & N. Sobari
Faculty of Economics and Business Universitas Indonesia, DKI Jakarta, Indonesia

ABSTRACT: Online Food Delivery (OFD) services transaction in Indonesia has been growing rapidly and become more popular compared to the pre-COVID-19 pandemic situation. Considering this issue, this study aims to empirically investigate consumers' attitude and repurchase intention towards OFD services in Indonesia by employing the technology acceptance model (TAM), a theory of planned behavior (TPB), and partial adoption of the extended information technology continuance model (ITCM). The study examined the structural relationship between hedonic motivation, price saving orientation, time saving orientation, prior online purchase experience, information fit-to-task, visual appeal, convenience motivation, postusage usefulness, attitude towards OFD services, and repurchase intention towards OFD services. There were 207 questionnaires collected to test the research model using structural equation modeling (SEM). It was found that eight of the sixteen proposed hypotheses were accepted. In addition, this study discusses its practical implication, limitation, and suggestion.

1 INTRODUCTION

Online Food Delivery (OFD) services transaction in Indonesia has increased significantly during the COVID-19 pandemic (Yuswohady, 2020). At such times, consumers are avoiding eating at a restaurant and prefer to order food and beverage (F&B) through OFD services. It shows that the consumers shift from ordering for indulgence to utility function and from occasional to repetitive activity. In this case, their motivations to order F&B using OFD services are invaluable. Several previous studies found several factors influencing consumers' attitude and repurchase intention of online shopping and OFD services, including hedonic motivation (HM), price saving orientation (PSO), time saving orientation (TSO), prior online purchase experience (POPE), information fit-to-task (IFT), visual appeal (VA), convenience motivation (CM), post-usage usefulness (PU), attitude (AOFDS), and repurchase intention towards OFD services (RIOFDS). Therefore, this study aims to examine the structural relationship between

HM, PSO, TSO, POPE, IFT, VA, CM, PU, AOFDS, and RIOFDS towards OFD services in Indonesia. It is expected that results could provide valuable inputs for OFD companies to develop strategies that could increase consumers repurchase intention towards OFD services.

2 LITERATURE REVIEW

This study is based on TAM (Davis, 1989), TPB (Ajzen, 1991), and partial adoption of the extended ITCM (Bhattacherjee et al., 2008). Hedonism is essential in online shopping as it links to shopping enjoyment (Ingham et al., 2015). It reveals that consumers' hedonic motivation positively affects convenience motivation and post-usage usefulness on OFD services (Yeo et al., 2017). In addition, price wars among online sellers provide convenience for consumers to compare prices (Quelch & Klein, 1996). A study revealed that consumers consider online purchase as a valuable transaction as

DOI 10.1201/9781003196013-40

they can compare offers from various retail before making purchase (Audrain-Pontevia et al., 2013) and beneficial to buyers who get products at lower costs (Chiu et al., 2014). Besides, that time saving provides convenience preferred by consumers. Thus, price saving orientation and time saving orientation positively affect convenience motivation and post-usage usefulness of OFD services (Yeo et al., 2017). Consumer online shopping experience is a determinant factor motivating the convenience of consumers in using service (Hernández-Ortega et al., 2008). Hence, consumers will gain accumulated knowledge from their shopping experience influencing their future decisions and usefulness perception (Lord & Maher, 1990).

Loiacono et al. (2017) explained that information fit-to-task simplifies website displays to show appropriate information for users to do a task. In this case, consumers will consider a website to be useful for shopping if it has complete information assisting them do tasks. In addition, visual appeal is an element seen on a website, which according to Tractinsky et al (2000) influences user perception on website's usefulness. In this case, Xiang et al. (2016) found that information fit-to-task and visual appeal positively influenced perceived usefulness. Davis (1989) stated that perceived ease of use refers to the extent to which a person believes the system will minimalize their effort in completing something. Perceived usefulness can be defined as the extent to which a person believes that using a system will improve the performance of an activity. Har & Eze (2011) stated that perceived ease of use and perceived usefulness positively affect repurchase intention. Yeo et al. (2017) supersedes perceived ease of use with convenience motivation and perceived usefulness with post-usage usefulness. Moreover, convenience motivation positively influenced post-usage usefulness, attitude, and behavioral intention on OFD services. On the other hand, post-usage usefulness positively influenced attitude and behavioral intention on OFD services. Limayem et al. (2006) found that attitude on online shopping is the factor determining the shopping intention.

3 RESEARCH METHOD

To test the proposed model (Figure 1), this study employed a quantitative and purposive sampling technique by involving 207 Go-Food or GrabFood users using the service on 9 March9 May 2020 as its sample. We used items from Yeo et al. (2017) to measure HM, PSO, TSO, POPE, CM, PU, AOFDS, and RIOFDS, Xiang et al. (2016) and Loiacono et al. (2007) to measure IFT, and Xiang et al. (2016) to measure VA. Moreover, a five-point Likert scale was adopted. The pre-test was distributed online to 30 respondents and fulfill validity and reliability requirement. Moreover, the main test used Structural Equation Modeling (SEM) to examine the relationship between variables, which consists of measurement and structural model.

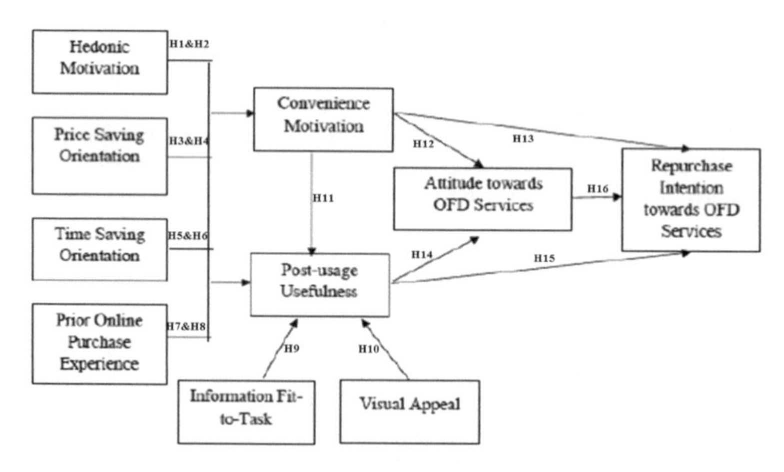

Figure 1. Research model.

4 RESULTS AND DISCUSSION

The respondents consisted of 76.33% female and 23.67% male. Most of them were 21–30 years old, had a monthly income between Rp 5,000,001-Rp10,000,000, private employees, have been using OFD services for more than two years, and recently used OFD services less than a week ago. It was found that after the COVID-19 pandemic ends, 28.50% respondents would increase their frequency of purchasing F&B through OFD services, 29.47% would decrease it, and the rest would remain the same. Based on confirmatory factor analysis, all indicators and variables were valid and reliable (SLF $\geq$ 0.50; t-value $\geq$ 1.645; CR $\geq$ 0.70; VE $\geq$ 0.50). The research model has ten good fit indexes (RMSEA, NNFI, NFI, RFI, IFI, CFI, normed chi-square, ECVI, AIC, and CAIC). In this study, only eight of sixteen hypotheses have t-value > 1.645 (H3, H5-H8, H11, H14-H15), which were accepted (Table 1). It was also found that: (1) PSO, TSO, and POPE had a significant positive influence on CM; (2) TSO, POPE, and CM had a significant positive influence on PU; and (3) PU had a significant positive influence on AOFDS and RIOFDS. These findings empirically support previous studies.

Table 1. Path diagram results for the structural model

H	Path	T-value	Evaluation	H	Path	T-value	Evaluation
1	HM-> CM	1.04	Not Supported	9	IFT-> PU	0.90	Not Supported
2	HM-> PU	0.57	Not Supported	10	VA-> PU	0.47	Not Supported
3	PSO-> CM	2.43	Supported	11	CM-> PU	1.99	Supported
4	PSO-> PU	0.76	Not Supported	12	CM -> AOFDS	0.18	Not Supported
5	TSO-> CM	2.92	Supported	13	CM-> RIOFDS	0.93	Not Supported
6	TSO-> PU	6.03	Supported	14	PU-> AOFDS	5.46	Supported
7	POPE-> CM	2.74	Supported	15	PU-> RIOFDS	5.84	Supported
8	POPE-> PU	2.54	Supported	16	AOFDS-> RIOFDS	−0.50	Not Supported

Source: Result of statistical data using LISREL

The OFD services were often used by consumers as a routine activity as they perceive the service convenient. In this case, consumers perceived that OFD services could give them benefits while fulfilling their daily needs (utility function) and not merely pleasure. According to Yuswohady (2020), consumers use OFD services to meet their utility and not for satisfying their indulgence. Therefore, H1 and H2 are rejected. The charge prices of Go-Food and GrabFood are 20%–25% higher on applications compared to when a customer buys it directly from the vendor. Even though the prices higher, customers were still willing to pay for the services. This finding supports Prabowo and Nugroho (2018) who found that price saving orientation did not affect the post-usage usefulness of Go-Food. Therefore, H4 is rejected. Information on OFD services may be considered as general information and respondents already had knowledge or experience regarding OFD services. In this case, the information listed may no longer be applied when determining the use of service. Images and designs displayed on the OFD application may be considered overly general and familiar for consumers because these images only showed the menu that does not affect consumers in perceiving the services as a useful platform. It supports Dewi et al. (2018) who found that visual appeal did not affect perceived usefulness. Thus, H9 and H10 are rejected.

The convenience motivation could be insignificant for the respondents because they were already familiar with OFD services. Pynoo et al. (2011) explained that ease of use is a significant predictor of attitude if consumers have no prior experience. The majority of respondents in this study were generation Y and Z (95.17%) who were already familiar with the internet, digitalization, and accustomed to high-tech products. Therefore, convenience motivation in using OFD services did not affect their repurchase intention. This finding supports Hussein (2017) who found that ease of use did not affect consumers' purchase intention because consumers already familiar with the system and they consider the technology to be relatively easy. Thus, H12 and H13 are rejected. Based on statistical analysis, the mean of attitude is higher than three. This finding indicates that respondents agreed that using OFD services is wise, good, sensible, and rewarding. However, almost 30% of the respondents stated they will decrease the purchase frequency through OFD services after the COVID-19 pandemic ends. Therefore, a positive attitude towards OFD services did not necessarily encourage consumers repurchase intention. According to Sentosa and Mat (2012), attitude did not influence behavioral intention. Therefore, H16 is rejected.

5 CONCLUSION AND SUGGESTIONS

Only eight hypotheses of this study accepted. It was found that: (1) PSO, TSO, and POPE had a significant positive influence on CM; (2) TSO, POPE, and CM had a significant positive influence on PU; and (3) PU had a significant positive influence on AOFDS and RIOFDS. Besides, consumers' attitude towards OFD services was positive and they would repurchase if the service benefits them. Thus, OFD companies must ensure their services are valuable and useful for consumers in every aspect (helping the consumer to shop quickly and effectively). Also, OFD companies must ensure transaction processes in the application is easy and clear, thus the OFD application must be user-friendly to speed up the F&B purchase process. In OFD services, the aforementioned speed is not just during the online transaction process. It starts from the F&B order process through the application, payment, and delivery processes.

OFD services companies must ensure the consumer shopping experience is positive (application, duration, driver and retailer quality, communication, etc.). It is also suggested that companies should update their promotions to attract price-sensitive consumers since they tend to choose a site that provides best value. This study is limited by its research model that did not differentiate Go-Food and GrabFood users. Therefore, it is suggested that further study would: (1) conduct differentiate analysis between Go-Food and GrabFood users; (2) add utilitarian motivation in the research model; (3) cultural and technology acceptance differences allow this study to be replicated in other countries; and (4) the technological difference acceptance between generations will allow this study to be tested based on respondent age.

REFERENCES

Ajzen, I. (1991). The theory of planned behavior. *Organizational behavior and human decision processes*, *50*(2).

Audrain-Pontevia, A. F., N'Goala, G., & Poncin, I. (2013). A good deal online: The Impacts of acquisition and transaction value on E-satisfaction and E-loyalty. *Journal of Retailing and Consumer Services*, *20*(5).

Bhattacherjee, A., Perols, J., & Sanford, C. (2008). Information technology continuance: A theoretic extension and empirical test. *Journal of Computer Information Systems*, *49*(1), 17–26.

Chiu, C. M., Wang, E. T., Fang, Y. H., & Huang, H. Y. (2014). Understanding customers' repeat purchase intentions in B2C e-commerce: the roles of utilitarian value, hedonic value, and perceived risk. *Information Systems Journal*, *24*(1), 85–114.

Davis, F. D. (1989). Perceived usefulness, perceived ease of use, and user acceptance of information technology. *MIS quarterly*.

Dewi, D. S., Sudiarno, A., Saputra, H., & Dewi, R. S. (2018). The effect of emotional design and online customer review on customer repeat purchase intention in online stores. In *Conference Series: Materials Science and Engineering* (Vol. 337, No. 1, pp. 1–9). Har, L. C., & Eze, U. C. (2011). Factors influence consumers' intention to repurchase online in Malaysia. *International Journal of Electronic Commerce Studies*, *2*(2), 157–164.

Hernández-Ortega, B., Jiménez-Martínez, J., & Martín-DeHoyos, M. J. (2008). Differences between potential, new and experienced e-customers. *Internet Research*.

Hussein, Z. (2017). Leading to intention: The role of attitude in relation to technology acceptance model in elearning. *Procedia Computer Science*, *105*, 159–164.

Ingham, J., Cadieux, J., & Berrada, A. M. (2015). e-Shopping acceptance: A qualitative and meta-analytic review. *Information & Management*, *52*(1), 44–60.
Limayem, M., Khalifa, M., & Frini, A. (2000). What makes consumers buy from Internet? A longitudinal study of online shopping. *IEEE Transactions on systems, man, and Cybernetics-Part A: Systems and Humans*, *30*(4), 421–432.
Loiacono, E. T., Watson, R. T., & Goodhue, D. L. (2007). WebQual: An instrument for consumer evaluation of web sites. *International Journal of Electronic Commerce*, *11*(3), 51–87.
Lord, R. G., & Maher, K. J. (1990). Alternative information-processing models and their implications for theory, research, and practice. *Academy of management review*, *15*(1), 9–28.
Quelch, J. A., & Klein, L. R. (1996). The Internet and international marketing. *MIT Sloan Management Review*, *37*(3).
Prabowo, G. T., & Nugroho, A. (2019). Factors that influence the attitude and behavioral intention of Indonesian users toward online food delivery service by the Go-Food application. In *12th International Conference on Business and Management Research (ICBMR 2018)*. Atlantis Press.
Pynoo, B., Devolder, P., Tondeur, J., Van Braak, J., Duyck, W., & Duyck, P. (2011). Predicting secondary school teachers' acceptance and use of a digital learning environment: A cross-sectional study. *Computers in Human behavior*, *27*(1), 568–575.
Sentosa, I., & Mat, N. K. N. (2012). Examining a theory of planned behavior (TPB) and technology acceptance model (TAM) in internetpurchasing using structural equation modeling. *Researchers World*, *3*(2 Part 2), 62.

Contemporary Research on Business and Management – Noviaristanti (Ed.)

ZMOT marketing strategy during the Covid-19 pandemic

M. Jannah
Trunojoyo Madura University, Bangkalan, Indonesia

ABSTRACT: During the Covid-19 pandemic, customers spend their time finding information online. The process of getting a new customer begins with ZMOT (zero moment of truth). This study aims to analyze the ZMOT marketing strategy applied in the Google search engine. The descriptive quantitative approach was used with spread questionnaires by an online survey. The sampling technique used for this study is purposive sampling. The result of this study showed that a majority of customers look for spot tourism on ZMOT Google. Customers started their process about 2–3 weeks before and spent a week making their purchase decision. Instagram is another source they use before confirming their decision.

1 INTRODUCTION

In recent times, customers have found it convenient to express their needs and wants. Digitalization gives more power to the customer and makes the businessman think about how to win the market. The businessman must explore the consumers' needs, wants, and motivations of buying, which is an important thing.

The moment of truth was started by A.G. Lafley, chairman and CEO of Procter & Gamble (2005). The first moment of truth is the customers' initial thoughts. The customers have the capability of knowing about a product in the first 3–7 seconds of seeing it. The first moment of truth is the moment customers choose the product instead of the competitor's product (Wikipedia.org). There are also the second moment of truth, the third moment of truth, and the zero moment of truth.

The zero moment of truth is a term coined by Google in 2011. Google introduced the term to represent winning the customers. If in the traditional model there are a stimulus, the first moment of truth, and second moment of truth in sequence before the purchase decision, then the modern model has the zero moment of truth before the first moment of truth.

The internet has an important role in ZMOT (Zero Moment OfTruth). ZMOT catch the ignorance of customer from digital life and prevail on all industry, goods, and services, and it is not just a store but also e-commerce too, B2C and C2C. It does not stop there. ZMOT relates to customer loyalty too.

Businessmen must be aware that the goal of relationship marketing is customer loyalty. When customers are satisfied, they give more advantage to the company (Anderson & Srinivasan, 2011). They help get new consumers through positive word of mouth, and some have a blog and review the products. Google, Bing, Yandex, CC Search, Swisscows, DuckduckGo, StartPage, Search Encrypt, Wiki.com are examples of search engines.

Google attracts a lot of users. It did even more during the Covid-19 Pandemic. Early in 2020, the world was shocked by the spread of Covid-19, which was later declared a pandemic. Covid-19 is a hot topic, especially in discussions regarding world economic growth. IMF stated that the world economy grew negatively by 4.9 percent in 2020 (money.compas.com). Being no exception, Indonesia was one of the countries that were severely affected by the Covid-19 pandemic. Badan Pusat Statistik (BPS) noted that in the second quarter period, Indonesian's economic growth contracted to 5.32 percent.

DOI 10.1201/9781003196013-41

Contraction of economic growth, both of world and Indonesia, had no effect on search engines, especially Google. During the pandemic, Google witnessed an increase in traffic. Their company generated a profit of 2.5 billion in March 2020 (tekno.compas.com). On the zero moment of truth, consumers have more control on finding information about their products on the internet. They can make their minds before purchase decision or going to shop.

In this era, buyers are getting smarter and more critical. They trust the public's recommendation more than sellers'. This research will analyze the ZMOT marketing strategy, describe findings obtained from Google users, and answer the questions such as when the customers decide to buy a product. Factors regarding the sites on Google were captured as ZMOT strategy marketing.

2 LITERATURE REVIEW

2.1 *Zero moment of truth*

Procter & Gamble observed that the first seven seconds was the critical moment before the purchase decision was made. A moment of truth can be described as a stimulus to the consumer to discover the brand in question (Ertemel & Basçı, 2015).

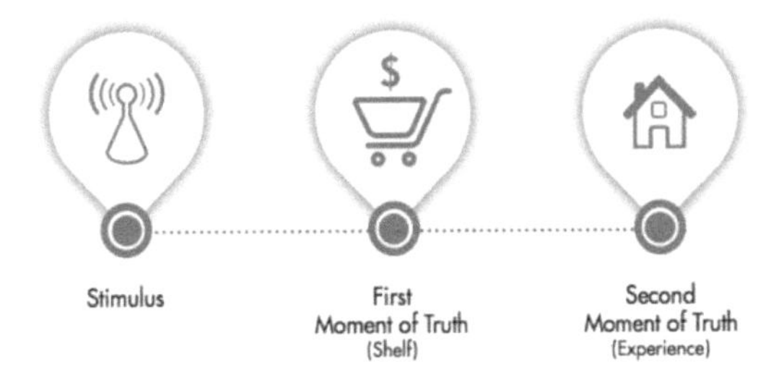

Figure 1. Traditional mental model of the decision-making process.
Source: Lecinski (2011).

Traditionally, the decision-making process started with a stimulus. Lafley, as the P&G CEO, described the opportunity of the company and brand to be the winner of a booth of truth, first and second moment. The first moment occurs at the store when consumers buy a product. The second occurs at their home. They try the product and decide whether they are satisfied or not. The new mental model of decision making is different. Digitalization and technology are making the consumer smarter. Consumers begin their path of decision making of a purchase in the digital world; it is called the zero moment of truth (ZMOT). They search for product information every time and everywhere. Consumers use the internet to get product and brand information before making a purchase decision (Venkatesan, et al. 2007).

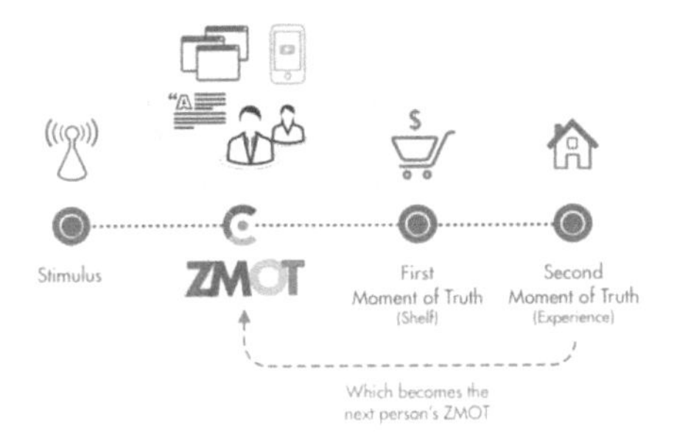

Figure 2. The new mental model of the decision-making process.
Source: Lecinski (2011).

3 RESEARCH MODEL

3.1 *Data collections*

The data were collected from online participants. For this exploration, non-probability technique sampling was used, wherein no all members of the population have an equal chance. Purposive sampling was used in this research. This research has some criteria on the sampling (Cooper & Schindler, 2011): (1) Google users; (2) browsing a product or brand during the Covid-19 pandemic; and (3) minimum 17 years old. The final sample consisted of 363 samples with 32% male and 68% female.

3.2 *The questionnaire*

The questionnaire captured demography, age, and gender of the respondents. Income and job were also captured by this questionnaire to describe findings obtained from Google users using the ZMOT marketing strategy.

The researcher gives more items on the questionnaire for capturing the factor of sites on a search engine Google as ZMOT marketing strategy. In this part, a five-point Likert type scale was used (1 = "strongly disagree", 5 = "strongly agree").

4 RESULT AND DISCUSSION

This research will cover Google users during the Covid-19 pandemic as a ZMOT marketing strategy. Most of the respondents in this research were female (68%), and 32% were male. This indicates that during the Covid-19 pandemic, females spend more time than males to get product and brand information from Google.

Table 1. Age of respondents.

Age	Total	%
17–25 years old	195	54%
26–35 years old	117	32%
36–45 years old	39	11%
>45 years old	12	3%
Total	363	100%

Table 1 shows that the majority of the participants in this study are the Z Generation, that is, those who are 17 to 25 years old (people born in 1995–2020). These findings are similar to those of prior research by Grencíková & Vojtovic (2017), who observed that Generation Z are spending more time Googling product information than X and Y generations.

This study also captures the factors as an antecedent of the sites on Google as a ZMOT marketing strategy. The researcher tries to get the factors of expected site participants. This study uses exploratory factor analysis, KMO MSA >0.5, and Barlett's test (<0.05) (Hair et al.,2010). The results are shown in Table 2.

Table 2. KMO and Bartlett's test.

Kaiser-Meyer-Olkin Measur Adequacy.	e of Sampling	.882
Bartlett's Test of Sphericity	Approx. Chi-Square df	2324.959 91
	Sig.	.000

Table 2 describes that KMO MSA 0.882 (>0.5) and the Bartlett's test sig. 0.000 (<0.05) fulfill the requirements of exploratory factor analysis.

This Study has 14 items of measured expected sites as ZMOT. From 14 items divided by 3 factors or variables can good ZMOT as a marketing strategy from Google. The 14 items are:

X1 Sites have a complete description of goods/ services
X2 Sites have a picture of the goods/services
X3 Sites show goods and services and people's comments
X4 Sites have no advertisement
X5 Sites have good privacy
X6 We can communicate directly with persons
X7 Site has a good design and layout
X8 Site provides accurate information
X9 Site uses the correct language
X10 Site provides interesting and relevant information
X11 Site has popular and paid domain extensions (eg. .com; .co.id; or .net)
X12 Site has a good speed
X13 Site is responsive
X14 Site is easy to understand

The 14 items are divided into 3 variables or factors. First, variables X1, X2, X7, X8, X9, X10, X12, X13, and X14 are site design; second, variables X3, X6, and X11 are site social, and the variables X3 and X4 are site privacy.

5 CONCLUSION

The big question is why tourism spots became the first option, although Indonesia had a lockdown policy. From the exploratory factor analysis, we get 3 variables from 14 items; they are site design, site social, and site privacy on the Google site as a ZMOT marketing strategy. From the result of this study, we know that a recommendation website as a ZMOT marketing strategy can be good for a startup business. The companies must emphasize several points before creating a website, keeping site design, site social, and site privacy in mind to connect with consumers.

REFERENCES

Anderson, R.E., & Srinivasan, S.S. (2003) 'E-satisfaction and e-loyalty: a contingency framework', *Psychology And Marketing*, Vol. 20 No. 2, pp. 123–138.

Cooper & Schindler. 2011. *Business Research Methods 11thed.* New York: McGraw-Hill

Ertemel, A.V., & Basçı, A. (2015) 'Effect of Zero Moment of Truth on Consumer Buying

Decision: An Exploratory Research in Turkey', *International Journal of Social Science and Education Research,* Vol. 1 No. 2, pp. 526–526.

Hair, J. F., William. C. B., Barry. J. B., & Rolph. E A. (2010) *Multivariate Data Analysis: A Global Perspective,* Seventh Ed, Pearson Education Inc. The United States of America.

Lecinksi, J. (2011). *Winning the zero moment of truth.* Zero Moment of Truth

Venkatesan, R., V. Kumar, & Nalono R., (2007), 'Multichannel Shopping: Causes and Consequences, *Journal of Marketing,* Vol. 71 No. 2, pp. 114–132. Wikipedia.org

money.compas.com

tekno.compas.com

Contemporary Research on Business and Management – Noviaristanti (Ed.)

Designing a marketing strategy with a consumer behavior approach at STIDKI Ar Rahmah Surabaya

M.H. Musyanto & G.C. Premananto
Magister Management of Economic and Business Faculty of Airlangga University Surabaya, Indonesia

ABSTRACT: Sekolah Tinggi Ilmu Dakwah dan Komunikasi Islam (STIDKI) Ar Rahmah Surabaya has been experiencing a significant decrease in the number of participants at the test stage in new student admissions since 2015; so this has caused losses because the chances of getting qualified student candidates are getting smaller, while the costs incurred are not small. This study seeks to find a solution in order to maintain the number of registrants or increase the number of PMB test participants by designing an effective marketing strategy with a consumer behavior approach. The research was conducted using qualitative methods, interviewing applicants who did not take the test and the students to obtain primary data and analyze new student admission documents as a secondary data source. From the data obtained, a focus group discussion was carried out with STIDKI management members who were directly involved in the admission of new students, and the result was that STIDKI Ar Rahmah would focus on various efforts to (1) become a favorite campus for memorizing the Quran, (2) increase the role of lecturers in institutional cooperation and capitalize on Quran lecturers who have international qualifications, (3) involve students in marketing to schools, especially their home schools, and (4) apply a registration fee at the time of new student admission.

1 INTRODUCTION

This research was conducted at the Sekolah Tinggi Ilmu Dakwah dan Komunikasi Islam (STIDKI) Ar Rahmah Surabaya, a higher education institution with a full scholarship that was founded in 2015 with a vision to become a campus for producing imams and professional mosque managers who memorize the 30 juz Quran and help build civilization through mosques.

Since 2015, there has been a decrease in the number of test takers at the time of new student admission (NSA) as follows:

Table 1. Attendance of new student admissions test takers

Year	Registrant	Test Taker	Percentage
2015	125	90	72%
2016	251	158	63%
2017	245	112	46%
2018	306	144	47%
2019	319	123	39%
2020	410	216	53%

The magnitude of the decrease in the number of test takers is certainly very detrimental, at least in two ways: (1) the loss of potential quality students and (2) the high cost of organizing NSA. Therefore, knowing the cause of the decline in NSA test participants is important in order to increase the number of registrants and maintain this number until the NSA test stage.

DOI 10.1201/9781003196013-42

1.1 *Research question/objective*

The problem focused on in this research is how to design an effective marketing strategy with consumer behavior approaches to increase the number of registrants and maintain it until the New Student Admissions test stage?

2 LITERATURE REVIEW

2.1 *College marketing*

The increasing need for higher education encourages each higher education institution to carry out the most effective marketing strategy to attract stakeholders. The stakeholders referred to in this case are high school graduates, parents of prospective students, funders, as well as graduates (Bialon, 2015). In order to survive and achieve the competitive advantage in a sustainable manner, universities must use a marketing framework (Hoyt and Brown, 2003) and must also be able to meet customer needs well (Kotler, 2016). Approaching the customer is very effective in identifying customer needs precisely; this can be done by using feedback from customers or prospective customers to control organizational change. Furthermore, the marketing mix (marketing mix) was effectively applied to influence the need for higher education services (Ivy, 2008).

2.2 *Consumer behavior*

Understanding consumer behavior is one of the important points in order to achieve marketing effectiveness. Understanding consumer behavior is essentially understanding “why consumers do what they do”. Marketing stimuli and environmental stimuli enter the consumer’s consciousness, and a series of psychological processes coupled with certain consumer characteristics internally influence the decision-making process and purchasing decisions. The marketer’s job is to understand what goes on in consumer consciousness from getting outside marketing stimuli to the final purchase decision. (Kotler, 2016).

2.3 *Stages of the consumer purchasing process*

The five-stage model of the consumer purchase process (Kotler, 2016) is the basis for discussing how consumers make decisions to buy products. There are five stages for a person to arrive at a decision to buy a product or service: (1) recognition of needs, (2) seeking information, (3) evaluation of alternatives, (4) purchasing decision, and (5) post-purchase behavior.

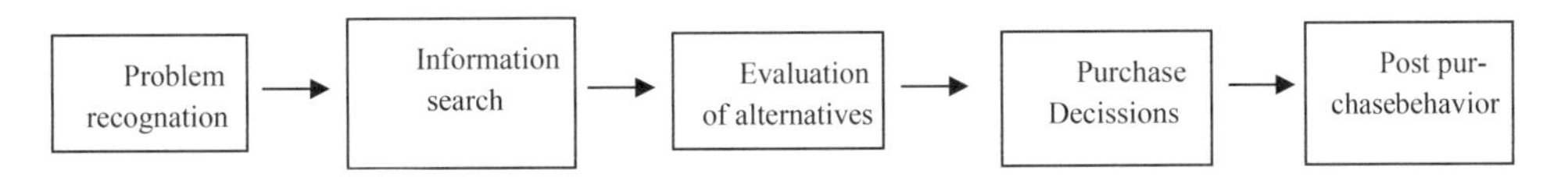

Figure 1. Stages of the consumer purchasing process (Kotler, 2016).

2.4 *Alternative evaluation and purchasing decisions*

One of the crucial stages of the five stages mentioned above is the alternatives evaluation stage. In this stage, consumers gather information regarding various preferences related to brands of the products they need and then select one. Then, the consumer intends to buy the most preferred product. There are five things the consumers consider when they are at the purchase intention stage, including the brand (dealer), amount (quantity), time of purchase (timing), and the payment method. In the context of higher education institutions such as STIDKI Ar Rahmah, what will be taken into consideration is the profile of STIDKI Ar Rahmah, the number of students accepted, the time required for the registration process, and the school fees. After the purchase intention is formed, there are two factors that the consumers consider before making a purchase decision:

a. The first factor is the attitude of other people. The extent to which the attitudes of others affect us will depend on two things: (1) the intensity of the negative attitudes of others toward alternatives that consumers like and (2) the motivation of consumers to obey others. The more intense the negative influence of a person on us and the closer we are to that person, the more we will take decisions according to that person's preferences and vice versa.
b. The second factor is the factor of unanticipated situations that can arise and change the purchase intention.

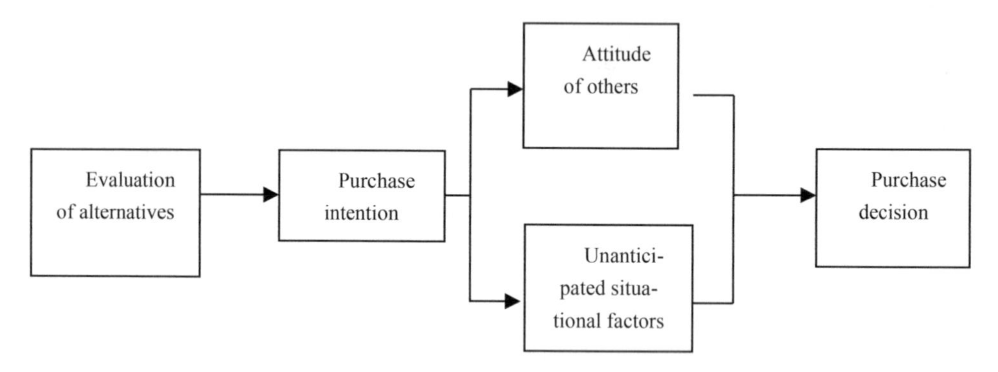

Figure 2. Stages between evaluation of alternatives and purchasing decisions (Kotler, 2016).

This research will be focused on the stages from identifying needs to purchase decisions to get an idea of how the student candidates for STIDKI Ar Rahmah take decisions about choosing schools.

3 RESULTS AND DISCUSSION

3.1 *Secondary data*

From the new student admissions document of STIDKI Ar Rahmah Surabaya, several important things related to the research theme are observed, as shown in tables below.

Table 2. The origin of the registrant school for STIDKI Ar Rahmah Surabaya.

	ORIGINAL SCHOOL		
	MA/SMAIT/SMKIT/PP (Islamic Senior High School)	SMA (Senior High School)	SMK (Vocational Senior High School)
Percentage	75%	17%	8%

Table 3. Data sources of information for new student admissions.

	INFORMATION SOURCE					
	Brochure	School	Media	Web	Friend/Family	Others
Percentage	5%	29%	16%	5%	38%	7%

3.2 *Primary data*

To obtain primary data, interviews were conducted with 10 registrants of STIDKI Ar Rahmah Surabaya who did not continue to the test stage and with 14 students of STIDKI Ar Rahmah Surabaya with the following results:

1. The reasons why registrants were not present at the test stage were as follows: (1) STIDKI Ar Rahmah was not their first choice; (2) they were tied to other institutions; (3) they did not meet the minimum requirements of memorizing 3 juz Al Quran; and (4) they were not allowed by parents.
2. Regarding the sources of information for new student admissions of STIDKI Ar Rahmah, both registrants and students of STIDKI Ar Rahmah answered the following: (1) they received information from acquaintances (parents, high school alumni, and teachers), and this was the most common answer; (2) they received information from social media such as WhatsApp and Facebook; 3) they received information from the STIDKI website.
3. The most motivating factors to register with STIDKI Ar Rahmah, according to the informants' answers, were in the following order: (1) tahfizh Quran 30 juz, (2) scholarship, and (3) mosque imam.

4 RESULTS OF THE FOCUS GROUP DISCUSSION

Furthermore, the data given above were used as the material in a focus group discussion (FGD), which was attended by the management team of STIDKI Ar Rahmah. The FGD decided on several marketing strategies, which were briefly described as follows:

1. STIDKI Ar Rahmah will try to become a favorite high school for memorizing the Quran.
2. Increasing the role of lecturers in institutional cooperation and capitalizing on Quran lecturers who have international qualifications.
3. Involving students in marketing the curriculum to schools, especially their home schools.
4. Applying a registration fee at the time of admission of new students.

REFERENCES

Alves, H dan Raposo, M (2010). The Influence of University Image on Student Behavior. *International Journal of Educational Management* vol. 24 No. 1 2010 pp. 73-85

Bialon, L. (2015). Creating Marketing Strategies for Higher education Institutions. *Minib Marketing Scientific And Research Organizations*. Vol. 18, Issue 4, p. 129–146

Hoyt, J.E. & Brown, A.B. 2003, "Identifying college choice factors to successfully market your institution", *college and university*, vol. 78, no.4, pp. 3–10

Ivy, J. 2008, "A New Higher Education Marketing mix: the 7Ps for MBA marketing", *The International Journal of Educational Management*, vol. 15, no. 6/7, pp. 276282

Kotler, P., Kartajaya H., & Setiawan I. (2010). Marketing 3.0 from products to costumers to the human spirit. Wiley

Kotler, P., Keller, K.L.(2016) *Marketing Management.* 15 global edition, Pearson

Moogan, Y.J., Baron, S., & Haris, K (1999). Decision making Behaviour of Potential Higher Education Students. *Higher education quarterly, 0951–5224 vol. 53.No. 3* p.211–228

Peter, J.P., Olson, J.C. (2010). *Consumer Behavior & Marketing Strategy*. 9th ed. McGraw-Hill

Schiffman, L.G., Wisenblit.J. (2015). *Consumer Behaviour*. 11th ed. Pearson

Zikmund,W.G., Babin, BJ., Carr, JC., & Griffin, M. (2013). *Business Research Methods*. 9th ed.South-Western Cangage Learning

Contemporary Research on Business and Management – Noviaristanti (Ed.)

The impact of board diversity on energy and mining firms in the emerging countries in South East Asia and their stock price crash risk in 2015–2019

M.A.Y. Setiawan & J. Jahja
Universitas Indonesia, Jakarta, Indonesia

ABSTRACT: This research aims to determine the impact of board diversity on energy and mining firms in the emerging countries in South East Asia and each firm's stock price crash risk in 2015–2019. The diversity in this research includes the board's background diversity and the board's quantity diversity. This research applied a cross-sectional data panel of ordinary least squares (OLS) to obtain its final data. The result of this research shows that board diversity does not influence the stock price crash risk. Significantly, the influence of the independent variable on the dependent variable was found only for the quality of board's education background and board's average tenure of the firm.

1 INTRODUCTION

This research's main objective is to determine the impact and influence of board diversity on energy and mining firms in the emerging countries around South East Asia and their stock price crash risk potential. Apparently, the striking difference between this research and its predecessor is the subject of the research, which is the energy and mining firms in the emerging countries around South East Asia. The energy and mining industry historically has been dominated by men, and one of the variables used for this research is the gender of the board members (Mayes and Pini, 2014). The relation between male-dominated firms and stock price crash risk of the firm in question is one of the benchmarks set for this research in order to determine the impact of the said diversity and the financial performance shown by the firm as the lasting effect of the former.

2 THEORETICAL BACKGROUND AND HYPOTHESIS DEVELOPMENT

Each type of board diversity is categorized into different groups, such as observation attribute and non-observation attribute. Observation attribute is deemed as the visible attribute, and demographic characteristics, such as age, ethnics, race, and gender, are covered. On the other hand, non-observation attribute includes education history, work tenure, professional background, and behavior (Kilduff et al., 2000). Observation attribute is the main focus of a majority of the research studies about board diversity (Erhardt et al., 2003). Non-observation attribute would show the clear differences between each firm in regard to the firm's problem and its interaction style.

The lack of information exchanged between manager and stakeholders would potentially cause significant stock price crash risk. In this kind of situation, the manager has a special authority to control the spread of the information to public. This special authority to control the spread of information to public leads to the possibility of the manager's discretion to hide negative news that is ready to be published at the stock market in order to keep the firm's interests intact (Kothari et al., 2009). However, the manager cannot hide that negative information forever; in the end,

DOI 10.1201/9781003196013-43

the information would be eventually published. This is one of the reasons behind the significant downturn on the firm's stock price.

Board diversity could very well have an impact on the stock price crash risk potential of a firm. Observation attribute (age and gender) is deemed to reduce the stock price crash risk. This statement is based on past researches that prove the tendency of a diverse board to increase its information efficiency (Gul et al., 2011) and decrease the asymmetric information on the stock market (Abad et al., 2017). Additionally, the board with gender diversity would tend to increase disclosure of stock price informativeness as a transparency tool for the firm. This affects the disclosure and transparency of stock price informativeness to public, which automatically reduces hiding of bad news and stock price crash risk significantly (Kaur and Singh, 2017).

2.1 *Hypothesis development*

Based on comprehensive facts obtained from a few past researches and the current development of board's diversity and firm's stock price crash risk, the hypothesis for this research was then developed. The first hypothesis is that the board's background diversity has a significant impact on stock price crash risk, and the second hypothesis is that the board's quantity diversity has a significant impact on stock price crash risk.

3 SAMPLE, VARIABLES, AND METHODS

The sample and data tabulated for this research consist of the firm's financial report, corporate governance report, and the historical stock price data of energy and mining firms in the emerging countries in South East Asia during the five-year period, starting from 2015 to 2019. The primary source of data from each firm for this research is compiled from Thomson Reuters. Based on these data, 19 firms from the energy and mining industry in four different emerging countries in South East Asia, namely, Indonesia, Malaysia, Thailand, and Philippines, are selected as the main object and sample of this research.

The dependent variable applied for this research is the representation of firm's stock price crash risk. Stock price crash risk is represented as NSKEW (Negative conditional return skewness) for its math equation (Chen et al., 2001). The equation is as follows:

$$NSKEW_{i,t} = -[n(n-1)^{3/2}\Sigma W_{i,t}^{3}]/[(n-1)(n-2)(\Sigma W_{i,t}^{2})^{3/2}] \quad (1)$$

The independent variable for this research includes a handful of variables that can be used to prove the hypothesis of the research, such as gender of the board, board size, independent board size, and number of board meetings held each year. The regression analysis was basically done to prove the impact of the independent variable on the dependent variable, T. The majority of independent variables for this research have been used before by Yeung and Lento (2018) and Jebran, et. al (2020).

4 METHOD

The data analysis used for this research is based on ordinary least squares (OLS) cross-sectional regression analysis and panel data with the EViews software. The main objective of this regression analysis is to investigate the impact of board diversity of the firm on its stock price crash risk. The movement of stock price on an annual basis for each firm was noted and included in the regression analysis. Historical data of all 19 firms were obtained from investing.com and Yahoo Finance. In addition to that, the independent variable-related data were acquired from Thomson Reuters via the ESG (Environmental Social Governance) report that represents the corporate governance practice run by each firm.

5 EMPIRICAL RESULTS

5.1 *Preliminary test*

5.1.1 *Chow test*

The Chow test is a mandatory test to determine the most appropriate model to be used for the research between the common effect model (CEM) and fixed effect model (FEM).

According to the Chow test, the probability of cross-section F from the regression (0.4442) is more than the significance value (5%). This finding indicates the common effect model (CEM) as the most appropriate model to be used for this research over fixed effect model (FEM).

5.1.2 *Lagrange multiplier test*

The Lagrange multiplier test is a mandatory test to determine the most appropriate model to be used for the research between the common effect model (CEM) and random effect model (REM). According to the Lagrange multiplier test, the probability of one-sided cross-section Breusch Pagan from the regression (0.6084) is more than the significance value (5%). This finding indicates the common effect model (CEM) as the most appropriate model to be used for this research over the fixed effect model (FEM).

5.1.3 *Board diversity and stock price crash risk*

After selecting the common effect model (CEM) as the main source of regression data, the relation between board diversity and stock price crash risk is deemed ready to be interpreted and further analyzed. The regression result according to CEM is as follows (Table 1).

Table 1. Regression using the Common Effect Model (CEM)

Variable	Coefficient Std.	Error	t-Statistic	Prob.
C	1.4923	1.2045	1.2389	0.2187
GENDER	0.1362	0.1324	1.0287	0.3064
EDUCATION	0.1964	0.0888	2.2114	0.0296
TENURE	−0.5335	0.2666	−2.0013	0.0484
SIZE	−1.0294	0.6320	−1.6288	0.1069
INDEPENDENT	−0.2232	0.4330	−0.5154	0.6076
MEETING	0.2229	0.2939	0.7585	0.4502
R-squared	0.1015			
Prob(F-statistic)	0.1412			

Using the 5% significance value for this research, the regression data given in table 4 are therefore fit enough to be interpreted and analyzed. According to regression result obtained via the common effect model (CEM), the probability (F-statistic) of the regression stands at 0.1412. This value is way higher than the significance value of this research (5%) and indicates the non-significant simultaneous impact of board diversity on the firm's stock price crash risk.

Additionally, the R-squared value from the regression mentioned above also explains the influence of all independent variables combined on the dependent variable. The R-squared value for this research stands at 0.1015, and this means that 10.15% of all possible influences that impacted stock price crash risk originated from all independent variables of this research.

5.2 *Hypothesis test*

For the first hypothesis, there are two independent variables that exist. These variables are board's education history and work tenure. The regression result indicated the significant impact of board's education background and work tenure on the firm's stock price crash risk. Additionally, a positive figure on education's coefficient means that the impact of improvement board's education quality would increase firm's stock price crash risk, while the coefficient of board's work tenure shows that a more experienced group of board would decrease the firm's stock price crash risk altogether.

The second hypothesis focuses on board's quantity diversity, such as board's gender, size, independent board size, and amount of board meetings. All four variables possess identical numbers on its regression probability. These variables' probability suggests that the board's quantity diversity does not significantly impact firm's stock price crash risk.

6 CONCLUSIONS

The result of this research concludes that board diversity simultaneously does not influence the stock price crash risk. Significantly, the influence of the independent variable on the dependent variable was found only for the quality of board's education background and board's average tenure of the firm. The findings prove that the increasing quality of board's education background tends to increase the firm's stock price crash risk, while a more experienced group of board is likely to decrease the potential stock price crash risk of a firm. However, the board's quantity diversity is deemed not to have a significant impact on the firm's stock price crash risk, and the board diversity has nearly non-existent influence on stock price crash risk.

REFERENCES

Chen, J., Hong, H.; Stein, J.C. (2001). Forecasting crashes: trading volume, past returns, and conditional skewness in stock prices. J. Finance Econ. 61 (3), 345–381.

Erhardt, N.L.; Werbel, J.D.; Shrader, C.B. (2003), Board of director diversity and firm financial performance, Corporate Governance: An International Review, Vol. 11 No. 2, pp. 102–111.

Hao, D.Y.; Qi, G.Y.; Wang, J. (2018), Corporate Social Responsibility, Internal Controls, and Stock Price Crash Risk: The Chinese Stock Market, Sustainability.

Indonesia Stock Exchange (2019). IDX Statistics.

Jebran, K.; Chen, S.; Zhang, R. (2020), Board Diversity and Stock Price Crash Risk, Research in International Business and Finance, Vol. 51, pp. 1–19.

Kilduff, M.; Angelmar, R.; Mehra, A. (2000), Top management-team diversity and firm performance: Examining the role of cognitions, Organization Science, Vol. 11 No. 1, pp. 21–34.

Kothari, S.P.; Shu, S.; Wysocki, P.D. (2009), Do managers withhold bad news?, J. Account. Res. 47 (1), 241–276.

Lee, K.-T.; Ooi, C.-A.; Hooy, C.-W. (2019). Corporate diversification, board diversity and stock-price crash risk: Evidence from publicly listed firms in Malaysia. International Journal of Economics and Management 13(2), pp. 273–289.

Mayes, R; Pini, B. (2014). The Australian mining industry and the ideal mining woman: Mobilizing a public business case for gender equality. Journal of Industrial Relation Volume: 56 issue: 4, page(s): 527–546.

Thomson Reuters (2020). ESG Report

Yeung, W.H.; Lento, C. (2018). Ownership structure, audit quality, board structure, and stock price crash risk: evidence from China. Glob. Finance J. 37, 1–24.91–121.

Contemporary Research on Business and Management – Noviaristanti (Ed.)

The influence of browsing motivation toward consumer impulse buying behavior on Indonesian M-commerce

Mufadhzil & Y. Alversia
Master of Management, Universitas Indonesia, Indonesia

ABSTRACT: The purpose of this study is to examine how browsing motivation influences consumers' urge to buy impulsively (UBI) during the COVID-19 pandemic. Based on the stimulus–organism–response (S-O-R) framework and motivation theory, this study investigated situation factor interpersonal influence (UBI), visual appeal (VA), portability (PB), and reaction factor in m-commerce to examine online impulsive buying. This study also investigated the impulse buying tendency personality trait as a moderating factor that affects the relationship between browsing motivation and urge to buy impulsively. We collected data from 368 respondents via an online questionnaire and carried out analysis using SEM-AMOS. The result indicates that situation factors affect browsing motivation differently. While utilitarian browsing (UB) has a significant and negative influence on consumers' urge to buy impulsively, hedonic browsing (HB) has a significant and positive influence on consumers' urge to buy impulsively. Further, impulsive buying tendency (IBT) significantly moderates the relationship between browsing motivation and consumers' urge to buy impulsively.

1 INTRODUCTION

At the end of 2019, the world was shocked by the spread of a disease called COVID-19 (coronavirus disease 2019). To stop a wider spread, the Indonesia government had imposed PSBB (large-scale social restrictions). During PSBB, there was a change in people's shopping behavior from offline to online by 14% (Kantar, 2020). The shift in patterns from offline to online also affects consumer shopping behavior, one of which is impulsive buying behavior (Verhagen and van Dolen, 2011).

According to X. Zheng et al. (2019), situation cues such as the visual appearance of m-commerce applications, interpersonal influence, and portability are considered to influence impulsive purchases from consumers. Meanwhile, the impulsive buying tendency (IBT) is also found to be an influencing factor toward impulsive purchase behavior (Chen & Yao., 2018).

This study expands the motivation theory and adopts hedonic browsing and utilitarian browsing as two motivation values to investigate the consumer's impulse buying behavior. Also, this study suggests that IBT as a personality trait moderates the relationship between browsing motivation and impulsive buying behavior.

The research questions of this study are: (1) Do situation cues affect browsing motivation on m-commerce during the COVID-19 pandemic?; (2) does browsing motivation affect consumers' urge to buy impulsively on m-commerce during the COVID-19 pandemic?; and (3) does the moderating role played by IBT affect the relationship between browsing motivation toward the urge to buy impulsively during the COVID-19 pandemic?

2 LITERATURE REVIEW

Based on previous literature reviews, family and friends are considered to be able to direct consumer purchasing behavior patterns in collectivist countries (Lee and Kacen, 2008). The research

 DOI 10.1201/9781003196013-44

conducted by X. Zheng et al. (2019) shows that the interpersonal influence positively affects hedonic browsing behavior. Meanwhile, consumers with utilitarian values, according to Ismagilova et al. (2019), tend to find information about a product that was previously provided by other consumers before shopping.

Visual appeal is closely related to the appearance of letters and other visual elements (such as graphics) as well as any efforts made to increase the visual appeal of the web as a whole (Parboteeah et al., 2009). Therefore, visual attractiveness is expected to encourage consumers to obtain hedonistic or utilitarian values by browsing m-commerce applications. This statement is supported by research from X. Zheng et al. (2019), which shows that visual appeal positively and significantly influences hedonic browsing or utilitarian browsing.

Portability is a key feature that concerns e-commerce users (Okazaki and Mendez, 2013) because consumers can connect with the m-commerce application anytime and anywhere so that it can increase the length of exposure time. According to the research by X. Zheng et al. (2019), portability positively and significantly influences utilitarian browsing behavior, but it is the opposite for hedonic browsing.

Browsing is a way consumers access information in an online environment, which is the first level in information retrieval and decision making (Rowley, 2002). Utilitarian and hedonic browsing influence impulsive buying (Madhavaram and Leverie, 2004). Research by X. Zheng et al. (2019) showed that utilitarian browsing is negatively associated with impulsive buying, while hedonic browsing has a positive effect.

IBT is a personality trait that refers to the extent to which an individual tends to make unwanted and direct purchases with little consideration or evaluation of the consequences (Flight et al., 2012). Previous research also found that IBT significantly and positively predicts impulsive buying (Liu et al., 2013).

3 RESEARCH METHODS

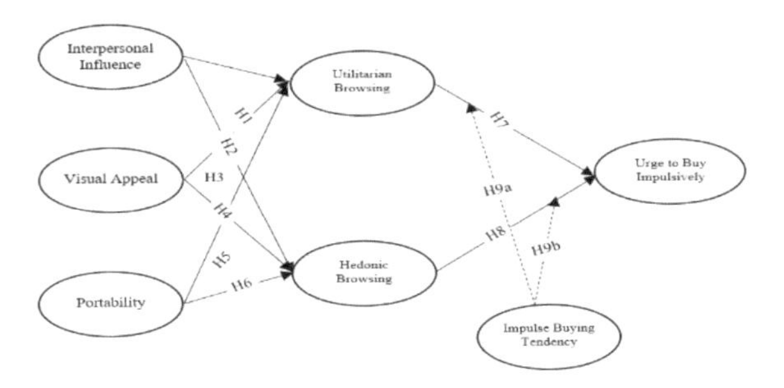

Figure 1. Research model.

We collected data via an online questionnaire to measure customer perception concerning the constructs of interest. Our target sample is consumers who made purchases in October 2020 through one of the m-commerce applications during the COVID-19 pandemic. We invited participants by placing a link on Instagram and WhatsApp, and 368 valid responses were obtained for data analysis.

All of the questionnaire items in this research were adapted from previous studies, with appropriate modification to fit the research context, and assessed on a six-point Likert scale ranging from 1 (strongly disagree) to 6 (strongly agree).

This study selected structural equation modeling (SEM) to evaluate the research models, and hypotheses were proposed. Therefore, we used SPSS 25 and AMOS 23 to conduct the structural equation modeling estimation.

4 RESULTS

We tested the proposed hypotheses with SEM-AMOS. The results show that most of the hypotheses support the relationship between constructs in the research model. The results of hypotheses testing are shown in Table 1.

Table 1. Hypothesis testing results.

Hypotheses	C.R	P	Result
H1: "Interpersonal influence" positively affects the "utilitarian browsing" by m-commerce application users during the COVID-19 pandemic	3.674	***	Supported
H2: "Interpersonal influence" positively affects the "hedonic browsing" by m-commerce application users during the COVID-19 pandemic	5.403	***	Supported
H3: "Visual appeal" positively affects the "utilitarian browsing" by m-commerce application users during the COVID-19 pandemic	4.836	***	Supported
H4: "Visual appeal" positively affects the "hedonic browsing" by m-commerce application users during the COVID-19 pandemic	7.713	***	Supported
H5: "Portability" positively affects the "utilitarian browsing" by m-commerce application users during the COVID-19 pandemic	6.648	***	Supported
H6: "Portability" positively affects the "hedonic browsing" by m-commerce application users during the COVID-19 pandemic	−.400	.689	Not Supported
H7: "Hedonic browsing" positively affects the "urge to buy impulsively" by m-commerce application users during the COVID-19 pandemic	−2.929	.003	Supported
H8: "Hedonic browsing" negatively affects the "urge to buy impulsively" by m-commerce application users during the COVID-19 pandemic	10.222	***	Supported

Hypothesis nine suggests that IBT moderates the relationship between browsing motivation and urge to buy impulsively. The results show that IBT is a significant predictor of the urge to buy impulsively. The results of hypothesis nine are shown in Table 2.

Table 2. Hypothesis testing results

Hypotheses	C.R	P	Result
H9a: "IBT" moderates the relationship between "utilitarian browsing" and the "urge to buy impulsively" during the COVID-19 pandemic.	0.007	***	Supported
H9b: "IBT" moderates the relationship between "hedonic browsing" and the "urge to buy impulsively" during the COVID-19 pandemic	0.005	***	Supported

5 CONCLUSION

First, interpersonal influence has a significant impact on both consumer browsing motivations. This result differs from a previous study by X. Zheng et al. (2019), who suggested that interpersonal influence has a significant impact on hedonic browsing but an insignificant impact on the consumer with utilitarian browsing motivation.

Second, the visual appeal has a significant effect on both consumer browsing motivations. The finding is consistent with the previous study by X. Zheng et al. (2019), who suggested that m-commerce does not only make consumer feel good but also provide information to improving efficiency in the consumer shopping experience in m-commerce.

Third, portability has a significant relationship with utilitarian browsing. However, on hedonic browsing, portability does not have a significant impact. This result is consistent with a previous study by X. Zheng et al. (2019), who suggested that consumers with hedonic motivation are just concerned with pleasure and enjoyment of shopping behavior, but for the consumer with utilitarian motivation, portability is about efficiency.

Fourth, utilitarian browsing has a significant and negative effect on the urge to buy impulsively. Meanwhile, hedonic browsing has a significant and positive impact. This finding is similar to the study conducted by Park et al. (2012), who suggested that people with utilitarian motivation are more concerned with the goal in shopping, while people with hedonic motivation are mostly concerned with fun and enjoyment.

Fifth, IBT moderates the relationship between browsing motivation toward the urge to buy impulsively. This finding proposes that a higher personality trait of IBT in both consumer utilitarian and hedonic browsing motivations will increase the intention to buy impulsively.

This study provides information on how situational factors affects consumer behavior toward the urge to buy impulsively on m-commerce, which might help m-commerce providers to plan their marketing strategies. For example, visual appeal and interpersonal influence, which have a significant effect on consumers with utilitarian and hedonic values, are a hint for the m-commerce retailers to design the visual appearance of their m-commerce application and also to provide interesting product information visually to fulfill utilitarian consumers' need.

This study focuses only on m-commerce customers and cross-sectional research because of limited resources and time. Future research could adopt experiment methods or a longitudinal approach to examine consumer buying behavior.

REFERENCES

Chen, C.C., & Yao, J.Y., (2018). What drives impulse buying behaviors in a mobile auction? The perspective of the S-O-R model. Telematics and Informatics, 35, 1249–1262.

Flight, R. L., Rountree, M. M., & Beatty, S. E. (2012). Feeling the urge: Affect in impulsive and compulsive buying. The Journal of Marketing Theory and Practice, 20, 453-466.

Ismagilova, E., Dwivedi, Y. K., & Slade, E. (2019). Perceived helpfulness of eWOM: Emotions, fairness and rationality. Journal of Retailing and Consumer Services. https://doi.org/10.1016/j.jretconser.2019.02.002.

Lee, J. A., & Kacen, J. J. (2008). Cultural influences on consumer satisfaction with impulse and planned purchase decisions. Journal of Business Research, 61(3), 265–272.

Madhavaram. S. & Laverie, D.A. (2004) Exploring Impulse Purchasing on the Internet. Advances in consumer research. Association for Consumer Research (U.S.), 31(1).

Okazaki, S., & Mendez, F. (2013). Exploring convenience in mobile commerce: Moderating effects of gender. Computers in Human Behavior, 29(3), 1234–1242.

Parboteeah, D. V., Valacich, J. S., & Wells, J. D. (2009). The influence of website characteristics on a consumer's urge to buy impulsively. Information Systems Research, 20, 60–78.

Park, E. J., Kim, E. Y., Funches, V. M., & Foxx, W. (2012). Apparel product attributes, web browsing, and e-impulse buying on shopping websites. Journal of Business Research, 65(11), 1583–1589.

Rowley, J. (2002). Window' shopping and browsing opportunities in cyberspace. Journal of Consumer Behaviour, 1(4), 369–378.

Verhagen, T. & van Dolen, W. (2011). The influence of online store beliefs on consumer online impulse buying: a model and empirical application. Inf. Manage, 48(8), 320–327.

X. Zheng, et al. (2019). Understanding impulse buying in mobile commerce: An investigation into hedonic and utilitarian browsing. International Journal of Information Management, 48 (2019), 151–160.

Contemporary Research on Business and Management – Noviaristanti (Ed.)

How uncertainty and firm characteristics determine the capital structure: Evidence from Indonesia's sharia- and non-sharia-compliant firms

M.L. Nurhakim & I. Gandakusuma
Master of Management, Universitas Indonesia, Jakarta, Indonesia

ABSTRACT: This paper inspects the determinants of the capital structure of Indonesia's sharia- and non-sharia-compliant firms. We employed a broader and more comprehensive uncertainty index than Economic Policy Uncertainty (EPU) called the World Uncertainty Index (WUI). This study derived results from a static panel data model that allows for interactions between the uncertainty index and firm characteristics. Our findings unveil that some WUI and firm characteristic interactions are statistically significant to determine the sharia-compliant firm capital structure. Interestingly, some sharia-compliant firm characteristics are insignificant individually, but significantly affect the capital structure when dealing with uncertainty. The results show that firm characteristics have a partial effect on the non-sharia-compliant firm capital structure. This study extends the empirical literature of the uncertainty index and becomes the pioneer in research, employing the newly introduced WUI in sharia firms and emerging markets.

1 INTRODUCTION

Capital structure plays an important role in fulfilling stakeholders' needs. Capital structure is also the main topic in corporate finance, and its study remains inconclusive (Vo, 2017; Yildirim et al., 2018). As sharia finance in Indonesia is growing, especially in the sharia-compliant firms, the urgency of capital structure studies in this type of firm is increases. Besides, Indonesia was also recognized as the best sharia capital market in 2019 and 2020. Sharia-compliant firms are listed in Indonesia's Financial Authority (OJK) list of sharia effect called Daftar Efek Syariah (DES). Sharia-compliant firms in Indonesia's Stock Exchange (BEI) share 50% of the capital market, and 63% of the stocks in BEI is sharia-compliant firms. Thus, the capital structure topic is becoming increasingly important for the stakeholders, as the sharia market grows.

In this study, we employed the World Uncertainty Index (WUI), introduced by Ahir, Bloom, and Furceri in 2018. As suggested by Li & Qiu (2021), we used the WUI with firm characteristics to extract its effect on the capital structure. We employed profitability, firm size, tangibility, and growth opportunity as the firm characteristics. The research questions are determinants of capital structure in both sharia-compliant firms and non-sharia-compliant firms. We also investigated the effect of the uncertainty index on the capital structure. The rest of the paper are as follows: section 2—literature review, section 3—methodology, section 4—results, and section 5—conclusions.

2 LITERATURE REVIEW

Sharia-compliant firms enlisted in DES are limited to firms that are stated in its articles of association bylaws about business conduct under sharia principles or are reporting certain levels of debt

DOI 10.1201/9781003196013-45

and non-halal revenue sources. These limitations may drive the capital structure approach of such firms in different ways. The main theories to explain the capital structure approach is trade-off and pecking order (Yildirim et al., 2018). Trade-off indicates that financial mix is to adjust the tax shield and various costs of issuing debt (Myers, 1984). Pecking order theory provides hierarchy options for financing decisions and firms issue the most secure external financing if required (Myers, 1984). Trade-off and pecking order are quite different. Trade-off tends to leverage the firm's capability, while pecking order opts for external financing only if necessary.

WUI is an uncertainty index covering political and economic uncertainty (Ahir et al., 2018). The previously used uncertainty index was the Economic Policy Uncertainty (EPU), which covered economic uncertainty and was available for developed countries only. The WUI is a development of this index and is now available for emerging countries. The WUI is correlated negatively with macroeconomics indicators (e.g., GDP) and is suggested by the developer to be used as an alternative macroeconomic variable (Ahir et al., 2018).

3 METHODOLOGY

3.1 *Sample design*

We employed the list of the sharia effect or called Daftar Efek Syariah (DES) published by OJK and Sharia National Board in the 2015–2019 period. Sharia-compliant firms' sample was collected by including firms that were always listed in DES in 2015–2019. Non-sharia-compliant firms' sample was collected by including all firms that were always out of the DES during the same period. The sample excluded the financial industry and all firms which were either inconsistently present in the list or not present at all during the period. This exclusion becomes the limitation of this study, as it may not explain all of the firms in the exchange. We collected 163 sharia-compliant firms and 51 non-sharia-compliant firms.

We employed book leverage and market leverage for the capital structure indicator. Profitability was measured by EBITDA to TA; firm size was measured by the natural logarithm of revenue; tangibility was measured by PPE to TA; and growth opportunity was measured by the market to book ratio. GDP was measured by its annual growth and WUI was measured by its natural logarithm. These variables were collected from Thomson Reuters data stream, world bank, and EPU.

3.2 *Model specification*

The model in this study is based on two main reasons as previous studies suggested. The first reason is mitigation of the endogeneity problem. The second reason is that Yildirim et al. (2018) stated a common belief in the studies in which firm characteristics affect capital structure decisions at least in the following year. Li & Qiu (2021) also stated that the uncertainty index or macroeconomic variables are likely to affect the leverage at least in the subsequent year. Incorporating the static panel regression and prior considerations, we estimate:

$$\begin{aligned} BLEW_{it} = {} & \alpha + \beta_1 PROF_{it-1} + \beta_2 SIZE_{it-1} + \beta_3 TANG_{it-1} + \beta_4 GOPP_{it-1} + \beta_5 LWUI_{it-1} \\ & + \beta_6 LWUI_{it-1}{}^{*} PROF_{it-1} + \beta_7 LWUI_{it-1}{}^{*} SIZE_{it-1} + \beta_8 LWUI_{it-1}{}^{*} TANG_{it-1} \\ & + \beta_9 LWUI_{it-1}{}^{*} GOPP_{it-1} + \beta_{10} GDPG_{it-1} + \mu_1 + \mu_t + \varepsilon_{it} \end{aligned} \tag{1}$$

$$\begin{aligned} MLEW_{it} = {} & \alpha + \beta_1 PROF_{it-1} + \beta_2 SIZE_{it-1} + \beta_3 TANG_{it-1} + \beta_4 GOPP_{it-1} + \beta_5 LWUI_{it-1} \\ & + \beta_6 LWUI_{it-1}{}^{*} PROF_{it-1} + \beta_7 LWUI_{it-1}{}^{*} SIZE_{it-1} + \beta_8 LWUI_{it-1}{}^{*} TANG_{it-1} \\ & + \beta_9 LWUI_{it-1}{}^{*} GOPP_{it-1} + \beta_{10} GDPG_{it-1} + \mu_i + \mu_t + \varepsilon_{it} \end{aligned} \tag{2}$$

4 RESULTS

4.1 *Results*

Before the estimation of the results, we conducted a model selection procedure for all 4 models (book leverage and market leverage for both sharia and non-sharia). The selection process also considered the validity of the regression after facing the autocorrelation and heteroskedasticity issue. As suggested by Schaffer and Stillman (2006), we tested for overidentifying to choose the model with robust-cluster standard error. Given below is the summary.

Table 1. Static panel regression results

	Book Leverage		Market Leverage	
	Sharia	Non-sharia	Sharia	Non-sharia
Variable	Coefficient	Coefficient	Coefficient	Coefficient
L1.PROF	0.0916	−0.2454	0.3750	−0.2310
L1.SIZE	0.0262**	0.0628**	0.0090	0.0320
L1.TANG	−0.1837	−0.0465	0.0860	0.1230
L1.GOPP	0.0031	−0.0173	−0.0560***	−0.0680**
L1.LWUI	−0.0538	0.0922	0.0090	0.0290
L1.LWUIXPROF	0.0840**	0.1569	0.1830**	0.7170
L1.LWUIXSIZE	0.0046	−0.0055	−0.0010	−0.0030
L1.LWUIXTANG	−0.0495	−0.0291	0.0090	0.0130
L1.LWUIXGOPP	−0.0018	−0.0073	−0.0090***	−0.0050
L1.GDPG	53.7200	92.0400	16.5370***	44.7900
Constant	−0.1598	−0.6873	−0.4940	−0.0470
R2	0.0823	0.1198	0.205	0.4271
F-stat	3.65***	7.02***	5.700***	83.49***
Observations	715	255	715	255
Model	Fixed effect	Fixed effect	Fixed effect	Random Effect

*, **, and *** denote significance at 10%, 5%, and 1% levels respectively.

From Table 1, we found consistency with previous studies in the importance of size and growth opportunity in firms' capital structure decision (Moradi & Paulet, 2019; Onofrei et al., 2015; Rokhayati et al., 2019; Vo, 2017; Yildirim et al., 2018; Zhang et al., 2018). Both size and growth's effect can be explained by the trade-off theory since both factors are indicating the firm's capability to endure the risk of leverage. From the size point of view, the positive relationship is the cause of increasing capability to endure more risk as the firm's getting bigger in size. Therefore, a firm with a bigger size tends to increase leverage. High growth opportunity indicates higher risk, and firms tend to lower the leverage as the growth opportunity increases.

Profitability is not significantly affecting leverage and is consistent with previous studies (Moosa & Li, 2012; Oktavina & Manalu, 2018; Rokhayati et al., 2019). We may argue that profitability may not indicate the needs for external financing, yet it shows the cash availability. Tangibility is also not significant and is consistent with previous studies (Fauzias et al., 2011; Moosa & Li, 2012; Serrasqueiro & Caetano, 2015; Yildirim et al., 2018). Fauzias et al. (2011) compared the tangibility between countries and found it insignificant due to the assumptions of firms in different countries regarding the importance of tangibility.

Comparing our study results with those reported by Li and Qiu (2021), we found that the uncertainty index indicated by WUI is not significant. Li and Qiu (2021) also found it insignificant in two of three uncertainty indexes they employed. We also confirm the significant interaction between firm characteristics (profitability and growth opportunity) with the uncertainty index, although we only found one firm characteristic that is partially significant through interaction, namely, the growth opportunity.

The effect of the interaction is also as expected, where the index's effect is the same as the firm characteristic it interacted with.

5 CONLUSION

In conclusion, we found that determinants of the capital structure of both sharia-compliant and non-sharia compliant firms are size and growth opportunity in terms of the firm's characteristics. Comparing the results of the two types of firms, the significant firm's characteristics are affecting the capital structure similarly. The uncertainty index effects on capital structure are seen through its interaction with growth opportunity and profitability only to sharia-compliant firms and are partially insignificant on both types. The interaction effects indicate that firms are considering external factors and internal factors when deciding financial mix. The indifference effects may indicate a similar managerial approach on both firm types. It may also concern the policymakers on how to regulate the different types of firms as they show quite a similar approach to capital structure decisions. Further, this may also raise a question for investors regarding the difference between sharia-compliant and non-sharia-compliant firms since sharia-compliant firms are facing more constraints than non-sharia-compliant firms. Results in this study also show that firms are rational when deciding their financing and the firm's risks. For further studies, the researcher may employ uncertainty index in another related topic or provide a comparative study between countries, since it is available for both emerging and developed countries.

REFERENCES

Ahir, H., Bloom, N., & Furceri, D. (2018). The World Uncertainty Index. *SSRN Electronic Journal*. https://doi.org/10.2139/ssrn.3275033.

Fauzias, M. N., Razali, H., Khairunisah, I., Izani, I., & Norazlan, A. (2011). Determinants of Target Capital Structure: Evidence on South East Asia Countries Nor, Haron, Ibrahim, Ibrahim & Alias. *Journal of Business and Policy Research, 6*(3), 39–61.

Li, X., & Qiu, M. (2021). Journal of International Money and Finance The joint effects of economic policy uncertainty and firm characteristics on capital structure?: Evidence from US firms. *Journal of International Money and Finance, 110*, 102279. https://doi.org/10.1016/j.jimonfin.2020.102279.

Moosa, I., & Li, L. (2012). Firm-specific factors as determinants of capital structure: Evidence from Indonesia. *Review of Pacific Basin Financial Markets and Policies, 15*(2), 1–17. https://doi.org/10.1142/S021909151150007X.

Moradi, A., & Paulet, E. (2019). The firm-specific determinants of capital structure – An empirical analysis of firms before and during the Euro Crisis. *Research in International Business and Finance, 47*(July 2018), 150–161. https://doi.org/10.1016/j.ribaf.2018.07.007.

Myers, S. C. (1984). The Capital Structure Puzzle. *The Journal of Finance, XXXIX*(3), 575–592.

Oktavina, M., & Manalu, S. (2018). Pecking Order and Trade-off Theory in Capital Structure Analysis of Family Firms in Indonesia. *Jurnal Keuangan Dan Perbankan, 22*(1), 73–82. https://doi.org/10.26905/jkdp.v22i1.1793.

Onofrei, M., Tudose, M. B., Durdureanu, C., & Anton, S. G. (2015). Determinant Factors of Firm Leverage: An Empirical Analysis at Iasi County Level. *Procedia Economics and Finance, 20*(15), 460–466. https://doi.org/10.1016/s2212-5671(15)00097-0

Rokhayati, I., Pramuka, B. A., & Sudarto. (2019). Optimal financial leverage determinants for smes capital structure decision making: Empirical evidence from Indonesia. *International Journal of Scientific and Technology Research, 8*(11), 1155–1161.

Schaffer, M., & Stillman, S. (2006). XTOVERID: Stata module to calculate tests of overidentifying restrictions after xtreg, xtivreg, xtivreg2, xthtaylor. *Statistical Software Components*.

Serrasqueiro, Z., & Caetano, A. (2015). Trade-Off Theory versus Pecking Order Theory: capital structure decisions in a peripheral region of Portugal. *J. Bus. Econ. Manage, 16*(2), 445–466.

Vo, X. V. (2017). Determinants of capital structure in emerging markets: Evidence from Vietnam. *Research in International Business and Finance, 40*, 105–113. https://doi.org/10.1016/j.ribaf.2016.12.001
Yildirim, R., Masih, M., & Bacha, O. I. (2018). Determinants of capital structure: evidence from Shari'ah compliant and non-compliant firms. *Pacific Basin Finance Journal*, 51(May), 198– 219. https://doi.org/10.1016/j.pacfin.2018.06.008
Zhang, Q., Saqib, Z. A., & Chen, Q. (2018). Determinants of Capital Structure: An Empirical Analysis of Fuel and Energy Sector of Pakistan. *2018 15th International Conference on Service Systems and Service Management*. https://doi.org/10.1109/ICSSSM.2018.8465061

Contemporary Research on Business and Management – Noviaristanti (Ed.)

Brand awareness of Crisp!! Enye-Enye

M. Astuti & M. Khoerunisa
Pembangunan Nasional Veteran Jakarta University, Indonesia

ABSTRACT: This research is motivated by the low brand awareness of Crisp!! Enye-Enye, as it was ranked 11th out of 14 popular cassava-processed product brands in Banten Province in 2018. Therefore, this research aims at evaluating the effect of advertising, direct marketing, and personal selling on brand awareness. A quantitative descriptive research design was employed with 80 Crisp!! Enye-Enye consumers in different regions as the respondents. The research data analysis was conducted with the assistance of smartPLS version 3.2.7. The research results revealed that advertising, direct marketing, and personal selling had quite a high effect on brand awareness, with the R-squared value of 74.1%. One of the three hypotheses was rejected; advertising and direct marketing were proven to affect brand awareness, while personal selling did not affect brand awareness.

1 INTRODUCTION

Micro, small, and medium enterprises (MSMEs) are very important for economic development in the community. Moreover, MSMEs have a strategic role in the national economic development. When the global economic crisis hit the world, MSMEs were still able to boost the national economy. Therefore, they contribute significantly to the gross domestic product (GDP) in Indonesia's economic growth (Sulistyo, 2020). Based on data from 2017 to 2019, MSMEs' contribution to Indonesia's GDP increased by 57.08%, 60%, and 60.34% for three consecutive years.

According to the Department of Agriculture and Plantation of Lebak Regency (2019), the GRDP percentage in Lebak Regency fluctuated from 2015 to 2019. From 2015 to 2016, the value decreased by 10%. Meanwhile, from 2017 to 2019, the value increased by 10%, 5%, and 4.9%, respectively. This can occur because the management of potential and productivity in the area has not been optimized, even though it has a high potential related to superior food.

Foodstuffs are a promising source of culinary potential. One of the foodstuffs in Lebak Regency is cassava. With a harvest area of 2,539 hectares and a production level of 45,230 tons, cassava productivity is 17.81 tons/ha (Suryana, 2019). Meanwhile, the ideal cassava productivity is 20–60 tons/ha, so the cassava productivity in Lebak Regency has not reached the expected ideal figure (Ammurabi, 2019).

Research conducted by Astuti, Sembiring, and Amanda (2018) showed that Kelompok Wanita Pagoda (KWP/Pagoda Women's Group) in Lebak Regency produced enye-enye, which is a ready-to-eat snack from cassava under the brand name "Crisp!! Enye-Enye". Even though the product marketing distribution covers three areas, namely, DKI Jakarta Province, Lebak Regency, and Tangerang Regency, the brand is not yet known by the public. The lack of understanding of the marketing strategies and knowledge in using social media is MSMEs' weakness (Mardiatmi and Pinem, 2016).

Based on the popularity of cassava-processed product brands in Banten Province, Crisp!! Enye-Enye was ranked 11th out of 14 in 2019. This indicates that the brand awareness was still low. The concept of a brand is not only for products from large-scale or multi-national companies. An MSME must also have a brand to encourage business progress and development (Lingga, 2019).

DOI 10.1201/9781003196013-46

2 RESEARCH METHODOLOGY

This research employed a descriptive quantitative research design. The population in this research was Crisp!! Enye-Enye consumers from Lebak Regency, Tangerang Regency, and DKI Jakarta Province. The sampling technique used was probability sampling with a random sampling design. 80 respondents were involved. The data were collected using primary data sources, namely, questionnaires. The data analysis technique was a descriptive analysis approach through the interpretation of the value of cross loading and loading factors (output partial least squares) and inferential analysis to test the hypothesis with the assistance of SmartPLS version 3.2.7 software.

3 RESULT AND DISCUSSION

3.1 *Descriptive analysis*

Based on the results of the analysis using SmartPLS version 3.2.7, the output factor loading and cross-loading for the variables used in the research are presented in Table 1.

Table 1. Descriptive analysis results.

Variables	Cross Loading Average	Loading Factor Average
Brand awareness (Y)	0.705	
Crisp!! Enye-Enye is the first product that I remember when I remember snack foods made from cassava (High)		
I know Crisp!! Enye-Enye products (Low)		
Advertising (X_1)		0.760
The concept of the advertisement displayed makes me curious (High)		
The advertising slogan is catchy to remember (Low)		
Direct Marketing (X_2)		0.766
I can find advertising via social media (WhatsApp) (High)		
The seller can provide detailed information about the product (Low)		
Personal Selling (X_3)		0.754
Sellers ask for reviews after purchase (High)		
Sellers maintain good relationships with consumers (Low)		

Source: Data Processed, 2020

Table 1 shows that the average cross-loading of brand awareness is 0.705, which infers that the consumers realize that product promotion in various media helps the products to be easily recognized, so that this can help increase Crisp!! Enye-Enye's visibility. The average loading factor for advertising is 0.760, which means that the consumers believe that advertising in the form of images, brand ambassadors, and slogans can attract consumers to remember the Crisp!! Enye-Enye brand. The average loading factor for direct marketing is 0.766, which infers that the consumers think that direct marketing through social media, telephone, or face-to-face can give a good impression to consumers. The sellers can provide a detailed explanation of the product and respond to questions asked by consumers quickly. Furthermore, the average loading factor for personal sales is 0.754, which indicates that in addition to direct advertising and marketing, the consumers also think that personal sales can raise consumer interest in the Crisp!! Enye-Enye products. In conclusion, the process of approaching, presentation and demonstration, overcoming objections, closing transactions, and follow-up had been done well.

3.2 *Inferential analysis*

R-squared was used to measure the extent of the model's ability to explain variations in the dependent variable (Ghozali, 2016, p. 97). The output of SmartPLS is presented in Table 2.

Table 2. Value of R-squared.

	R-squared
Brand Awareness	0.741

Source: SmartPLS Output

Table 2 shows that the independent variables of advertising, direct marketing, and personal selling are able to increase brand awareness by 74.1%.

The $t_{\text{!"\#\$\%}}$ value in this research is 1.992 with a degree of error of 5%. The results of data processing for significance testing (t-test) obtained are presented in Table 3.

Table 3. T-statistical test results.

	Original Sample (O)	T Statistic (|O/STDEV|)	P-Values
Advertising–Brand awareness	0.351	2.456	0.014
Direct marketing–Brand awareness	0.416	2.682	0.008
Personal selling–Brand awareness	0.174	0.962	0.336

Source: SmartPLS Output

The results of hypothesis testing show that advertising positively and significantly affects brand awareness with a path coefficient value of 0.351. In other words, if advertising decreases in terms of quality of the product concept and appearance, then brand awareness also decreases. The indicators used in advertising include drawing attention, attractive appearance, and increase of desire and action that can increase the Crisp!! Enye-Enye brand awareness. The results are also in line with the research studies conducted by Dewi and Sugandi (2019), Putri et al. (2019), and Septiningrum and Sudrajat (2019), who stated that advertising can increase brand awareness.

In addition, direct marketing positively and significantly affected brand awareness with a path coefficient value of 0.416, which indicates that when direct marketing increases by one unit, brand awareness also increases by 0.416. The results are supported by Shintarani (2017).

Furthermore, personal selling has no significant effect, but has a positive direction on brand awareness with a path coefficient of 0.174, which indicates that personal sales do not affect the Crisp!! Enye-Enye brand awareness. The results are not in line with the research conducted by Dewi and Magdalena (2017) and Shintarani (2018), who state that personal selling can increase brand awareness.

The difference in the research results could be affected by different research objects. Previous research used more large-scale company objects, while in this research, MSMEs with unknown product results were used as the research object. In addition, the criteria for sampling were different. The research conducted by Shintarani (2017) concluded that personal sales in companies were carried out by sales professionals. Meanwhile, at the Crisp!! Enye-Enye as one of MSMEs, the marketing was still lacking in terms of good marketing strategy practices. The follow-up and maintenance are needed to be carried out in personal sales in order to ensure customer satisfaction and repeat customers (Kotler & Armstrong, 2016). Meanwhile, The Crisp!! Enye-Enye sellers do not follow up, such as asking for reviews and maintaining good relationships with consumers.

4 CONCLUSION

This research concluded that advertising increased the Crisp!! Enye-Enye brand awareness. Furthermore, direct marketing also increased the Crisp!! Enye-Enye brand awareness. However, personal sales neither increased nor decreased the Crisp!! Enye-Enye brand awareness.

REFERENCES

Ammurabi, S. D. (2019, December 2). *Tekan Impor Gandum, MSI Usul Pengembangan 1 Juta Ha Singkong*. Gatra. https://www.gatra.com/detail/news/459799/ekonomi/tekan-imporgandum-msi-usul-pengembangan-1-juta-ha-singkong.

Astuti, M., Sembiring, R., & Amanda, A. R. (2018). Program Kemitraan Komunitas pada 'EnyeEnye' di Desa Mekar Agung, Kabupaten Cibadak, Lebak – Banten. *Kumawula: Jurnal Pengabdian Kepada Masyarakat, 1(3)*, 131–149. doi:10.24198/kumawula.v1i3.19196.

Dewi, L. & Magdalena, F. (2017). Pengaruh Personal Selling Dan Word Of Mouth Terhadap Brand Awareness Bisnis Mahasiswa Universitas Ciputra. *Jurnal Eksekutif, 14(2)*, 253261.

Dewi, Y. L. and Sugandi, M. S. (2019). Pengaruh Iklan Web Series SPace # "Kenapa Belum Nikah?" Terhadap Brand Awareness JD.ID. *Profetik: Jurnal Komunikasi, 12(1)*, 141–148. doi:10.14421/pjk.v12i1.1556.

Kotler, P., & Armstrong, G. (2016). *Principles of Marketing Global Edition* (Sixteenth). Harlow: Pearson.

Lingga, M. A. (2019, February 26). *Meski Produk UMKM, Ternyata Sangat Penting Miliki Brand dan Hak Cipta*. Kompas. https://ekonomi.kompas.com/read/2019/02/26/174108026/meski-produk-umkm-ternyatasangat-penting-miliki-brand-dan-hak-cipta.

Mardiatmi, B. D., & Pinem, D. (2016). The Study on the Marketing Mix Development Strategy Analysis of Creative Industry Sme-Based in Depok West Java. *International Journal of Business & Commerce, 5(6)*, 91–104.

Putri, A. N., Handayani, T., & Astuti, M. (2019). Pengaruh iklan, Selebriti Pendukung dan Pemasaran dari Mulut ke Mulut Terhadap Kesadaran Merek pada Produk Mie Sedaap. *Jurnal Manajemen, 11(1)*, 24–34. doi:10.29264/jmmn.v11i1.4524.

Septiningrum, W., & Sudrajat, R. H. (2019). Pengaruh Jingle Iklan Susu Koperasi Peternakan Bandung Selatan Pengalengan Terhadap Brand Awareness. *e-Proceeding of Management*, 6(2), pp. 5107–5117.

Shintarani, E. D. (2018). Pengaruh Promosi Produk Seafoodking terhadap Brand Awareness Produk. *Nyimak (Journal of Communication)*, 1(2), pp. 209–220. doi:10.31000/nyimak.v1i2.484.

Sulistyo, C. B. (2020, January 4) *Pemberdayaan UMKM Menuju Go International*. Investor Daily. https://investor.id/opinion/pemberdayaan-umkm-menuju-go-international.

Suryana, M. (2019, November 23) *Produksi ubi kayu di Lebak tembus 45.230 ton dengan luas tanam 2.539 hektare*. Antara News Biro Banten. https://banten.antaranews.com/berita/74984/produksi-ubi-kayu-di-lebak-tembus-45230ton-dengan-luas-tanam-2539-hektare

Contemporary Research on Business and Management – Noviaristanti (Ed.)

Maturity analysis of E-government in West Java public service websites

N. Hanifa & S. Noviaristanti
Telkom University, West Java, Indonesia

ABSTRACT: The electronics-based government system (SPBE, Sistem Pemerintahan Berbasis Elektronik)) is a new chapter of governance in Indonesia, especially in the West Java province. The West Java government has the vision to accelerate the implementation of the digital government by implementing a digitalization system to support public services. This study aims to test the E-government maturity model by using best practices in 27 public service portals implemented in West Java, Indonesia. There are 17 cities and regencies classified into stage 1 (presence) and 8 cities and regencies classified into stage 2 (interaction) of the maturity model.

1 INTRODUCTION

Transformation is mandatory for regions that are pioneering smart cities. The application of digital government requires the government to implement a digitalization system to support the administrations of public services (Pikiran Rakyat, 2019), for instance, in the forms of E-government application in government agencies' official websites and the availability of integrated services through online systems (Dhevina, 2018). The Governor of West Java, Ridwan Kamil, aims to implement digitalization in West Java to increase the region's potential and accelerate bureaucratic services (Kompas, 2019). The Governor mentioned four theories to accelerate the intended development: the Government 3.0 Theory, Pentahelix, the Eight Door Budget, and Digital Government.

From the regency to city level, the West Java government provides websites for public services. This research seeks to produce a framework using four stages of maturity stage, i.e., presence, interaction, transaction, and integration, with best practices for E-government development maturity model in 27 regencies and cities in West Java. This research aims to compose a maturity model to guide the local government of West Java in developing effective E-government, especially for its regencies and cities. The E-government model can function as the main portal for information and news search that supports the government's vision in realizing digitalization.

2 LITERATURE REVIEW

2.1 *E-government and maturity model*

According to Fath-Allah et al. (2014), the maturity model of the E-government website is a series of stages (from basic to advanced) that determine the E-government website's maturity. The stages are divided as follows:

1. The presence stage presents information and content to citizens, including news, law, publications, databases, and interactive maps. In this research, the stage will be mapped based on the web content category's best practices, as the focuses are both information and portal content.
2. The interaction stage enables interaction between the government and the public. In this research, the stage will be mapped based on the external category's best practices, since the model focuses on interaction and communication with citizens.

DOI 10.1201/9781003196013-47

3. The transaction stage enables citizens to make transactions by executing some services. This stage also includes best practices from other categories (including back-end and front-end web design, web content, and external category), producing sophisticated services or transactions.
4. The integration or transformation stage is the integration between the government agencies, i.e., joining the government to exchange information between agencies from the same or different jurisdictions. The methodology was adopted from the work by Fath-Allah et al. (2016).

2.2 *Best practice of maturity model E-government website*

A comparative study was conducted based on the work by Fath-Allah et al. (2016) in the four categories of maturity, i.e., presence, interaction, transaction, and integration. The study is divided into four categories of best practice:

1. The back-end category includes best practices that run in the background and are usually invisible to the user: customer centricity, interoperability, use of standards, modularity, security, privacy, single sign-on, delegation, e-participation, payments, workflows, and responsiveness.
2. Front-end web design categories include best practices that users usually observe and interact with. The front-end web design also relates to portal design aspects, namely, one-stop-shop, ease of navigation, social network, personalization, user forms, industrialization, and structuration.
3. Front-end web content categories include best practices that users typically see and interact with. They are related to portal information and its contents, namely, relevancy, accessibility, search engines, periodical change, rich content, interactive games, mobile apps, statements, transactions, and understandability.
4. The external category includes best practices loosely integrated with the technical aspects of the portal. This category mostly relates to the integration of citizens in the E-government process and portal marketing aspects, which are advertising, referencing, incentives, contest, and reusability.

3 METHODOLOGY

3.1 *Content analysis*

This research employs a quantitative approach and Martono's (2014) content analysis framework. According to Martono (2014), the content analysis stages are divided into data coding, data entering, data output, and data analysis. In the data coding stage, the data were prepared in a coding sheet, where the data were inputted and coded. The data from 27 regency and city websites in West Java were processed by converting the data from website displays to numerical data based on the coding guidelines. Then, in the data entering stage, the numeric data were entered into a Microsoft Excel sheet before processed by SPSS version 21.

The next stage was data output, which, according to Martono (2014), is a display of data in a frequency distribution table. Lastly, in the data analysis stage, the data were collected from observation of 27 websites, which were presented in tables, graphs, or pie charts.

Table 1. Regency and city improvement.

No	Regency/City	Percentage	Stage
1	Bogor Regency	45.35%	1 (Presence)
2	Sukabumi Regency	49.2%	2 (Interaction)
3	Cianjur Regency	51.77%	2 (Interaction)
4	Bandung Regency	49.50%	2 (Interaction)
5	Garut Regency	45.50%	2 (Interaction)

(*conitnued*)

Table 1. Continued.

No	Regency/City	Percentage	Stage
6	Tasikmalaya Regency	37.50%	1 (Presence)
7	Ciamis Regency	39.25%	1 (Presence)
8	Kuningan Regency	45.50%	2 (Interaction)
9	Cirebon Regency	53.52%	2 (Interaction)
10	Majalengka Regency	35.27%	1 (Presence)
11	Sumedang Regency	33.17%	1 (Presence)
12	Indramayu Regency	43.27%	1 (Presence)
13	Subang Regency	39.42%	1 (Presence)
14	Purwakarta Regency	39.27%	1 (Presence)
15	Karawang Regency	35.27%	1 (Presence)
16	Bekasi Regency	51.92%	2 (Interaction)
17	West Bandung Regency	26.92%	1 (Presence)
18	Pangandaran Regency	37.32%	1 (Presence)
19	Bogor city	45.35%	2 (Interaction)
20	Sukabumi City	43.27%	1 (Presence)
21	Bandung City	22.77%	0
22	Cirebon City	22.45%	0
23	Bekasi city	41.35%	1 (Presence)
24	Depok City	41.35%	1 (Presence)
25	Cimahi City	37.17%	1 (Presence)
26	Tasikmalaya City	41.50%	1 (Presence)
27	Banjar City	22.77%	0

4 CONCLUSION

This study aims to test the E-government maturity model using best practices in 27 public service portals used in West Java province. The research found 17 cities and regencies classified into stage 1 (presence) and 8 cities and regencies classified into stage 2 (interaction) of the maturity model. The average results for the two stages are then evaluated to improve the public service quality. This study surveys the implementation of E-government by best practice in West Java and formulates solutions to improve its maturity stages. This paper portrays the government portal maturity model by identifying the best practices and providing the justification of the mapping and best practice models.

REFERENCES

Dhevina, I. 2018. *E-government: Innovations in Communication Strategies*. Ministry of State Secretariat of the Republic of Indonesia.

Fath Allah, A., Cheikhi, L., Al-Qutaish, R., & Idri, A. 2014. E-government Maturity Models: A Comparative Study. *International Journal of Software Engineering and Application*, 71–91.

Fath Allah, A., Cheikhi, L., Al-Qutaish, R., & Adri, A. 2016. A Mapping Between a BP model and an E-government Portals' Maturity Model. *Electrical and Information Technology*.

Kompas. 2019. Advancing the West Java Digitizing System, Kang Emil Received International Award. *Kompas.com*. Bandung.

Martono, N. 2014. Quantitative Research Methods: Content Analysis and Secondary Data Analysis. *PT Raja Grafindo Persada*. Jakarta.

Pikiranrakyat. 2019. E-government and E-Governance. *Pikiranrakyat.com*. Bandung.

Contemporary Research on Business and Management – Noviaristanti (Ed.)

Employee engagement and organizational performance in the new normal era: A case study of companies with millennial generation employees

B. Rachman, N.K. Darmasetiawan, & J.L.E. Nugroho
University of Surabaya, Surabaya, Indonesia

ABSTRACT: This research aims to analyze the effects of employee engagement on employee engagement behavior, tangible performance, and intangible assets and on shareholder values in companies with millennial employees in the new normal era. The research applied mixed approach methods, combining both quantitative and qualitative methods. The variables were tested based on the dimensions expressed. Data collection was done by distributing questionnaires to managers who work at a company with millennial employees. Based on the research findings, it can be concluded that there was a significant influence of employee engagement behavior, tangible performance, and intangible asset on the shareholder value in the companies with millennial employees in the new normal era.

1 INTRODUCTION

The COVID-19 pandemic has changed the fabric of life. People must face various new challenges, including the way to carry out daily life and at work. In the new normal era, everyone is expected to prepare and accept all the changes that occur mentally. Everyone is asked to have a healthy lifestyle and pay attention to health protocols, while the demands to continue working productively also remain.

Furthermore, Macey, Schneider, Barbera, and Young in 2009 scrutinized a model of employee engagement on organizational performance, manifested in shareholder value. The research pinpoints that a high-performance work environment affects employee engagement feelings and employee engagement feelings affect employee engagement behavior. Furthermore, it is said that employee engagement behavior affects tangible performance outcomes, which include enhanced productivity and intangible assets, which include brand equity, customer satisfaction and loyalty, innovation, and lower risk, which will generate shareholder value.

According to NI Ardiansyah and NK Darmasetiawan (2019), generation Y individuals tend to be more confident. Moreover, it is also known that millennial employees, who have more technological capabilities, are more competitive in using digital technology. Hart (in the work by Tolbize, 2008) argued that the millennial generation has grown up with technology and made technology a part of their life. Apart from demanding new ways of working that apply strict health protocols, the new normal condition also demands some restrictions on working hours, which forces companies' employees who usually work from the office to work from home (WFH) alternately.

This study aims to investigate the effects of employee engagement on the organizational performance of the companies with millennial employees in the new normal era. Thus, the research questions are the following: (1) How does employee engagement behavior affect tangible performance outcomes and intangible assets in companies with millennial generation employees in the new normal era? and (2) How can tangible performance outcomes and intangible assets generate shareholder value in companies with millennial generation employees in the new normal era?

DOI 10.1201/9781003196013-48

2 LITERATURE REVIEW

2.1 *Employee engagement behavior on tangible performance outcomes and intangible assets*

Organizational performance can be in the form of intangible assets that include brand equity, customer satisfaction and loyalty, innovation, lower risk, and tangible performance outcomes, which include enhanced productivity. Employee engagement behavior is a manifestation of employee engagement, which is an essential contributor to a company's success (Ayers, 2006, and Pillai, 2013). In fact, an organization's existence is influenced by employee engagement, which is one of the factors that affect organizational performance (Bersin, 2014). Furthermore, Robinson et al. (2004) also revealed that employees who are firmly tied to the company would increase their performance and provide optimal efforts for its progress. This signifies that the performance of employees who are tied to their work tends to be positive; hence, it positively affects company performance as the company's performance is achieved by its employees' performance (Antonaco Poulou, 2000).

Suan et al. (2013), who studied the effects of employee engagement on organizations in Malaysian manufacturing companies, found that employee engagement influenced organizational sustainability. Then, Anita (2014) affirmed that employee engagement had a significant relationship with employee performance. Besides, employee engagement also affects team performance, as explained by Uddin et al. (2018). In their research, Hallberg et al. (2006), Saks (2006), Schaufeli et al. (2004), and Suan et al. (2013) revealed that employee engagement is manifested in employee engagement behavior, which has a relationship with positive work results such as low attrition, high performance, and positive business results. Furthermore, Efendy (2002) stated that employee performance is performance behavior that is also a result of work produced by employees or real behavior demonstrated by their role in the organization. Therefore, the following hypothesis can be built:

H1: Employee engagement behavior affects tangible performance outcomes and intangible assets in companies with millennial generation employees in the new normal era.

2.2 *Tangible performance outcomes and intangible assets on shareholder value*

Organizational performance is an overview of how the organization works to achieve its goals, which will definitely be influenced by the organization's resources. Organizational performance can be evaluated from financial and non-financial aspects, or it can also consist of tangible performance and intangible assets.

Tangible performance outcomes are often indicated by immediately visible things, for example, an increase in profit and productivity (Macey et al., 2009). Meanwhile, intangible assets include brand equity, customer satisfaction and loyalty, innovation, and lower risk.

Adopted from previous research conducted by Pandey (2005), this study analyzes the effects of fundamental factors (profitability, growth, firm size, and leverage) on the market to book value. Shareholder value is measured using the market to book value ratio, so that shareholder value will be created if the company's market value exceeds its book value. In addition to tangible performance, intangible assets are needed by the company. According to Luis and Biromo (2007), intangible assets can be shown if a product or service produced by a company has value for consumers, so that both tangible performance outcomes and intangible assets will create shareholder value.

H2: Tangible performance outcomes and intangible assets can generate shareholder value in companies with millennial generation employees in the new normal era.

3 RESEARCH METHODS

This research focused on employee engagement and organizational performance. The research applied mixed methods, combining both quantitative and qualitative methods. It is a combination of explanatory and basic research to understand the effects of employee engagement on employee engagement behavior, tangible performance, and intangible assets and on shareholder values in companies with millennial employees in the new normal era. The phenomenological approach and qualitative descriptive research were also employed, where the researchers know the position at the time of data collection in the field and become the data interpreters (Moleong, 2011) to analyze this topic through the experience, opinion, thoughts, and feelings of PT. A and PT. B employees. The qualitative data analysis was supported by various employee engagement behavior theories in employee engagement, tangible performance outcomes, and intangible assets on organizational performance, and their effects on shareholder value.

4 RESULTS AND DISCUSSION

From the results of quantitative research that has been done, it has been proven that (1) employee engagement behavior affected tangible performance outcomes and intangible assets in companies with millennial generation employees in the new normal era, which is positively significant and (2) tangible performance outcomes and intangible assets were able to generate shareholder value in companies with millennial generation employees in the new normal era, which is also positively significant.

From the results of the qualitative analysis, it was found that according to a PT. A senior manager who is a millennial employee, in this new normal era, he tries to maintain organizational performance, even though adjustments on work methods need to be done to break the chain of COVID-19 transmission. Furthermore, he said that he was involved in adjusting his company's business strategy, so that the company could survive and continue to grow in this new normal era.

Furthermore, the company opted to increase its competitive advantage, as it is essential to know that when planning to start a business. This is inseparable from economic principles, namely, running the company's operations smoothly, minimizing all costs, and maximizing profits.

According to Bersin (2014), this can be categorized as positive employee behavior because after all, they still think about organizational performance. The senior manager encourages employees to try their best to win consumers' hearts as part of a marketing strategy because marketing activities are to create short-term transactions and establish long-term relationships with customers, distributors, and suppliers. To implement this strategy, these managers feel that they cannot work alone, as they need help or work with other people to achieve company profitability and create stakeholder value. Judging from employees' responses in implementing the company strategy, employees who are classified as millennials show enthusiasm and high initiative and care deeply about company concerns through creative ideas put forward by employees for the progress of PT. A. This is also supported by the results of qualitative research on PT. A's millennial employees, where they show enthusiasm, strive to achieve success in their roles, show persistence, are proactive, and adapt to the new normal era.

At PT B, almost all of the millennial employees who work admit that they are ready to work in this new normal era, as they are flexible in adjusting their workplace policies and feel capable of nurturing a comfortable working atmosphere even if they only meet and coordinate virtually because it is a new work atmosphere that can make them more enthusiastic about working. Additionally, they feel that their office can facilitate working remotely by providing clear performance targets, flexible time, and workplace, making them work more productively. Millennial employees were used to living with digital convenience; thus, the support of current facilities from the company they work for makes millennial employees show more engagement behavior, which encourages performance, and this performance is directed at achieving the company's tangible and intangible outcomes, for example, in just a few moments, millennial employees have been able to adjust the way they work to serve customers through the digital platform developed by the company in a short time.

The company needs to develop tangible performance and intangible assets. According to Macey (2009), tangible performance outcomes are indicated by immediately visible things, for example, an increase in profit and productivity and intangible assets that include brand equity, customer satisfaction and loyalty, innovation, and lower risk.

5 CONCLUSION AND IMPLICATION

From the abovementioned discussion, it can be concluded that (1) employee engagement behavior had a positive and significant effect on tangible performance outcomes and intangible assets in companies with millennial generation employees in the new normal era and (2) tangible performance outcomes and intangible assets were able to generate shareholder value in companies with millennial generation employees in the new normal era, which are positively significant.

Meanwhile, from the results of quantitative data analysis supported by the results of qualitative analysis, it can be concluded that in this new normal era, companies with millennial generation employees who have engagement behavior can adapt more quickly to maintain organizational performance, both in the form of tangible performance outcomes and intangible assets. Both tangible performance outcomes and intangible assets are aimed at creating shareholder value.

REFERENCES

A.P. Prasetyanto, N.K. Darmasetiawan, J.L.E. Nugroho (2020). The high-performance work environment and employee engagement on millennials generation working at green business organization. *The 4th International Conference on Family Business and Entrepreneurship.*

Antonacopoulou, E.P. (2000). Employee Development through Self-Development in Three Retail Banks. *Personal Review, 29, 491–508.*

Bersin, Josh. (2014). *Predictions for 2014: Building A Strong Talent Pipeline for The Global Economic Recovery-Time for Innovative and Integrated Talent and HR Strategies.*

Macey, Schneider, Barbera, Young. 2009. *Employee Engagement, Tools for Analysis, Practice, and Competitive Advantage.* John Wiley & Sons, Chichester, West Sussex, UK., p:8.

Moleong. L. J. (2011). Metode Penelitian Kualitatif. PT. Remaja Rosdakarya, Bandung.

Myers, K., & Sadaghiani, K. 2010. Millenials in the workplace: A communication perspective on millennials' organizational relationships and performance. *Journal of Business and Psychology,* 25(2): 225–238 June 2010.

N.I, Ardiansyah & N.K., Darmasetiawan. (2019). Psychological well-being and work place relations gaps on generational differences, *Advances in Social Science and Humanities Research Volume 308.* Atlantis Press: Proceeding of *16th International Symposium on Management (INSYMA 2019).*

Nusantria, Sandi and Suharnomo. (2011). *Employee Engagement: Anteseden dan Konsekuensi Studi pada Unit CS PT. Telkom Indonesia Semarang. Undergraduate thesis*, Universitas Diponegoro.

Pillai, Anandan. (2013). Role of content strategy in social media brand communities: A case of higher education institutes in India. *Journal of Product & Brand Management.* Emerald Group Publishing Limited.

Tolbize. (2008). *Generation differences in the workplace research and training center on community living,* University of Minnesota.

Thomas. (2007). A New Measurement Scale for Employee Engagement: Scale Development, Pilot Test, and Replication. *Academy of Management Proceeding*

Verbeeten, Frank H.M. (2008). Performance Management Practices in Public Sector Organizations: Impact on Performance. *Accounting Auditing & Accountability Journal 21 (April): 427454.*

Wong, M., Gardiner, E., Lang, W., & Coulon, L. (2008). Generational differences in personality and motivation. *Journal of Managerial Psychology. 23 (8), 878–890.*

Contemporary Research on Business and Management – Noviaristanti (Ed.)

Discovering big in apparel: Redesigning the logo for "Big Apparel House"

R. Adiputra & Arviansyah
Faculty of Economics and Business, Universitas Indonesia, Indonesia

ABSTRACT: This study employs a business coaching technique in Big Apparel House (BAH), an MSME that focuses on large-sized men's clothing. We initially examine the MSME using the business internal and external analysis methods. The data are collected through field observations and interviews. We find two primary problems: inconsistent logo and impulsive unplanned marketing activities. This study focuses on assisting BAH in redesigning the logo and aligning all the company's marketing tools. We follow Merrilees' guidelines to explore and develop the BAH redesign strategy. The redesigned logo and its implementation have shown substantial improvement in BAH's brand and marketing.

1 INTRODUCTION

Competition in the digital world requires companies to have competitive advantages; various aspects can be an advantage; however, brands are a significant advantage in the marketplace (Mizik, 2014). Two critical elements that are the keys to a brand are personality (Keller, 2013) and logo (Cian, Krishna, and Elder, 2014). A study of the relationship between the two explains that a brand with a good logo and personality can affect sales and valuation (Luffarelli et al. l, 2019). Big Apparel House does not have a fixed identity. A logo, an identity element (Keller, 2013), has differences in various types of products and campaigns; the difference between product and marketplace logos can be seen in Figure 1. Inconsistency in identity makes promotional actions lack a definite plan. According to Merrilees (2005), brand orientation occurs when a strategy is built around the brand, and all organization parts are tied to the brand. Thus, building a brand vision, brand orientation, and brand implementation strategy (Merrilees, 2005) is the step needed to carry out BAH transformation to increase promotional activities' effectiveness.

Figure 1. Logo of Big Apparel House.

2 LITERATURE REVIEW

Rebranding attempts to adjust its vision and orientation to the elements embedded in its brand (Bedbury, 2002). One of the rebranding goals is to keep the brand fresh and relevant to the market

 DOI 10.1201/9781003196013-49

(Aaker, 1991). Kapferer (2000) says that a rebranding is necessary; a brand can maintain the quality of its aspirations only if it continues to change and reinvent itself.

In this study, the rebranding process will refer to the framework suggested by Merrilees (2005). Three stages are suggested in carrying out the rebranding process, namely, brand visioning, lining brand orientation, and brand strategy implementation; this stage aims to plan rebranding activities that are structured according to the company's needs and desires. Various strategy implementations can be carried out, starting from changing elements to brand names if necessary. However, in the case of Big Apparel House, the rebranding that will be done is creating a logo and sales support tools to change the company's image that previously experienced inconsistencies with their identity.

3 METHODOLOGY

This study utilizes a qualitative approach in the form of business coaching. Both primary and secondary data are acquired from interviews, observations, and BAH's internal data. This business coaching is implemented through four steps: (1) internal and external analysis; (2) TOWS, gaps analysis, and Pareto analysis; (3) alternative solutions and decision making; and (4) implementation and monitoring.

4 RESULT AND DISCUSSIONS

Analysis of MSME conditions was carried out based on internship sessions and interviews with Mr. Adji and consumers. After conducting internal and external analyses, we found two main problems: the inconsistency of the Big Apparel House brand association's logo and promotional activities carried out solely based on the impact of the decline in sales. After discussing with MSME owners and employees, we agreed to plan to fix these two problems compared to other problems.

We apply a rebranding strategy to the BAH. Rebranding carried out at MSMEs will refer to the case study research by Bill Merrilees; there are three primary stages carried out, namely, establishing a vision, orientation, and implementation strategy for the brand. The journal's application will be translated according to the company's vision and mission, setting orientation, and implementing planned changes.

4.1 *Alignment of the company's vision and mission*

According to Merrilees, alignment of the vision and mission needs to be done to carry out rebranding. The brand association that is formed needs to describe the company's ideals and goals. At first, BAH did not have a clear goal; they just tried and carried out the routine.

However, after further investigation, the two owners desired to change BAH to become "the main choice that is complete in meeting the needs of large men." Two missions were published with this vision, namely, creating comfortable products for customers and creating a database for customer needs.

4.2 *Setting orientation*

BAH's market is a specific niche market, where their target customers are large men. Brand orientation is a guideline for the company direction in determining the marketing strategy (Merrilees, 2005); it can be concluded that brand orientation forms brand characteristics needed to be instilled in consumers. According to BAH sales data, clothing designs that are in demand by consumers have a simple color orientation, not too bright but comfortable to look at. Then, the brand orientation will be taken according to the interests of consumers, namely, simple, comfortable, and large.

4.3 *Implementation of planned changes*

In this stage, planning for a logo change will begin, starting with three previously analyzed orientations, namely, simple, comfortable, and large. Thus, the logo sketch can be formed; we sketched with Mr. Adji and looked at several logos from other companies such as Progetto Moda Shop, an MSME from France that sells unique bags for women. From this logo, we take an idea of the inner lettering. This concept can represent size explicitly by only depicting the inner part of the letter. According to the old logo, we maintain the three-letter stacks because this logo remains a historic one that has accompanied BAH's four-year journey. According to Mr. Pipit, the main reason why this logo is not used in the product is the complexity of embroidery or screen printing. With these two references, we sketched up to express ideas based on the logo reference.

After finishing sketching, we asked both owners to choose the logo they think is best. The selection results are then elaborated again through color selection. According to the initial orientation, we agreed to choose the blue color that symbolizes broadness, comfort, and calm as the logo's color, besides black, as an affirmation. So, it was agreed that the new Big Apparel House logo would be as shown in Figure 2.

Figure 2. The new logo of Big Apparel House.

The selected logo is then implemented to all marketing tools used by BAH, such as employee clothes, product wrappers, business cards, cloth bags, and paper bags, which are given in Appendix A. This latest element is then applied to social media, especially in the product photo section. We surveyed 35 BAH consumers to find out consumer responses to logo changes; after we compared the latest logo with the previous one, 90.9% of respondents thought that the latest logo was better than the previous logo with various indicators offered in the survey.

The most significant change when implementing logos on social media occurs in product photos; currently, this marketing tool has two main elements: the product and the logo itself. This action is done to infuse the brand's elements into consumers' minds to differentiate them from competitors.

5 CONCLUSION

The redesigned logo of BAH is a change based on the problem, namely, the logo's inconsistency. Previously, MSMEs had four different logos as markers in their online shops and social media. Logo changes are carried out by building a company philosophy from the ground up, starting with establishing a vision and mission and orientation to their brand.

From the philosophy development process, a description of the company is found. This philosophy is outlined in a sketch tailored to the tastes of the owner and the target market. With this change, BAH has experienced an increase in social media interactions and an increase in the number of followers on their Instagram account. When this implementation was carried out, products with the latest photos had sales greater than the average sales during the pandemic, which had a direct positive impact on MSMEs.

REFERENCES

Aaker, D. A. (1991). Managing brand equity: Capitalizing on the value of a brand name. Free
Press; Maxwell Macmillan Canada; Maxwell Macmillan International.

Bakhtieva, E. (2017). B2B digital marketing strategy: A framework for assessing digital touchpoints and increasing customer loyalty based on Austrian companies from heating, ventilation and air conditioning industry. Oeconomia Copernicana, 8(3), 463?478–463?478. https://doi.org/10.24136/oc.v8i3.29

Buchanan, L., Kelly, B., & Yeatman, H. (2017). Exposure to digital marketing enhances young adults' interest in energy drinks: An exploratory investigation. PLOS ONE, 12(2), e0171226. https://doi.org/10.1371/journal.pone.0171226

Cian, L., Krishna, A., & Elder, R. S. (2014). This logo moves me: Dynamic imagery from static images. Journal of Marketing Research, 51(2), 184–197. https://doi.org/10.1509/jmr.13.0023 Keller, K. L. (2013). Strategic brand management: Building, measuring, and managing brand equity (4th ed). Pearson.

Kotler, P. (2003). Marketing management (11th ed). Prentice Hall.

Luffarelli, J., Stamatogiannakis, A., & Yang, H. (2019). The visual asymmetry effect: An interplay of logo design and brand personality on brand equity. Journal of Marketing Research, 56(1), 89–103. https://doi.org/10.1177/0022243718820548

Merrilees, B. (2005). Radical brand evolution: A case-based framework. Journal of Advertising Research, 45(2), 201–210. https://doi.org/10.1017/S0021849905050221.

Mizik, N. (2014). Assessing the total financial performance impact of brand equity with limited time-series data. Journal of Marketing Research, 51(6), 691–706. https://doi.org/10.1509/jmr.13.0431.

Mullins, J. W., & Walker, O. C. (2008). Marketing management: A strategic decision-making approach (6th ed). McGraw-Hill.

Osterwalder, A., Pigneur, Y., & Clark, T. (2010). Business model generation: A handbook for visionaries, game changers, and challengers. Wiley.

Contemporary Research on Business and Management – Noviaristanti (Ed.)

The effect of transformational leadership on innovative work behavior with psychological capital as the mediating variable

R. Febita
Universitas Indonesia, Central Jakarta, DKI Jakarta, Indonesia

P.M. Desiana
Universitas Indonesia, Depok, West Java, Indonesia

ABSTRACT: Innovation is very important for the sustainability and success of the company. Therefore, organizations should consider improving the employees' innovative work behavior. This study aims to examine the relationship between transformational leadership and the innovative work behavior of employees. This study used a mediation process to uncover the mediating impact of psychological capital in the proposed model. Furthermore, it used cross-sectional design and administered a questionnaire to 285 employees working at a major information and communication technology company in Indonesia (PT XYZ) to assess the proposed relationships. SPSS 25 and Lisrel 8.8 were used for statistical analysis. The result showed that psychological capital mediated the relationship between transformational leadership and innovative work behavior. Thus, the study shows that transformational leaders improved their employees' innovative work behavior by improving their psychological capital.

1 INTRODUCTION

Innovation is the most important factor for many organizations in the business environment. Correspondingly, previous studies show that organizational innovation exists only when employees have innovative work behavior. Organizational innovation is built on new ideas, which are generated, promoted, and implemented by employees. Therefore, organizational innovation depends on employee innovative work behavior (Amankwaa et al. 2019).

PT XYZ is a major information and communication technology (ICT) company in Indonesia. PT XYZ is demanded to provide quality and fast services, so that innovation continues to be carried out in terms of both internal and external processes, especially in the field of digital telecommunications. So far, PT XYZ's digital business is considered not enough to compete with other competitors, especially with foreign companies.

Psychological capital has a positive and significant effect on creativity and innovative behavior (Lei et al. 2020). Psychological capital is a person's positive psychological state, which is categorized by self-efficacy, optimism, hope, and resilience (Luthans et al. 2007). Individuals who have high psychological capital will try to come up with creative ways to achieve their goals (Abbas & Raja 2015).

Transformational leadership is one of the most effective types of leadership and has a significant effect on the performance of subordinates (Le & Lei 2018). Transformational leadership characteristics are the right driver to improve the four factors of psychological capital (Schuckert et al. 2018). Transformational leaders encourage company innovation through innovative behavior of employees to create innovative ideas and implement these innovative ideas into new products or processes (Birasnav et al. 2011). However, empirical studies show different results regarding its effects on innovative behavior; some show a significant relationship (Afsar & Umrani 2019;

DOI 10.1201/9781003196013-50

Amankwaa et al. 2019), others an insignificant one (Pundt 2015; Bednall et al. 2018). Leadership is very important aspect in PT XYZ. Therefore, a lot of training programs related to leadership have been carried out.

There are a few studies on the relationship of psychological capital with innovative work behavior (Schuckert et al. 2018; Lei et al. 2020). Based on the aforementioned explanation, the authors are interested in conducting research on the innovative work behavior of PT XYZ employees to examine whether transformational leadership can directly and indirectly increase innovative work behavior, with psychological capital as the mediating variable.

2 LITERATURE REVIEW

Transformational leadership characteristics are identifying and communicating vision, taking action that should be used as an example, encouraging acceptance of team goals, improving team performance, providing support for employees, and stimulating employee intelligence (Carless et al. 2000). Meanwhile, innovative work behavior consists of three different behavioral tasks for innovation, namely, idea generation, idea promotion, and idea realization. The first task is idea generation, which is creating novel and useful ideas. The next task is idea promotion, which is engaging in social activities to build supporters who provide the power to realize the ideas. The final task is idea realization, which is producing a model or prototype of the innovation (Janssen 2000). Transformational leaders build enthusiasm among organizational members to be more creative, as well as to create new ideas and solutions related to organizational structure, processes, and implementation (Afsar & Umrani 2019). Schuckert et al. (2018) showed that transformational leadership has a positive and significant effect on innovative work behavior. Thus, in this study, the first hypothesis is as follows:

H1: Transformational leadership has a positive and significant effect on innovative work behavior of employees.

Psychological capital is a person's positive psychological condition, which is categorized by four factors. These factors are self-efficacy, optimism, hope, and resilience (Luthans et al. 2007). Logically, the characteristics of transformational leadership are the right drivers to improve the four factors of psychological capital (Schuckert et al. 2018). Transformational leaders can affect the psychological capital of their subordinates because they can influence the beliefs, values, and goals of their subordinates (Wang et al. 2018). Schuckert et al. (2018) proved that transformational leadership has a positive and significant effect on psychological capital. Thus, the second hypothesis of this study is as follows:

H2: Transformational leadership has a positive and significant effect on psychological capital of employees.

Individuals who have high self-efficacy will have high intrinsic motivation, self-confidence, and intelligence, and they dare to face challenging and creative situations (Mishra et al. 2017). Optimistic individuals will not give up easily, and they have a positive view in stressful conditions. Thus, they will look for creative ways to solve problems and take advantage of every opportunity (Rego et al. 2012). Individuals who have high hope see problems and opportunities from different perspectives and look for creative ways to solve problems (Mishra et al. 2017). Resilience is also important when they are undergoing activities that require long-term effort, while maintaining high efficiency (Wojtczuk-Turek & Turek 2015), such as innovation activities. Mishra et al. (2017) showed that psychological capital has a positive and significant effect on innovative work behavior. Thus, the third hypothesis is as follows:

H3: Psychological capital has a positive and significant effect on innovative work behavior of employees.

In this study, the authors intended to test the mediating effect of psychological capital on the relationship between transformational leadership and innovative work behavior. Thus, the fourth hypothesis is as follows:

H4: Psychological capital mediates the relationship of transformational leadership and innovative work behavior.

3 RESEARCH METHODS

The participants of this study are 285 employees working for a major information and communication technology (ICT) company in Indonesia. The authors informed potential participants that the questionnaire was confidential and anonymous. The participants were asked to rate on their supervisors' transformational leadership characteristics, psychological capital, and their innovative work behavior. From 285 participants, 53.33% were male, 64.56% completed a bachelor's degree, 41.40% have been working for four up to six years in this company, and 55.49% have staff positions. 35.09% of the participants were from sales and marketing, 34.74% were from operations, and the rest were from other departments.

The authors used a six-point Likert scale ranging from (1) "strongly disagree" to (6) "strongly agree" for transformational leadership and psychological capital indicators and (1) "never" to (6) "very often" for innovative work behavior indicators. A seven-item scale reported by Carless et al. (2000) was used to measure transformational leadership behavior. Meanwhile, a scale reported by Luthans et al. (2007) was used to measure employee's psychological capital. The dimensions were self-efficacy, optimism, hope, and resilience; each dimension has six items, so there are 24 items. Innovative work behavior was rated using Janssen (2000)'s nine-item measure. The dimensions include idea generation, idea promotion, and idea realization; each dimension has three items.

4 RESULTS

Data were analyzed using the structural equation modeling (SEM) approach with Lisrel 8.8. The authors dropped three items from psychological capital because they did not meet the validity requirements. Thus, the authors only used 21 scales for psychological capital measurement. The results of data collection that have met the standards of validity and reliability were used for data processing.

The direct relationships among the core constructs in the theoretical model used one-tailed test method with a 95% confidence level. Transformational leadership ($\beta = 0.07$, $p < 0.05$) does not have a significant effect on innovative work behavior; so H1 was rejected. In addition, transformational leadership ($\beta = 0.42$, $p < 0.05$) has a positive and significant effect on psychological capital; so H2 was supported. Psychological capital ($\beta = 0.71$, $p < 0.05$) has a positive and significant effect on innovative work behavior; so H3 was also supported.

The value of the mediating effect (0.37) is greater than the direct effect (0.07), indicating that psychological capital strengthens the relationship between transformational leadership and innovative work behavior. The authors also used Sobel (1982) Z-test to test the mediation effect. Z-value for the mediating effect of psychological capital on transformational leadership and innovative work behavior relationship is 6.28 with a p-value of zero. This result shows that psychological capital mediates the relationship between transformational leadership and innovative work behavior, and so H4 was supported.

5 CONCLUSION

In conclusion, our study clarifies the mechanism by which transformational leadership influences innovative work behavior with psychological capital as the mediating variable. Transformational leaders can convince employees to think positively in terms of new visions and ideas. Thus, employees are not only more confident in their abilities but also more optimistic in their actions and in thinking about the future. Therefore, they can motivate employees to be more innovative (Lei et al. 2020). In other words, the characteristics in transformational leadership are the right drivers for increasing psychological capital, which is also a driver for innovative behavior (Schuckert et al. 2018).

Our study suggests several practical implications from strategic and operational levels to increase innovative work behavior of employees. From the strategic level, management should consider creating a pleasant work environment to create positive employee emotions, so that psychological capital of employees can be improved. Management also should consider using transformational leadership measurement in training and promotion programs. From the operational level, supervisors and managers should preach what they have taught to their subordinates. In addition, they should instill pride and respect in others and inspire their subordinates to be more competent.

This study has certain limitations that can be fixed in the future research. First, as our data came from only one company, the result can be different if applied on the industrial level. A sample of employees on the industry level may extend our findings. Second, the research data were distributed through an online questionnaire. This resulted in the decision to fill out the questionnaire depending on the willingness of the respondent, and so the distribution of the results of this study was not even for each group of respondents. For the next research, it is better to distribute the questionnaires offline.

REFERENCES

Abbas, M., & Raja, U. 2015. Impact of psychological capital on innovative performance and job stress. Canadian Journal of Administrative Sciences/Revue Canadienne des Sciences de l'Administration 32(2): 128–138.

Afsar, B., & Umrani, W. A. 2019. Does thriving and trust in the leader explain the link between transformational leadership and innovative work behaviour? A cross-sectional survey. Journal of Research in Nursing 25(1): 37–51.

Amankwaa, A., Gyensare, M. A., & Susomrith, P. 2019. Transformational leadership with innovative behaviour. Leadership & Organization Development Journal 40(4): 402–420.

Bednall, T. C., E. Rafferty, A., Shipton, H., Sanders, K., & J. Jackson, C. 2018. Innovative behaviour: how much transformational leadership do you need?. British Journal of Management 29(4): 796–816.

Birasnav, M., Rangnekar, S., & Dalpati, A. 2011. Transformational leadership and human capital benefits: The role of knowledge management. Leadership & Organization Development Journal 32(2): 106–126.

Carless, S. A., Wearing, A. J., & Mann, L. 2000. A short measure of transformational leadership. Journal of business and psychology 14(3): 389–405.

Janssen, O. 2000. Job demands, perceptions of effort-reward fairness and innovative work behaviour. Journal of Occupational and Organizational Psychology 73(3): 287–302.

Le, P. B., & Lei, H. 2018. The mediating role of trust in stimulating the relationship between transformational leadership and knowledge sharing processes. Journal of Knowledge Management 22(3): 521–537.

Lei, H., Leaungkhamma, L., & Le, P. B. 2020. How transformational leadership facilitates innovation capability: the mediating role of employees' psychological capital. Leadership & Organization Development Journal 41(4): 481–499.

Luthans, F., Youssef, C.M. & Avolio, B.J. 2007. Psychological Capital: Developing the Human Competitive Edge. New York: Oxford University Press Oxford.

Mishra, P., Bhatnagar, J., Gupta, R., & Wadsworth, S. M. 2017. How work-family enrichment influence innovative work behavior: Role of psychological capital and supervisory support. Journal of Management and Organization 25(1): 58–80.

Pundt, A. 2015. The relationship between humorous leadership and innovative behavior. Journal of Managerial Psychology 30(8): 878–893.

Rego, A., Sousa, F., Marques, C., & e Cunha, M. P. 2012. Authentic leadership promoting employees' psychological capital and creativity. Journal of business research 65(3): 429–437.

Schuckert, M., Kim, T. T., Paek, S., & Lee, G. 2018. Motivate to innovate How authentic and transformational leaders influence employees' psychological capital and service innovation behavior. International Journal of Contemporary Hospitality Management 30(2): 776–796.
Sobel, M. E. 1982. Asymptotic confidence intervals for indirect effects in structural equation models. Sociological methodology 13: 290–312.
Wang, Y., Zheng, Y., & Zhu, Y. 2018. How transformational leadership influences employee voice behavior: The roles of psychological capital and organizational identification. Social Behavior and Personality: an international journal 46(2): 313–321.
Wojtczuk-Turek, A., & Turek, D. 2015. Innovative behaviour in the workplace. European Journal of Innovation Management 18(3): 397–419.

Contemporary Research on Business and Management – Noviaristanti (Ed.)

Transformational leadership style as a driving factor of organizational citizenship behaviors mediated by work engagement and transfer of training: A case study on start-up companies in Indonesia

R.A. Safaati & P.M. Desiana
Universitas Indonesia, Depok, West Java, Indonesia

ABSTRACT: This study aims at understanding the effect of the transformational leadership style on organizational citizenship behavior (OCB) by taking work engagement and transfer of training into account as mediating factors. The data were collected using self-administered surveys taken by 240 employees who had received formal training or informal learning and have worked in start-up companies in Indonesia. Structural equation modeling analysis revealed that transformational leadership had a direct effect on OCB. The result also indicated that the casual relationship between transformational leadership and OCB was fully mediated by work engagement and transfer of training. To summarize, this study disclosed the importance of transfer of training implementation within start-ups as well as courses of action to boost the OCB of the employees.

1 INTRODUCTION

The increasing number of failed start-ups has indicated the complexities of developing a start-up company. It also demonstrates that start-ups, as organizations, are designed to seek repeatable and scalable technology-based business models using the platform (Blank, 2013). To grow sustainably in a dynamic environment, the companies need to organize and pay attention to internal social aspects, such as the implementation of training programs as policies in human resource development. However, the main problem is although the company is willing to invest more in training, there is still a challenge to ensure employees' transfer training process in the workplace (Sookhai and Budworth, 2010).

Transformational leadership is one of the leadership styles that can streamline training transfer and positive work engagement, particularly in start-ups (Tucker and Lewis, 2004). In this case, Robbins and Judge (2015) found that a company with employees who are willing to do tasks beyond their job descriptions are predicted to have better performance than other organizations in terms of efficiency. Furthermore, regardless of the difference between this study and previous studies, there is no comprehensive study on the effect of transformational leadership on organizational citizenship behavior (OCB) mediated by work engagement and transfer of training. Based on the investigation, this study was conducted by incorporating transformational leadership, work engagement, transfer of training, and OCB into a complete model of study. The originality of this paper is that it investigated the mediation effect of work engagement and transfer of training on the effect of transformational leadership on OCB.

2 LITERATURE REVIEW

Bass (1993) explained that there are five factors of transformational leadership. The first factor that can help followers achieve higher performance is being excellent role models (ideal influence; II). The second factor focuses on how they communicate their expectations and goals (inspirational

DOI 10.1201/9781003196013-51

motivation; IM). The third factor is based on intelligence and rationality (intellectual stimulation; IS). The fourth factor focuses on personal attention (individualized consideration; IC) to the followers. In addition, an alternative view was developed, which considers work engagement as a unique stand-alone concept, indicated by a positive and satisfied state of mind related to work, such as vigor and dedication to be active in absorption (Schaufeli et. al., 2006). Transfer of training occurs when the behavior that employees learn from their learning experiences through training is generalized to their work context to form sustainable behaviors that are maintained over time (Baldwin and Ford, 1988). In this case, employee performance consists of task performance or role performance (in-role) and contextual performance or extra role performance. On the other hand, OCB is defined as discretionary individual behavior that refers to the act of extra roles outside of work and judged in the aggregate to promote effective organizational functioning (Organ, 1988). Meanwhile, Graham (1991) argues that there are three types of organizational citizenship behavior, including organizational obedience, loyalty to the organization, and organizational participation.

3 RESEARCH METHODOLOGY

3.1 *Instrument and measures*

The data were measured using a self-report approach with a five-point Likert-type scale from strongly disagree (1) to strongly agree (5). To measure transformational leadership (TFL), seven items were used (Carless et al., 2000). In addition, in measuring work engagement (WE), nine items were used (Schaufeli and Bakker, 2006). Transfer of training (TOT) was measured with thirteen items. Meanwhile, organizational citizenship behavior (OCB) was measured with sixteen items (Lee and Allen, 2002). It was found that only one item of OCB was invalid. On the other hand, the rest were reliable as CR $\geq$ 0.7 and AVE $\geq$ 0.5.

3.2 *Participants*

The data were collected using a non-probability sampling method with a purposive sampling technique. Participants in this study were employees who had received formal training or informal learning and had worked in start-ups in Indonesia.

3.3 *Sample size*

For a minimum sample size, this study required up to five times the number of indicators based on structural equation modeling (SEM) requirement on LISREL 8.80 software (Hair et al., 2009). Therefore, a minimum of 230 respondents were required. This study gathered a total of 240 respondents.

4 RESULTS

4.1 *Descriptive analysis*

The questionnaires were given to 157 women (65.41%) and 83 men (34.58%). The majority of respondents were 21–25 years old (65.41%). It was found that 76.25% of the respondents in this study had worked in the start-up industry for three years, and a majority of them worked as staff (73.75%). The respondents were concentrated in the JABODETABEK area, where most of them worked in business development, operations, and product development unit. Around 66.25% of the respondents had received re-skilling or up-skilling training, and as much as 42.5% had received team training.

4.2 *The measurement model (confirmatory factor analysis)*

Almost all indicators in this study were valid and reliable. It was found that only one indicator, which is organizational citizenship behaviors, was below the desirable cut-off of SLF; thus it was omitted. According to the findings, all of the items in this study were regarded as a set of reliable measurements.

The model fitted to the data as nine of goodness-of-fit indices (GFI $0.90 \geq 0.90$; RMR $0.039 \leq 0.08$; RMSEA $0.05 \leq 0.063 \leq 0.08$; AGFI $0.87 \geq 0.90$; NFI $0.97 \geq 0.90$; NNFI $0.98 \geq 0.90$; CFI $0.98 \geq 0.90$; IFI $0.98 \geq 0.90$; and RFI $0.96 \geq 0.90$) resulted in good fit. Three variables included in this study (work engagement, transfer training, and OCB) are multidimensional. Therefore, an assessment of the second-order measurement model was done. The result showed that the dimensions within the variables were valid and reliable. However, the goodness of fit was decreased slightly. Therefore, the model was refined to improve the overall model fit.

4.3 *The structural model*

The overall structural model had a good fit for the data. After ensuring the model fitted the data, the hypotheses developed for this study were tested. In this model, the result of all variables was found to support all the hypotheses. Transformational leadership was found to have a positive and significant effect on work engagement (SLF = 0.48; t-value = 7.37), transfer of training (SLF = 0.53; t-value = 7.99), and organizational citizenship behaviors (SLF = 0.18; t-value = 2.36). On the other hand, work engagement was found to influence OCB positively and significantly (SLF = 0.43; t-value = 6.80) as well as the transfer of training (SLF = 0.12; t-values = 1.77). In this study, work engagement and transfer of training were found to fully mediate the effect of transformational leadership on OCB due to the significant direct effect of transformational leadership on OCB. It was indicated that because H1 to H5 was supported, H6 and H7 were also supported.

5 DISCUSSION

Results from the analysis of start-up companies in Indonesia showed that the implementation of effective transformational leadership and transfer of training that was combined with a high level of work engagement had a positive significant effect on OCB. This finding also supports the proposed mediation model. In addition, transformational leadership was found to boost work engagement, transfer of training, and OCB significantly. It indicates that employees who work in start-up companies, particularly in Indonesia, were willing to exhibit OCB with strong persuasion for the power to evenly distributed by the direct influence of the leader as a role model.

These findings are in line with several previous studies (e.g., Aboramadan and Dahleez, 2019; Winokur and Sperandio, 2017; Gemeda and Lee, 2020). It was assumed that transformational leadership can increase employees' motivation and contains an important exchange process (between leaders and followers), more specifically when the leader delegates vision and encourages them to unleash their full potential performance to achieve the desired objectives. Transformational leadership was found to be beneficial to enhance work engagement and OCB indirectly.

The results suggested that the transformational leadership style would somehow increase the level of employees' willingness to perform extra roles. However, it would somehow not immediately increase involvement. Rather, the transformational leadership style would strengthen the follower's energy and dedication to their work. It can be seen that work engagement was a significant factor as a mediator in terms of employee performance.

As an illustration, some studies found that work engagement had a significant part in mediating the role between transformational leadership and OCB (Purvanova et al., 2006). Hence, in situations where an effective leadership style has been implemented, there will be a higher likelihood that employees will be engaged in their work, who exhibit higher levels of dedication and extra role behavior. This indicates that a high culture of involvement is required in start-ups to increase work efficiency and effectiveness, reducing costs for operational errors, improving the quality of work outcomes, and increasing sales and profits.

Moreover, the fact that transfer of training was proven to directly increase the employees' tendency to exhibit OCB shows the significance of the supervisor's role in the transfer system. Therefore, through the process of acquiring new competencies as a result of training, employees can work more efficiently. Furthermore, this stimulates the members in a workplace to respect and support each other.

6 CONCLUSION

This study contributes to an understanding of factors that affect OCB by investigating the effect of transformational leadership on work engagement and transfer of training. The findings also imply that the transfer of training in organizations is crucial to strengthen work engagement to receive the impact of the transfer. Therefore, the transfer of training can increase the fully mediated effect of transformational leadership on OCB.

REFERENCES

Aboramadan, M., & Dahleez, K. A. (2020). Leadership styles and employees' work outcomes in nonprofit organizations: the role of work engagement. Journal of Management Development.

Bass, B., M. & Avolio J. B. (1993). Transformational leadership and organizational culture. Public Administration Quarterly, spring. 112–121.

Blank, Steve. (2013). Corporate Acquisitions of Startups: Why Do They Fail? [Website page]. Accessed from https://steveblank.com/2014/04/23/corporate-acquisitions-of-startups-why-do-they-fail/.

Gemeda, H. K., & Lee, J. (2020). Leadership styles, work engagement and outcomes among information and communications technology professionals: A cross-national study. Heliyon, 6(4), e03699.

Graham, J.W. (1991) An Essay on Organizational Citizenship Behavior. Employee Responsibilities and Rights Journal, 4, 249–270.

Hair, J. F., Anderson, R. E., Tatham, R. L., dan Black, W.C. (2010). Multivariate data analysis. (5th Ed.). UK: Pretince Hall International.

Organ, D. W. (1988). Issues in organization and management series.Organizational citizenship behavior: The good soldier syndrome. Lexington Books/D. C. Heath and Com.

Purvanova, R. K., Bono, J. E., & Dzieweczynski, J. (2006). Transformational leadership, job characteristics, and organizational citizenship performance. Human performance, 19(1), 1–22.

Robbins dan Judge. (2015). Perilaku Organisasi Edisi 16. Jakarta (ID): Salemba Empat.

Schaufeli, W. B., Bakker, A. B., dan Salanova, M. (2006). The measurement of work engagement with a short questionnaire: A cross-national study. Educational and Psychological Measurement, 66(4), 701–716.

Sookhai, F., dan Budworth, M. H. (2010). The trainee in context: Examining the relationship between self-efficacy and transfer climate for transfer of training. Human Resource Development Quarterly, 21(3), 257–272.

Tucker, L. R., dan Lewis, C. (2004). The Influence of the Transformasional Leader. Journal of Leadership and Organizational Studies, 10(4), 2004.

Winokur, I. K., & Sperandio, J. (2017). Leadership for effective teacher training transfer in Kuwaiti secondary schools. Teacher Development, 21(2), 192–207.

Contemporary Research on Business and Management – Noviaristanti (Ed.)

The bribery perceptions of Indonesian business sectors: Sand the wheels or grease the wheels?

R. Rokhim, N. Wulandary & R.E.F. Nasution
Universitas Indonesia, Indonesia

W. Thohary
Transparency International Indonesia, Indonesia

W. Jatmiko
Durham University, UK

ABSTRACT: This paper aims at testing two main hypotheses debated among corruption researchers: "sand the wheels" and "grease the wheels" in the Indonesian business sectors. The "Sand the wheels" hypothesis argues that corruption has a negative impact, while the "grease the wheels" hypothesis argues that corruption has a positive impact. Using the data from the *Bribery Risk Perception According to the Business Sector* survey conducted by Transparency International Indonesia and the provincial Gross Regional Domestic Product (GRDP), this study provides a descriptive overview of the business sector bribery perceptions and maps it against the provincial GRDP of 11 sample cities. The result shows that the larger the sector was, the higher the bribery perceptions that the sector had. Sectors dealing with natural resources such as mining, oil and gas, and forestry were perceived to have the highest bribery potential across all the sample cities. In terms of the two corruption impact hypotheses, the bribery cases in Indonesia confirmed the "grease the wheels" hypothesis since they occurred in sectors with higher provincial GRDP.

1 INTRODUCTION

Corruption is one of the most challenging problems faced by developing countries, including Indonesia. According to the Gallup World Poll (2013), 88% of the Indonesian people consider corruption practices to be prevalent in the government sector. However, not only the public sector, corruption has plagued the private sector. One of the most widespread corruption practices in the private sector is bribery. Bribery in the private sector will certainly impede overall economic growth. Several studies have proven that corruption could impede a company's growth and ultimately impact their economy in the aggregate. This hypothesis is known as "sand the wheels". On the other hand, further studies discover that corruption has a positive impact. This hypothesis is called "grease the wheels". According to this hypothesis, corruption helps in improving efficiency when a government has a poor bureaucratic system.

Based on the two hypotheses, this paper aims at providing a descriptive overview of the business sector bribery perception and mapping it against the GRDP generated by each sector. To give an overview of the effect of corruption in Indonesia, the Gross Regional Domestic Product (GRDP) of business sectors in the 11 sample cities was used as a variable. When the GRDP of a sector is high and its level of bribery is also high, the "grease the wheels" hypothesis is accepted. On the contrary, when the level of bribery is high, while the GRDP is low, it indicates the "sand the wheels" hypothesis effect.

DOI 10.1201/9781003196013-52

2 LITERATURE REVIEW

The impacts of corruption practices are often two sides of the same coin. According to the "grease the wheels" hypothesis, corruption improves efficiency in a business process.

Several studies support this hypothesis, such as Leff (1964), Leys (1965), and Huntington (1968). These studies reveal that in a country where bureaucracy is inefficient (highly centralized administration, excessive regulatory burden, and weak bureaucracy system), corruption practices could boost efficiency in business processes by minimizing lead time. Another study by Lui (1985) shows that bribery could remove constraints caused by a lengthy administration process and shortens queuing time in public facilities. Dreher & Gassebener (2007) show that the many procedures and the minimum amount of required capital negatively impact the launch of a new company into the industry, but through corruption, the negative effect of these regulations can be softened.

Contrary to the "grease the wheels" hypothesis, Mauro (1995) found that corruption negatively impacted investment and ultimately impeded its growth. Kaufmann & Wei (1998) tested the hypothesis using the company-level data and found the companies that paying bribes spent substantial time negotiating with other countries' representatives. Svensson, in another study, finds that bribery practices typically arise when a company has to deal with the government, especially when dealing with export, import, and infrastructure service acquisition (Svensson 2000). A study by Kimuyu (2007) on manufacturing companies in Kenya proves that a company, on average, pays 14.2% of its total contract value to government officers. On the other hand, Pham & Takayama (2015) investigated the relationship between corruption and a company's efficiency in Vietnam and found that a company that practices bribery from early on usually has to pay higher amounts of bribes over time.

3 METHODOLOGY

This study employed primary data from the 2015 Corruption Perception Index conducted by Transparency International Indonesia. More specifically, it employed Bribery Risk Perception According to Business Sector Survey. The respondents of this survey were

1,067 manufacturers from 11 cities: Manado, Bandung, Banjarmasin, North Jakarta, Makassar, Medan, Padang, Pekanbaru, Pontianak, Semarang, and Surabaya. They were provided with questions about their perception of bribery risk in other sectors, using a Likert scale questionnaire. Scale 0 shows that bribery practice is "Very Common", while the scale 5 "Very Uncommon". The questions were divided into four subcategories: (i) Potential for Bribery to Win a Business Competition, (ii) Potential for Bribery to Expedite an Administrative Process, (iii) Potential for Bribery in the form of a Political Donation, and (iv) Potential for Bribery in Other Private Sectors. The answers to the four-category questions were calculated to get the average bribery perception score of each business sector in 11 sample cities. Then, the average bribery perception score of each business sector in each city was mapped against the city's GRDP by province.

4 RESULTS AND DISCUSSIONS

4.1 *Bribery risk according to business sector*

The results show that the potential for bribery practices is perceived to be very high in the construction and mining sectors, both having an average bribery score of 3.41, followed by the oil and gas sector. According to Stansbury's data (2005), at least four reasons are contributing to the fact that the construction sector is prone to corruption: uniqueness, the complexity of transaction chains, the high level of official bureaucracy, and the large scale of investment. Since every project is unique, construction companies have the freedom to set higher costs, and this will go undetected as none of the projects is similar enough for comparison.

4.2 *Percentage of bribery to production cost*

Based on the perception of the percentage of bribery to production cost, the construction sector had the highest portion of bribes. According to the data from the Global Infrastructure Anti-Corruption Centre, companies in the construction sector have spent sizeable money on their production costs because massive investments are involved. In this case, bribes can easily go undetected. Following the construction sector was the service sector in the second place that covers many subsectors, such as education and health.

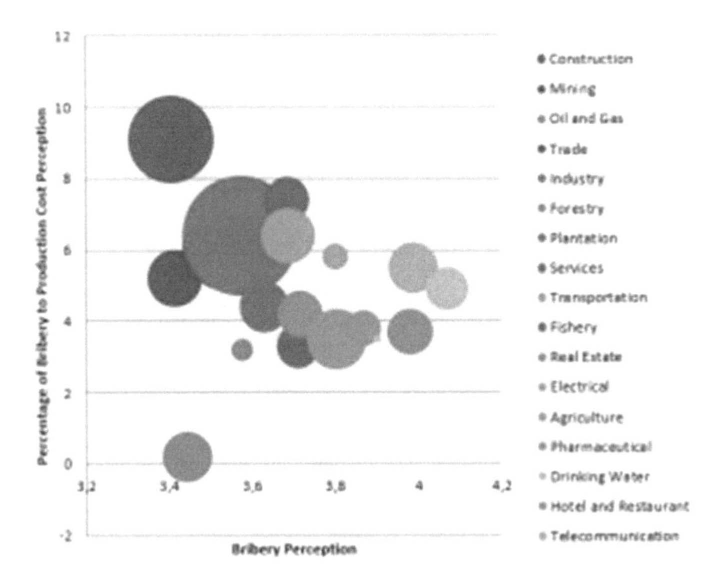

Figure 1. Indonesia's bribery perception, percentage of bribery to production cost, and Gross Domestic Product (GDP) perception according to business sector.

Figure 1 represents the aggregate bribery perception scores of the 11 sample cities according to their business sectors. Then, the results were mapped against the percentage of bribery to production cost. The size of each circle indicates the GDP generated by each business sector. Based on the mapping above, it is shown that the sectors most perceived as having higher levels of bribery are construction, oil and gas, mining, and trade. The construction sector has the highest percentage of bribery to production cost and has high bribery perception despite its major contribution to the GDP. This result is consistent with Stansbury's statement (2005) who found that construction is one of the sectors prone to corruption practices.

4.3 *Mapping of business sector bribery perception against the GDRP*

4.3.1 *Manado*

The construction sector was the largest contributor to the provincial GRDP of North Sulawesi, followed by the transportation sector in second place. Construction came first with the support from President Joko Widodo through his ambitious infrastructure development programs, especially outside Java Island. The data mapping shows that several sectors were perceived to have a higher potential for bribery. These sectors deal with natural resources, such as mining, oil and gas, and forestry.

4.3.2 *Bandung*

The business sectors perceived to have a higher potential for bribery in Bandung were related to natural resources: mining, forestry, and oil and gas. Bribery practice was also perceived to be prevalent in the construction sector. It was the second-largest contributor to the provincial GRDP of West Java after the trade sector.

4.3.3 *Banjarmasin*

The largest contributor to the provincial GRDP of South Kalimantan was the mining sector. In terms of topography, the island has a wealth of mining sites and forests. Despite the numerous mining sites in Banjarmasin and its status as the largest contributor to the provincial GRDP of South Kalimantan, the sector had a relatively low bribery perception. There were far more corruption cases in the forestry sector, especially illegal logging. It was exposed to the public and as a result, the sector became the top of mind of bribes in Banjarmasin.

4.3.4 *North Jakarta*

The largest contributor to the provincial GRDP of DKI Jakarta was the construction sector. Based on the results in other previous cities, construction was among the sectors with the highest bribery perception. For North Jakarta, however, bribery practice was perceived to be less ubiquitous thanks to the city's new administration led by Governor Basuki Tjahaja Purnama (Ahok), who had been determined to bring greater efficiency in governance, such as in business permit and licensing.

4.3.5 *Makassar*

The provincial GRDP of South Sulawesi most substantially came from the construction sector. On the other hand, the second-largest contributor was the trade sector, with the agricultural and service sectors in third and fourth places respectively. The plantation was perceived to be the most potential sector for bribery practice as Makassar has a great deal of plantation sector corruption cases.

4.3.6 *Medan*

The sector that was perceived to have the highest potential for and prevalence of bribery was mining, followed by agriculture and plantation in second and third places respectively. In terms of the provincial GRDP of North Sumatra, the greatest portion came from the sectors that deal with natural resources. Many corruption cases involving the city's plantation sector have been revealed to the public.

4.3.7 *Padang*

The survey found that the oil and gas, construction, plantation, and mining sectors were perceived to have a high potential for bribery. Other sectors that were perceived to have a high potential for bribery and among the top contributors to the provincial GRDP of West Sumatra include plantation and construction.

4.3.8 *Pekanbaru*

It was revealed that the sectors dealing with natural resource utilization were among the top contributors to the provincial GRDP. In terms of business sector bribery perception, the six sectors that had the highest score were those dealing with natural resources, such as forestry, agriculture, fishery, mining, plantation, and oil and gas.

4.3.9 *Pontianak*

The top provincial GRDP contributors were the construction, transportation, plantation, and service sectors. Meanwhile, in terms of bribery perception, transportation, telecommunication, and trade sectors had the highest potential for bribery practice.

4.3.10 *Semarang*

The provincial GRDP of Central Java largely derived from the trade sector. It was perfectly reasonable since Semarang—like Surabaya—is a port city. In addition to trade, the construction sector also contributed to the provincial GRDP of Central Java. In terms of bribery perception, the plantation sector was perceived to have the highest potential for bribery.

4.3.11 *Surabaya*

The provincial GRDP of East Java mostly derived from the trade sector, while the second major contributor was the construction sector. Surabaya is also the largest city in Indonesia, which means a greater construction intensity takes place in the city. The next largest contributors to the provincial GRDP were the service and agricultural sectors. As expected, the oil and gas, mining, and forestry sectors were perceived to have a high potential for bribery. One of the reasons was the considerable time required for obtaining permits and licenses, which prodding the business people to resort to bribery to smooth out their businesses.

5 CONCLUSION

This study provides an overview of the bribery perception of the business sector and maps it against the provincial GRDP generated by each sector. According to the mapping, bribery perception greatly depended on the major business sectors a city has. The larger the sector was, the higher the bribery perception people had of the sector. Likewise, sectors that deal with natural resources, such as mining, oil and gas, and forestry, were also perceived to have a higher bribery potential across all sample cities. This finding is consistent with Stansbury (2005) and Martini (2012) who conclude that the construction sector and any sectors dealing with natural resources are prone to corruption. Meanwhile, in terms of the impact of corruption, the mapping result is rather inconclusive to determine whether the findings are more supportive of the "sand the wheels" or "grease the wheels" hypotheses.

REFERENCES

Dreher, A., & Gassebner, M. (2007). Greasing the wheels of entrepreneurship? The impact of regulations and corruption on firm entry. Cesifo Working Paper No. 2013.

Fisman, R., Svensson, J. (2007). Are corruption and taxation really harmful to growth? Firm level evidence. Journal of Development Economics, 83 (2007), 63–75.

Huntington, S.P. (1968). Political order in changing societies. New Haven, CT: Yale University Press.

Kaufmann, D., Wei, S., 1998. Does grease money speed up the wheels of commerce? Mimeo, World Bank.

Leff, N. (1964). Economic development through bureaucratic corruption. The American Behavioural Scientist, 8(2), 8–14.

Leys, C. (1965). What is the problem about corruption? Journal of Modern African Studies, 3, 215–230, Reprinted in A.J. Heidenheimer, M.Johnston, & V.T. LeVine (Eds.) (1989), Political corruption: Ahandbook (pp. 51–66). Oxford: Transaction Books.

Lui, F. (1985). An equilibrium queuing model of bribery. Journal of Political Economy, 93(4), 760–781.

Mauro, P. (1995). Corruption and growth. Quarterly Journal of Economics, 110, 681–712.

Peter Kimuyu (2007),"Corruption, firm growth and export propensity in Kenya", International Journal of Social Economics, Vol. 34 Iss 3 pp. 197–217.

Pham, H.T., Takayama, S. (2015). Revisiting the missing middle: Production and corruption. Working Paper Series from Centre for Efficiency and Productivity Analysis, School of Economics, University of Queensland, Australia.

Stansbury, N. (2005). Exposing the foundations of corruption in construction. In Corruption in Practice, Transparency International Global Corruption Report 2005. Transparency International.

Svensson, J. (2001). The cost of doing business: Ugandan firms experiences with corruption. In: Svenson, J. (2005). Eight questions about corruption. Journal of Economic Perspectives, 19 (3), 19–42.

Transparency International. (2011). Bribe Payers Index Report.

Contemporary Research on Business and Management – Noviaristanti (Ed.)

Business coaching: Implementation of promotion mix and establishment of loyalty programs for MSMEs in Indonesia during the COVID-19 pandemic

S. Fiddiny & H. Suhaimi
Faculty of Economics and Business, University of Indonesia, Central Jakarta, Indonesia

ABSTRACT: A number of MSMEs in Indonesia have experienced a decline in income due to the COVID-19 pandemic, forcing MSMEs to work harder to survive the pandemic. This study aims to help MSMEs in the culinary industry by conducting the business coaching method through qualitative research, in which the data were obtained from observations, surveys, and in-depth interviews. The solution taken for this problem was done by designing a promotion mix and establishing loyalty programs. The implemented promotion mix elements were sales promotion and digital/internet marketing designed in the form of discounted offline promotions and online through the MSME Instagram platform. In addition, the establishment of the loyalty program was done by collecting a customer database. The implementation results showed an increase in MSME sales during the COVID-19 pandemic.

1 INTRODUCTION

According to data from the Indonesian Young Entrepreneurs Association, the pandemic has decreased the culinary industry's turnover by up to 30% (Kompas.com, 2020). Warung Sop Ayam Mbah Min is one of the MSMEs in the culinary industry offering a product named Klaten chicken soup, which uses a family recipe passed down from generation to generation to maintain its quality. The COVID-19 pandemic also had a negative impact on this business, as reflected in the significant difference in the sales before the COVID-19 pandemic occurred and during the pandemic. In addition, government regulation had implemented Large-Scale Social Restrictions (PSBB, *Pembatasan Sosial Berskala Besar*), where all businesses and offices were not allowed to operate. Due to this, Warung Sop Ayam Mbah Min had to close its shop for approximately three months.

The following characteristics were obtained based on the results of the questionnaire distributed to 103 respondents:

Table 1. Profile respondents.

Gender	
Men	52%
Woman	48%
Age	
17–25 year	39%
26–35 year	30%
36–45 year	27%
46–55 year	3%
>55 year	1%

(*continued*)

DOI 10.1201/9781003196013-53

Table 1. Continued.

Gender	
Occupation	
College Students	25%
BUMN Employees	22%
Private Employees	35%
Civil Servants	8%
Others	10%
Monthly Expenses	
>Rp700,000	9%
Rp700,000–Rp1,000,000	8%
Rp1,000,001–Rp2,000,000	28%
Rp2,000,001–Rp3,000,000	22%
>Rp3,000,000	33%

The survey aims to determine the respondents' brand awareness, purchase intention, and customer loyalty toward Warung Sop Ayam Mbah Min. The survey results were gathered using an online survey method. From the results of this study, it was known that the brand awareness of Warung Sop Ayam Mbah Min was relatively high: there were 60.3% of respondents who knew the MSME, but only 50.8% of the respondents had made purchases, and 49.2% had never made purchases. The questionnaire results also demonstrated the respondents' intention to make purchases at Warung Sop Ayam Mbah Min. It is known that the average value of purchase intention was only 3.07 on a scale of 5. This value is small, considering the average capability of the respondents to make purchases. Thus, Warung Sop Ayam Mbah Min needs to make innovations to be able to attract potential customers to want to make purchases.

On the other hand, the loyalty possessed by customers of Warung Sop Ayam Mbah Min has the value of 3.25. According to Evanschitzky and Wunderlich (2006), there are three phases of attitudinal loyalty, namely, cognitive loyalty, affective loyalty, and conative loyalty. Based on the value of the loyalty, customers of Warung Sop Ayam Mbah Min were still in the second phase, in which customers were satisfied with the services provided but have not entered into the third phase, where customers wanted to make repeated purchases of the products offered.

Based on the results of this survey, the MSME has not been implementing promotional activities to attract customers and establishing relationships with customers. With this background, this study aims to (1) implement the promotion mix for the MSME; and (2) create loyalty programs for the MSME.

2 LITERATURE REVIEW

According to Belch and Belch (2017), promotion is the coordination of the efforts made by sellers to prepare information channels and persuasion in the sales of goods and services or in promoting ideas. The basic tool that can be used to achieve organizational communication or promotion objectives is named a promotion mix. There are six elements of the promotion mix, which are advertising, direct marketing, digital/internet marketing, sales promotion, publicity/public relations, and personal selling.

According to Mullins and Walker (2013), maintaining customer loyalty is significant for business profits, especially in a mature market because loyal customers are more profitable. Companies can avoid spending high costs to get new customers and obtain loyal customers who tend to make continuous purchases and provide reviews or recommendations through positive word of mouth and are willing to pay high prices to get the desired product. Companies need to measure customer satisfaction regularly because a dissatisfied customer is unlikely to remain loyal over time. The main step in measuring customer satisfaction is examining customer behavior, such as frequency of purchases, the percentage of total customer purchases, and the company's ability to resolve complaints of any problems given by customers.

Loyalty programs are a powerful way for companies to increase customer retention and the volume or frequency of purchases (Bijmolt, Dorotic, & Verhoef, 2011; Bolton, Kannan, & Bramlett, 2000). The loyalty program is also a medium for service companies to start and maintain relationships with their customers so that they become long-term relationships (Bijmotl et al., 2011). An important element of a loyalty program is the reward structure and how it can change the purchasing and behavior of its members (Melancon, Noble, & Noble, 2011; Zhang & Breugelmans, 2012).

3 RESEARCH METHODOLOGY

This research is a qualitative study using a case study where the object of research is one of the MSMEs in Indonesia named Warung Sop Ayam Mbah Min, which was done by using primary data and secondary data.

Primary data were obtained by conducting the following: (1) in-depth interviews; (2) observations; and (3) surveys. In-depth interviews were conducted with the owners of Warung Sop Ayam Mbah Min to determine the business processes, including the internal and external problems that occurred. Observations were made during the operating hours of Warung Sop Ayam Mbah Min to determine the customers' behaviors. An online survey of 103 respondents was then conducted to determine customers' opinions regarding the strengths and weaknesses of Warung Sop Ayam Mbah Min. Secondary data used in this study were the related research journals discussing promotion and loyalty programs. In addition, the data obtained were analyzed to identify gap and Pareto analysis, which were then used to determine the alternative solutions and implement these solutions.

4 RESULT AND DISCUSSION

4.1 *Implementation of promotion activities*

Sales promotions were implemented to increase sales during the COVID-19 pandemic through online and offline activities. In online sales promotion, Warung Sop Ayam Mbah Min collaborated with an online food delivery application where customers were given free delivery fees and discounted prices on selected menus. The collaboration was carried out because the COVID-19 pandemic limits people's mobility; hence, many opted to purchase food through online food delivery. The implemented promotion was customer-oriented, where sales promotions targeted customers by giving discounts. Meanwhile, in offline sales promotion, Warung Sop Ayam Mbah Min applied four promotions as follows: (1) "MIN" Promo, where customers could eat for free if their name contains "min"; (2) Birthday Promo, where customers who were born in a certain month got 50% discount; (3) Rp5,000 Discount Promo, where customers got Rp 5,000 discount in the fourth week of each month; and (4) Independence Day Promo, where customers got a 'buy one get one' for those who made purchase on August 17, 2020. These promos were printed on a tent card placed on each table at the shop so that interested customers could find out about ongoing promotions.

Digital/internet marketing was used to complement sales promotion by communicating ongoing promotions on social media. Offline promotions carried out by Warung Sop Ayam Mbah Min needed to be advertised, so that the target audience could find out information about the promo. The social media utilized for this was Instagram, using an account owned by Warung Sop Ayam Mbah Min.

4.2 *Loyalty program implementation*

The application of the loyalty program was proposed by the authors as a solution to the problems and desires of the owner, considering that Warung Sop Ayam Mbah Min MSME had difficulty in maintaining long-term relationships with customers and increasing sales. The loyalty program's implementation begins with collecting the customer database, which was then entered in the system owned by MSME. The customer database was gathered by filling in the criticism and suggestion forms and is carried out in conjunction with the ongoing promotions at Warung Sop Ayam Mbah Min. The criticism and suggestion form was the preferred by the owner to find out the strengths and weaknesses of the shop from the customers' point of view. Customers who filled out the form were given free drinks.

The customer database was entered into the cashier system owned by Warung Sop Ayam Mbah Min. Customers who had registered were be given points for each arrival to determine the frequency of visits. The collection of points was carried out by the customer, and it could be exchanged for vouchers or a discounted price.

5 CONCLUSION

The problems faced by MSMEs during the COVID-19 pandemic were due to not carrying out promotional activities and not establishing long-term relationships with their customers. Hence, the solutions to these problems are to carry out the promotion mix and establish loyalty programs. These solutions carried out by MSMEs during the COVID-19 pandemic resulted in an increase in income by 17% compared to the previous month. Thus, it can be concluded that the implementation of these solutions can help MSMEs in increasing sales during the COVID-19 pandemic.

The limitations of this research are that it covers only the marketing aspects and can still be developed into other managerial aspects, and it involved only Warung Sop Ayam Mbah Min, the MSME being researched.

REFERENCES

Belch, G. E., & Belch, M. A. (2017). Advertising and Promotion: An Integrated Marketing Communications Perspective (11th Ed.). McGraw Hill Education.

Bijmolt, T. H. A., Dorotic, M., & Verhoef, P. C. (2011). Loyalty programs: Generalizations on their adoption, effectiveness and design. Foundation and Trends in Marketing, 5(4), 197–258.

Bolton, R. N., Kannan, P. K., & Bramlett, M. D. (2000). Implications of loyalty programs and service experiences for customer retention and value. Journal of the Academy of Marketing Science, 28(1), 95–108.

Dorotic, M., Verhoef, P. C., Fok, D., & Bijmolt, T. H. A. (2014). Reward redemption effects in a loyalty program when customers choose how much and when to redeem. International Journal of Research in Marketing, 31, 339–355.

Melancon, J. P., Noble, S. M., & Noble, C. H. (2011). Managing rewards to enhance relational worth. Journal of the Academy of Marketing Science, 39(3), 341–362.

Mullins, J. W. (John W., & Walker, O. C. (2012). Marketing management: a strategic decisionmaking approach (8th Ed.). McGraw-Hill.

Contemporary Research on Business and Management – Noviaristanti (Ed.)

The effect of person–organization fit and workplace fun on intention to stay with work engagement as a mediating variable

S. Apriyanti & E.S. Pusparini
Faculty of Economics and Business, Universitas Indonesia, Jakarta, Indonesia

ABSTRACT: The purpose of this research is to examine the effect of person–organization fit and workplace fun on work engagement and intention to stay and to examine work engagement as a mediating variable between the effects of person–organization fit and workplace fun on intention to stay. This research is a quantitative study with a cross-sectional design. An online questionnaire was used to collect data. In total, 342 employees from one of the leading companies in the retail industry in Indonesia participated in this research. The data analyzed by using SEM show that person–organization fit and workplace fun has a significant effect on work engagement, but there is no significant effect of workplace fun on intention to stay. Furthermore, the results of this research underscore the role of work engagement, which acts as a mediator between the effects of person–organization fit and workplace fun on intention to stay.

1 INTRODUCTION

The topic of employee turnover is still considered as an interesting topic for researchers and practitioners for several years (Hom, Shaw, Lee, & Hausknecht, 2017). Robbins and Judge (2017) stated that a high turnover rate can disrupt the efficiency of the organization. Furthermore, Ghosh, Satyawadi, Joshi, and Shadman (2012) stated that an employee leaving an organization is psychologically painful for the organization and also other employees because it does not only cause setbacks in the professional field but also disrupt the social life in the organization. On the other hand, Ghosh, Satyawadi, Joshi, and Shadman (2012) stated that today, organizations have realized that it is important to retain employees and build a stable workforce and then focus on the formulation and enforcement of a strong retention strategy to effectively lessen turnover. On the contrary to turnover intention, intention to stay refers to the willingness of an individual who consciously and deliberately wants to stay with the organization (Tett & Meyer, 1993). In simple terms, Kim, Price, Mueller, and Watson (1996) stated that intention to stay is the opposite of turnover intention.

2 LITERATURE REVIEW

Robbins and Judge (2017) stated that person–organization fit is a theory that explains an individual's interest in working in an organization and the individual is chosen by the organization to work because the values of the individual match with the organizational values, so the individual will leave the organization when he feels there is no match between them. One of the factors that causes high engagement is the conformity between individual values and organizational values. In line with this, Memon et al. (2018) in their research found that person–organization fit is a strong predictor of work engagement.

Person–organization fit is considered important because it is associated with a more positive work attitude, lower task performance, turnover intention, and suppressed stress of organizational behavior (Kinicki & Fugate, 2016). Bakker and Leiter (2010) defined that work engagement is a satisfied and positive state of mind linked to work characterized by dedication, vigor, and

DOI 10.1201/9781003196013-54

absorption. A research conducted by Rai, Ghosh, and Dutta (2019) found that engagement has a positive influence on intention to stay where the level of engagement assessed or perceived by individuals with their work tends to determine intention to stay.

Memon et al. (2018) stated that when individuals assess the conformity of the values they have with the values of the organization, these individuals will also feel that the organization is able to meet their needs well, and having the same attributes as other employees will cause individuals to be highly engaged. One of the main reasons that are keeping employees connected to the organization is the high engagement and emotional affection. Becker (2012) stated that a fun work environment has been used as a way to build an organizational culture that can support increased work engagement, employee performance, and employee retention. In addition, Fluegge (2008) defined workplace fun as interpersonal, social, or any task that is fun or humorous that can provide entertainment, pleasure, and a feeling of enjoyment to individuals. Consequently, workplace fun will effectively reduce turnover intention if the individuals feel engaged with the company.

3 RESEARCH METHODOLOGY

This research used a quantitative method with a convenience sampling technique and was conducted by distributing online questionnaires to the employees of one of the leading companies in the retail industry in Indonesia. The number of respondents was 342. We adopted items from the report by Jung and Yoon (2013) to measure person–organization fit, scale developed by Tews, Michel, and Allen (2014) to measure workplace fun, and UWES-9 developed by Schaufeli et al. (2006) to measure work engagement and items from Jung and Yoon (2013) to measure intention to stay. In this study, the researchers decided to continue using the scale developed by Jung and Yoon (2013) by changing negative statements to positive ones without changing the existing meaning. In line with that, Gupta and Singh (2018) also measured the intention to stay variable through a reversal of the turnover intention scale developed by Mobley et al. (1979) by changing negative statements to positive ones without changing the meaning. All measures used a 7-point Likert scale. The data were analyzed using structural equation modeling with LISREL to test the proposed model (Figure 1).

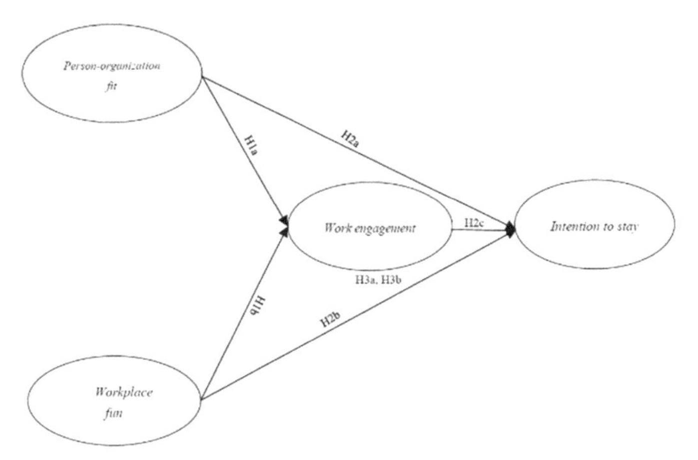

Figure 1. Research model

4 RESULTS

The questionnaires were filled by 281 women (82.16%) and 61 men (17.84%). The majority of respondents belonged to the 20–29 years age group (65.27%). Almost 81% of the respondents worked in PT X for 1–10 years. In this study, there are also ten values categorized in the good fit category: GFI (0.90), RMSEA (0.073), RMR (0.044), ECVI, TLI (0.97), NFI (0.97), RFI (0.96), IFI, (0.98), CFI (0.98), and CAIC. Hair et al. (2014) stated that using three to four fit indices provides evidence of adequate model fit. Therefore, the proposed research model can be declared to have met the good fit criterion. After the research model met the criterion of goodness of fit, the process was continued with a test of direct effect between variables. This study applied a one-tailed hypothesis test; therefore hypotheses will be accepted if the t-value is ≥ 1.645 (Lind, Marchal, & Wathen, 2017). In addition, it tested the mediating role of work engagement between person–organization fit and workplace fun to intention to stay. The findings of this research can be seen as shown below.

Table 1. Results of hypothesis testing.

Hypotheses	SLF	t-*values*	Decision
H1a: POF WE	0.52	8.21	Supported
H1b: WF–WE	0.34	5.51	Supported
Hypotheses	SLF	t-*values*	Decision
H2a: POF–ITS	0.45	4.77	Supported
H2b: WF–ITS	0.06	0.71	Not Supported
H2c: WE–ITS	0.24	2.82	Supported
H3a: POF–WE–ITS		2.695	Supported
H3b: WF–WE–ITS		2.533	Supported

5 DISCUSSION AND CONCLUSION

Based on Table 1, person–organization fit has a significant and positive effect on work engagement. In line with that, a research conducted by Memon et al. (2018) also found that person–organization fit is a strong predictor of work engagement in professional workers in the oil and gas industry. In addition, the relationship between workplace fun and work engagement is also significant and positive. This result is in line with the research conducted by Fluegge-Woolf (2014), who found that when workplace fun increases, there is also an increase in work engagement.

Person–organization fit has a direct and significant positive effect on intention to stay. In line with that, Demir, Demir, and Nield (2014) said that the higher the individual's assessment of person–organization fit, the higher the individual's desire to stay in the organization. However, the t-value on the relationship between the workplace fun on intention to stay in this study is <1.645, which is equal to 0.71, so it can be stated that it has no significant effect. Further, the result of this research proves that work engagement has a direct and significant positive effect on intention to stay. Kim and Yoo (2018) also found that work engagement has an effect on intention to stay.

The significance of the role of work engagement as a mediating variable was tested using the Sobel Test. Based on that, it can be concluded that work engagement has a good role as a mediator, which proves that the influence of person–organization fit maximizes the intention to stay if employees have a good level of engagement. In addition, the role of work engagement in mediating the influence of workplace fun on intention to stay shows that with work engagement, it can further increase the influence of workplace fun and the intention to stay. Sakr, Zotti, and Khaddage-Soboh (2019) stated that workplace fun has proven to be a key element in reducing work stress, absenteeism, and turnover as well as maintaining great customer service, employee retention, and attracting talent. However, everything is related to one essential concept, namely, engagement. Becker (2012) also stated that a fun work environment has been used as a way to build an organizational culture that can support increased work engagement, employee performance, and employee retention. Tsaur, Hsu, and Lin (2019) stated that if the organization can build and develop a fun workplace, it will help employees boost their social relationship and give social support to overcome sources of pressure, thus enabling employees to be more dedicated to their jobs.

In conclusion, this study reveals that the effect of person–organization fit and workplace fun on intention to stay is mediated by work engagement. This shows that work engagement is a very important factor in increasing intention to stay in order to increase employee work engagement. Therefore, the company should increase workplace fun and person–organization fit.

REFERENCES

Becker, F.W. (2012), The Impact of Fun in The Workplace on Experienced Fun, Work

Engagement, Constituent Attachment, and Turnover Among Entry-Level Service Employees, Disertation, Pennsylvania State University.

Demir, M., Demir, S. S., & Nield, K. (2014), "The relationship between person-organization fit, organizational identification and work outcomes", Journal of Business Economics and Management", Vol. 16 No. 2, pp. 369–386.

Fluegge-Woolf, E.R. (2014), "Play hard, work hard: fun at work and job performance", Management Research Review, Vol. 37 No. 8, pp. 682–705.

Ghosh, P., Satyawadi, R., Joshi, J. P. & Shadman, M. (2012), "Who stays with you? factors predicting employees' intention to stay", International Journal of Organizational Analysis, Vol. 21 No. 3, pp. 288–312.

Gupta, A. & Singh, V. (2018), "Enhancing intention to stay among software professionals", Academia Revista Latinoamericana de Administracion, Vol. 31 No. 3, pp. 569–584.

Hair, J.F., Black, W.C., Babin, B.J., & Anderson, R.E. (2014). Multivariate Data Analysis 7*th* Edition. Pearson New International Edition: London.

Hom, P.W., Shaw, J.D., Lee, T.W., & Hausknecht, J.P. (2017), "One hundred years of employee turnover theory and research", Journal of Applied Psychology,Vol. 102 No. 3,pp. 530–545.

Jung, H.Y & Yoon, H.H. (2013), "The effects of organizational service orientation on person – organization fit and turnover intent", The Service Industries Journal, Vol. 33 No. 1, pp. 7–29.

Kim, S.W., Price, J. L., Mueller, C. W., & Watson, T. W. (1996). The determinants of career intent among physicians at a U.S. Air Force Hospital. Human Relations, 49(7), 947–976.

Kim, K. J., & Yoo, M. S. (2018). The Influence of Psychological Capital and Work Engagement on Intention to Remain of New Graduate Nurses. JONA: The Journal of Nursing Administration, 48(9), 459–465.

Kinicki, A. & Fugate, M. (2016), Organizational Behavior: A Practical, Problem-Solving Approach, New York: McGraw-Hill

Lind, D.A., Marchal, W.G., & Wathen., (2017). Statistical Techniques in Business & Economics 17th Edition. McGrawHill: New York.

Memon, M.A., Salleh, R., Nordin, S.M., Cheah, J.H., Ting, H. & Chuah, F. (2018), "Personorganisation fit and turnover intention: the mediating role of work engagement", Journal of Management Development, Vol. 37 No. 3, pp. 285–298.

Rai, A.and Ghosh, P. & Dutta, T. (2019), "Total rewards to enhancee employees'intention to stay: does perception of justice play any role?", Evidence-based HRM, Vol. 7 No. 3, pp. 262–280.

Robbins, S.P & Judge, T. A. (2017), Organizational Behaviour Seventeenth Edition, London: Pearson Education.

Sakr, C., Zotti, R., & Khaddage-Soboh, N. (2019). The impact of implementing fun activities on employee's engagement: The case of Lebanese financial institutions, International Journal of Organizational Analysis, 27 (5), 1317–1335.

Schaufeli, W.B., Bakker, A.B., & Salanova, M. (2006), "The measurement of work engagement with a short questionnaire a cross-national study", Educational and Psychological Measurement, Vol. 66 No. 4, pp. 701–716.

Tett, R.P. & Meyer, J.P. (1993), "Job satisfaction, organizational commitment, turnover intention, and turnover: path analyses based on meta-analytic findings", Personnel Psychology, Vol. 46, pp. 259–293.

Tews, M.J., Michel, J.W., & Allen, D.G. (2014), "Does fun pay? the impact of workplace fun on employee turnover and performance", Cornell Hospitality Quarterly, Vol. 54 No. 4, pp. 370–382.

Tsaur, S.H., Hsu, F.S., & Lin, H. (2019), "Workplace fun and work engagement in tourism and hospitality: the role of psychological capital", International Journal of Hospitality Management, Vol. 81 No. 2019, pp. 131–140.

Contemporary Research on Business and Management – Noviaristanti (Ed.)

Backpackers' travel decision across generations and countries of origin: An empirical study in Indonesia

M.A. Stephanie, S. Wijaya & H. Semuel
Faculty of Business and Economics, Petra Christian University, Surabaya, Indonesia

ABSTRACT: In the past few decades, backpacking tourism has become more popular not only among young travelers but also among adults and elderly. Taking the trend into account, this study aims to investigate the variance of consideration of international backpackers across generational cohorts and origin countries when they backpack to Indonesia. In this study, travel perceived risk was analyzed as a covariate variable. In addition, the study had surveyed 156 international backpackers who have never traveled to Indonesia. This study employed descriptive statistics analysis, ANOVA, linear regression, and MANCOVA tests to examine the primary data. The results reveal a significant difference between baby boomers and generation Y backpackers during the pre-trip phase, especially in evaluating alternatives. In addition, the study also found a significant difference between Asian and non-Asian backpackers during the pre-travel phase, especially during the travel necessity introduction. Further study also reveals that travel perceived risk negatively influenced backpackers' consideration when deciding to visit Indonesia.

1 INTRODUCTION

Traveling has evolved into a must-do activity, representing a modern lifestyle followed across the globe to seek leisure and recreation. Indonesia has experienced significant growth in its tourism industry, with 15.81 million international travelers visiting the country, which is a 12.58% increase compared with 2017 (BPS, 2019). To attract more international visitors, the Ministry of Tourism in its strategic planning has prioritized the development of 10 tourism spots called "the new Balis" (Kemenpar, 2018). Most of these ten new developed Balis have the characteristics of natural tourist attractions closely related to adventure tourism. These adventure tourism destinations are expected to attract more international backpackers to visit Indonesia.

Several studies concerning backpacking have been done with various focuses. However, not a single study focused on backpackers' demographic factors that could influence their travel decision. In addition, it is still rare to find an empirical study that analyzes international backpackers' behaviors who visit Indonesia. Several studies show that baby boomers and X and Y generations have different perspectives and behavior in seeking information and making decisions (Williams & Page, 2011; Agosi & Pakdeejirakul, 2013; Rahulan, Troynikov & Watson, 2014; Paakkari, 2016). Before deciding their destination and activities, travelers will assess several possible risks. Travel perceived risk can affect the choice of destination (Garg, 2015). Therefore, this study examined the role of demographic factors, in particular generational cohorts and countries of origin, as a differentiator of backpacker's travel decisions by incorporating travel perceived risk as a covariate variable.

2 LITERATURE REVIEW

Previous studies revealed that age and generation groups can influence the decision-making process (Williams & Page, 2011; Agosi & Pakdeejirakul, 2013; Rahulan et al., 2014; Paakkari, 2016).

 DOI 10.1201/9781003196013-55

It was found that age and generation differences would influence behavioral change, perspective, and methods to reach a decision. In general, baby boomers are inclined to be more courageous in making decisions with careful considerations. On the other hand, generation Y generally tends to follow the trend and secondary information source in deciding a purchase, such as from social media (Rahulan et al., 2014). This finding indicates that each generation is influenced by different marketing approaches that cater to the needs of each generation. Based on the preceding discussion, the following hypothesis is proposed:

H1: There are differences in the travel decision consideration among backpackers across generations in Indonesia during the pre-travel and the mid-travel phases.

Demographic factors indirectly influence a person's travel decision. Pizam and Sussmann (1995) found that tourism behavior is closely related to the country of origin and cultural background. Meanwhile, McCleary, Weaver, and Hsu (2006) revealed a significant difference related to the perceived value of travel services experienced by international tourists from seven different countries in Hong Kong that influenced their willingness to revisit the country. Based on the preceding discussion, the following hypothesis is formulated:

H2: There are differences in the travel decision consideration among backpackers from various countries of origin in Indonesia during the pre-travel and mid-travel phases.

Traveling to a new destination outside travelers' daily environment could entail risks that have to be considered (Leep & Gibson, 2003). In this case, travelers are gathering information as much they could to minimize risks, and information on alternative destinations will be evaluated based on several criteria (Garg, 2015). After they evaluated the alternative destinations, the perceived risk will be the basis of the evaluation that will be considered. Moreover, it may even be possible for an alternative destination to be eliminated during the evaluation (Garg, 2015). Based on the discussion, the following hypothesis is formulated: H_3: Travel perceived risk negatively influences the travel decision consideration among backpackers in Indonesia during the pre-travel and mid-travel phases.

3 METHOD

The sample was purposively selected, considering the following: (1) travelers aged between 23 and 71 years at the time of the survey and (2) international travelers who have backpacked to Indonesia at least twice or those who were backpacking in Indonesia during the survey. The primary data collection was done using online and offline surveys that employed questionnaires. In this case, the offline survey was conducted in various tourist spots in Surabaya, Yogyakarta, Malang, and Bali. On the other hand, the online survey was conducted on backpacker's communities on Facebook. This study employed descriptive statistics analysis, ANOVA, linear regression, and MANCOVA tests to analyze the primary data.

4 RESULTS AND DISCUSSION

A total of 194 respondents participated in the survey, but only 156 questionnaires matched the criteria. Therefore, the response rate is 80.41%. It was found that balanced participation was received from both male and female respondents. In terms of age, the respondents were mostly 23–37 years old, which is a part of generation Y. It was also revealed that the majority of the respondents came from non-Asian countries and were currently working as employees. Half of them had the estimated travel costs of less than USD$ 1,000 and preferred to stay at a backpacker hostel/hotel. Additionally, most of them were traveling alone. It was also revealed that they have backpacked three times in the last 3 years and stayed for at least 1 month.

Table 1 illustrates that no significant differences can be found in the travel decision consideration among the international backpackers across generations during pre-travel. On the other hand, significant differences can be found in their travel decision consideration, specifically concerning alternatives during pre-travel. Furthermore, post-hoc ANOVA analysis revealed significant differences between the two generations' travel decision considerations when they visited Indonesia. They were generation Y and baby boomers (significance rate of 10%). It was also found that significant differences can be seen in alternative evaluation during pre-travel.

Table 1. ANOVA test of travel decision across generational cohorts.

Dependent Variable	Sum of Squares	df	Mean Square	F	Sig.
NEED Between Groups	.059	2	.030	.038	.963
RECOGNITION Within Groups	120.172	153	.785		
INFORMATION Between Groups	.681	2	.340	.249	.780
SEARCH Within Groups	208.928	153	1.366		
EVALUATION OF Between Groups	4.718	2	2.359 .828	2.848	.061
ALTERNATIVES Within Groups	126.718	153			

Concerning hypothesis 1, it can be seen that there were differences in the international backpackers' travel decision consideration across generations in Indonesia during the pre-travel and mid-travel phases. At the "*Need of Recognition*" phase, most Generation Y respondents stated that they want to travel because of the desire or motivation for experiencing something new. Most of them also admitted that they were backpacking to answer their curiosity about Indonesia. On the other hand, respondents aged 38–52 years (gen X), like the average people in the productive age, revealed that they were backpacking to spend their free time.

With regard to hypothesis 2, differences can be found regarding travel decision consideration of international backpackers from various countries of origin in Indonesia during the pre-travel and mid-travel. In the "*Need for Recognition*" phase, respondents from non-Asian countries travel to satisfy their needs of exploring something new. It may be influenced by differences between Indonesia and other non-Asian countries, especially their natural resources, culture, and weather. Therefore, non-Asian travelers preferred to visit Indonesia for a longer period. This tendency is supported by the official website of the Ministry of Tourism (2018), which shows that the average stay time of international tourists from non-Asian countries when they are visiting Indonesia is 12–14 days.

With regard to hypothesis 3, it was revealed that travel perceived risk negatively influenced the travel decision consideration of the international backpackers in Indonesia during pre-travel and mid-travel. It shows travel risks had a negative influence on their travel decision. However, due to the insignificance of its value, this influence did not affect respondents' travel decision consideration. Furthermore, it was also revealed that the covariate variable significantly and negatively influenced the travel decision consideration in the evaluation of alternatives. It may be caused by a travel risk indicator that has no direct relationship with the other two steps. It means that even if the respondents sense a risk, their pattern of needs or information source choice will not be considerably influenced.

5 CONCLUSION AND RECOMMENDATIONS

The main findings of the study can be summarized in several parts. First, a significant difference in travel decision consideration between generation Y and baby boomer backpackers in Indonesia during the pre-travel phase was found, especially on the evaluation of alternatives. Second, a significant difference in the travel decision consideration of non-Asian and ASEAN backpackers during pre-travel, especially in the "*Need Recognition*" phase, can also be found. Third, travel perceived risk negatively influenced travel decision consideration of the international backpackers in Indonesia during the pre-travel phase, especially in the evaluation of alternatives. It means that the higher the travel risk perceived by the international backpackers, the lower the probability of these groups to choose Indonesia as their preferred destination. In addition, it could also influence the "*Evaluation of the Alternatives*" phase.

These findings have several managerial implications. First, the local and national destination management organizations must develop adventure tourism events that could expose the natural and cultural diversity of Indonesia. It is suggested to develop a campaign that could work well with backpackers and independent traveler communities by making videos showing fun backpacker activities that blend well with Indonesia's nature potential. It also needs to show that it can be enjoyed at a relatively affordable cost for backpackers. Considering the findings, it would be valuable to attract more international backpackers to visit Indonesia. It is suggested that further studies apply the variables examined in this study to investigate other aspects such as cultural, religious, or sports tourism to enhance the understanding of tourist behavior from a more diverse perspective.

REFERENCES

Agosi, M. & Pakdeejirakul, W. (2013). Consumer selection and decision-making process: A comparative study of Swedish generation Y decision-making style between high involvement and low involvement products. Sweden: Master's Thesis, Malarden University, School of Business, Society, and Engineering.

Backpacker Guide NZ. (2017). Top seven water sports activities to do in New Zealand. Retrieved September 13, 2017, from http://www.backpackerguide.nz/top-7-water-sports-activities-to-do-in-newzealand/

Garg, A. (2015). Travel risks vs tourist decision making: A tourist perspective. *International Journal of Hospitality & Tourism Systems*, 8(1), 1–9.

McCleary, K.W., Weaver, P.A., Hsu, C.H. (2006). The relationship between international leisure travelers' origin country and product satisfaction, value, service quality, and intent to return. *Journal of Travel & Tourism Marketing, 21* (3), 117–130.

Paakkari, A. (2016). Customer journey of generation z in fashion purchases. Finland: Lahti University, Faculty of Business and Hospitality Management.

Pizam, A., & Sussmann, S. (1995). Does nationality affect tourist behavior? *Annals of Tourism Research, 22*(4), 901–917.

Rahulan, M., Troynikov, O., & Watson, C. (2014).Consumer behavior of generational cohorts for compression sportswear. *Journal of Fashion Marketing and Management, 19*(1), 87–104.

Sorensen, A. (2003). Backpacker ethnography. *Annals of Tourism Research, 30*(4). 847–867.

Sumarwan, U. (2002). *Perilaku konsumen*. Bogor: PT Ghalia Indonesia.

Swarbrooke, J., Horner, S. (2007). *Consumer behaviour in tourism*. Oxford: Elsevier Ltd.

Williams, K. C., & Page, R. A. (2011). Marketing to the generations. *Journal of Behavioral Studies in Business, 3*(1), 37–53.

Contemporary Research on Business and Management – Noviaristanti (Ed.)

Quest for entrepreneurial opportunity from the analytics education (course) perspective

S.G. Zahirah & G. Ramantoko
Faculty of Management and Business, Telkom University, Bandung, Indonesia

ABSTRACT: This study aims at describing the benefits of studying "data analytics" in understanding the objectives of training students to become entrepreneurs. It signifies that learning analytics has become one part of the special interest domain that encourages the finding of entrepreneurial opportunities. This study exemplifies the perceptions of business students on how the implementation of big data analytics would matter. The data were collected through questionnaires. The finding revealed some important variables based on overall students' perception, gender, and type of business of interest to assist in identifying the entrepreneurial opportunity process. The indicators were distributed normally using normal distribution. As shown by the results, all variables that were involved in this research can be considered for everyone who are interested in building independent businesses in the era of digital transformation.

1 INTRODUCTION

An entrepreneur must have prior knowledge that drives the establishment of business in seeking business opportunities. Prior knowledge is significant in the search for knowledge, as it can lead to entrepreneurial opportunities. Before technology changes the process of entrepreneurial exploitation, entrepreneurs must look for opportunities by utilizing the use the technology (Shane, 2000). Prior knowledge on the special interest domain is owned by a self-taught person who spends time and effort to learn the necessary skills deeply (Ardichvili and Ray, 2003).

Big data and data analytics are not new, but due to their rapid adoption, they appear to be new (Theobald, 2017). Various fields have used big data, for instance, accountancy (Brink and Stoel, 2018), innovation (Marshall, *et al.*, 2015), procurement (Handfield, *et al.*, 2019), academic librarians (Ahmad, *et al.*, 2019), and competitor analysis (Guo, *et al.*, 2015). Big data is also studied because of the benefits of its implementation. Moreover, it gives opportunities to "big analysis" to become "big opportunities", which has solved a lot of business problems and increased the quality of life (Ishwarappa and Anuradha, 2015).

The analytics course has become a way to obtain knowledge about big data and data analytics. By utilizing the analytics course, people can find relevant information anywhere. It is used to complement the existing information and support the finding of entrepreneurial opportunities. Besides, it can become a potential innovation tool for entrepreneurship (Uremovic and Marxt, 2018). Sedkaoui (2018) justifies the benefits of big data analytics to entrepreneurship. It supports decision making, decreases expenses, imparts knowledge to the consumer, and is able to use open data. Big data analytics can help entrepreneurs to get a lot of information relevant to their business. This information helps entrepreneurs make a better decision, since the operating cost of making a decision is high. That way, they can use big data analytics and reduce the probable loss for their company. This study exemplifies the perceptions of business students on how the implementation of big data analytics would matter. It describes the benefits of studying "data analytics" in understanding the ultimate objectives of training students to become entrepreneurs.

 DOI 10.1201/9781003196013-56

2 RESEARCH METHOD

The number of respondents participating in this study, sampled randomly, is 122. They were undergraduate students of Business School, Telkom University, Bandung, Indonesia, who had taken big data analytics as well as entrepreneurship courses in the 2016–2017 academic year. They were asked to fill out a questionnaire related to students' perceptions regarding the implementation of the big data analytics course to identify entrepreneurial opportunities.

The questionnaires, adapted from Sedkaoui (2018), were distributed online. He pointed out that the use of IT tools and ability to analyze data can improve skills and develop new business models. His study found that the learning program plays a vital role in changing the possibility into opportunity to become new entrepreneurs. In identifying entrepreneurial opportunities, the further learning of big data analytics is required by involving self-regulated learning. In addition, this research also adopted the concept proposed by Eom (2012), who stated that self-regulated learning demands the learners to study autodidactically to make the learning process active. Variables proposed by Sedkaoui (2018) are "use of IT tools" (indicators: easy to access, provide information, information to develop, privacy, security issue, communication tool, communication network, and dependency), "ability to analyze data" (indicators: understanding, manipulating, interpreting, applying, and benefit), "improve skill" (indicators: reduce risk, identify data error, solving problem, decision making, and efficiency), "develop new business model" (initiative, creativity, goal setting, understand to develop business, ability to analyze the situation, and effectiveness), and "opportunity to be a new entrepreneur" (self-efficacy and optimism); while that proposed by Eom (2012) is "self-regulated learning" (indicators: self-directed, time setting, discipline, target setting, initiative, managing time, on time, elaboration, organization, rehearsal, control, and planning). All variables had satisfied the Cronbach's Alpha reliability test.

3 RESULT AND DISCUSSION

3.1 *Demographics profile of respondents*

From 122 students participating in this study, there were 38 female respondents (68.03%) and 39 male respondents (31.97%). Based on the type of business of interest, most respondents (66.39%) were interested in building an independent business, 25.41% chose to buy a franchise business, while only 8.20% were interested in buying an existing business. Based on gender, both male and female were most interested in building a business, then franchising, and buying an existing business.

3.2 *Relationship of each indicators*

From 703 combinations of correlation between indicators, all values were below 80%. The correlation table shows 203 relationships at the very low level (correlation $\leq$ 19%), 343 relationships at the low level (19% <correlation $\leq$ 39%), 135 relationships at the medium level (39% < correlation $\leq$ 59%), and 22 relationships at the strong level (59% < correlation $\leq$ 79%). It assures that the relationship between indicators will not lead to multicolinearity. The highest correlation was between "discipline" and "time setting". "Discipline" leads the student to do the assignment, while "time setting" leads to setting the time to study. It may cause miss-perception because students think doing the assignment and studying are the same thing. Therefore, "discipline" and "time setting" could have the strong relationship.

3.3 *Overall view of students' perception*

The average score of the "use of IT tools" variable based on the perception of students interested in "building an independent business" showed a value of 4.3. It is in accordance with the work by Guo et al. (2015) which revealed that the "use of IT tools" could make the benefits of big data analytics feel better because the "use of IT tools" is used to collect data to produce valuable information. Therefore, students agreed that the "use IT tools" was the most necessary variable to build an independent business. It is consistent with the previous research that "use of IT tools" had led to a new way of communication, information retrieval, and technology dependency (Akcam, 2015). Moreover, "improve skill" of analytics was also above average ($\bar{x} = 4.1$). It showed that analytics skill was needed to help students implement data analytics in finding the business opportunity. "Self-regulated learning" ($\bar{x} = 3.7$), "ability to analyze data" ($\bar{x} = 3.7$), "develop new business" ($\bar{x} = 3.8$), and "opportunity to be new entrepreneur" ($\bar{x} = 3.6$) were moderately used in identifying the entrepreneurial opportunity process by business students

Students' perception to "manipulating" skill was varied (standard deviation 32%). "Manipulating" is a skill to sorting and cleaning up useless data to get the desired result. The process is time-consuming; therefore, students may think that the "manipulating" process can be bypassed to reduce time (Tole, 2013). In addition, the mode on the "manipulating" indicator was neutral. Therefore, students neither disagreed nor agreed to have the "manipulating" skill to help them identify business opportunities. In contrast to the result of the majority of indicators, it was agreed. Indicators "dependency", "communication for networking", "provide information", and "easy to access" were indicators with a strongly agree assessment mode. These indicators were representing variable "use of IT tools". This ensures that the "use of IT tools" was the most agreed upon thing to be able to assist in the process of identifying entrepreneurial opportunities.

3.4 *Students' perception based on gender*

When compared by gender, female students showed a higher "use of IT tools" ($\bar{x}_{female} = 44$, $\bar{x}_{male} = 43$) and "self-regulated learning" ($\bar{x}_{female} = 38$, $\bar{x}_{male} = 35$). Meanwhile, male students showed a higher "ability to analyze data" ($\bar{x}_{female} = 37$, $\bar{x}_{male} = 39$) and the "develop new business" variable ($\bar{x}_{female} = 37$, $\bar{x}_{male} = 3{,}9$). In the "improve skill" variable ($\bar{x} = 41$) and "opportunity to be a new entrepreneur" ($\bar{x} = 37$), both male and female students showed the same average score. It indicated what male and female students preferred in identifying opportunity strategies. Female students preferred using "use of IT tools" and "self-regulated learning" whereas male students preferred having the "ability to analyze data" and the ability to "develop new business" to identify entrepreneurial opportunities. Both male and female students believed that the variables "improve skill" and "opportunity to be a new entrepreneur" are needed in the process of identifying entrepreneurial opportunities

3.5 *Students' perception based on the type of business of interest*

The pattern that obtained by those were interested in building independent business was in the agree area. This means that the students felt that each indicator used in this study helped them in identifying the entrepreneurial opportunity to build independent businesses. It is not in accordance with the students who were interested in buying an existing business and buying a franchise business. The results showed that a diversity of perspectives resulted in less specific patterns being formed. However, students who were interested in buying an existing business showed a positive pattern by agreeing the "use of IT tools" variable. So, "use of IT tools" did help not only the students who were interested in building an independent business, but also the ones who interested in buying an existing business.

"Identify data error", "reduce risk", "problem solving", "decision making", "efficiency", "control", and "understanding" were antecedents that had consequences for "independent business". It means that each association of these indicators had a probability of influence of 40% to build "independent business". This result was considered quite significant because it was represented by 60% of respondents' perceptions. Of the seven indicators, the corresponding variable "improve skill" dominated the antecedent to the consequences. "Control" is an indicator that represents the variable "self-regulated learning", and "understanding" represents the variable "ability to analyze data".

All indicators involved are normally distributed approached by normal distribution even divided by "type of business interest". Despite this, the result would show the differences on the level of data variation. The "privacy" indicator, for the students who were interested to "build an independent business", had higher data variations compared to those who were interested to "buy an existing business". However, by looking at beta distribution, the "privacy" indicator had high probability to be a predictor. Hence, those who were identifying entrepreneurial opportunities should notice the "privacy" of "using the IT tools". Moreover, "creativity" was also distributed normally, even though the variation of the data was high. However, based on the beta distribution result, this indicator also had the probability to help in identifying entrepreneurial opportunities.

4 CONCLUSION

It can be concluded the indicators that were used in this research can be considered for those who were interested in building independent businesses in the today's digital transformation era. Sedkaoui (2018) stated that entrepreneurs in the future must have expertise in managing analytic data by understanding how and when to use analytic approaches. In addition, the result of this study has strengthened Sedkaoui's (2018) concepts that utilize the data and data analytics to open a new entrepreneurship area. Moreover, students, as the future entrepreneurs, should explore the entrepreneurial opportunity by integrating the big data analytics course to enrich their understanding and improve the effectiveness. Eom's (2012) concept states that self-regulated learning will lead the student to get a higher level of satisfaction. Thus, the result of this study revealed that "self-regulated learning" is one of the variables that help the students to identify the entrepreneurial opportunity. According to descriptive calculation results, the indicators used in this study could sufficiently interpret each of the corresponding variables. These indicators could also be used for further research using quantitative measurement methods to get more measurable results.

REFERENCES

Ahmad, K., JianMing, Z., & Rafi, M. (2019). An Analysis of Academic Librarians Competencies and Skills for Implementation of Big Data Analytics in Libraries: A Correlational Study. Data Technologies and Application.

Akcam, B. K., Hekim, H., & Guler, A. (2015). Exploring Business Student Perception of Information and Technology. Procedia – Social and Behavioral Sciences, 195, 182–191. https://doi.org/10.1016/j.sbspro.2015.06.347

Ardichvili, A., Cardozo, R., & Ray, S. (2003). A Theory of Entrepreneurial Opportunity Identification and Development. Journal of Business Venturing, 105–123.

Brink, W. D., & Stoel, M. D. (2018). Analytics Knowledge, Skills, and Abilities for Accounting Graduates. Advances in Accounting Education: Teaching and Curriculum Innovations, 23–43.

Eom, S.B. (2012) "Effects of LMS, Self-Efficacy, and Self-Regulated Learning On LMS Effectiveness in Business Education", Journal of International Education in Business, Vol. 5 Issue: 2, pp.129–144.

Guo, L., Sharma, R., Yin, L., Lu, R., & Rong, K. (2015). Automated competitor analysis using big data analytics. Business Process Management Journal, 735–762.

Handfield, R., Jeong, S., Choi, T. (2019). International Journal of Physical Distribution & Logistics Management.

Ishwarappa, dan Anuradha. (2015). A Brief Introduction on Big Data 5Vs Characteristics and Hadoop Technlogy. ICCC-2015. Procedia Computer Science 48, pp 319–324.

Marshall, A., Mueck, S., & Shockley, R. (2015). How Leading Organizations Use Big Data and Analytics to Innovate. Strategy and Leadership, 32–39.

Sedkaoui, S. (2018). How Data Analytics is Changing Enrepreneurial Opportunity. Internation Journal of Innovation Science, pp. 274–294.

Shane, S. (2000). Prior Knowledge and the Discovery of Entrepreneurial Opportunities. Organization Science. Vol.11 No.4 pp. 448–469.

Theobald, O. (2017). Data Analytics for Absolute Beginners. Kindel edition.

Tole, A, A. (2013). Big Data Challenges. Database Systens Journal Vol. IV No. 3, pp. 31–40.

Uremovic, Z. Z., & Marxt, C. (2018). Cognitive Process of Entrepreneurial Opportunity Identification: Toward a Holistic Understandingof the Micro-mechanisms. Cognitive and Innovation, 95–123.

Contemporary Research on Business and Management – Noviaristanti (Ed.)

Root cause analysis fulfillment upon the implication of Fatty Acid Methyl Ester (FAME) as biofuel for sustainable energy in Eastern Java and Bali Nusa Tenggara (MOR V)

TSH. Sagala & S. Hartini
Airlangga University, Surabaya, Indonesia

HF. Prasetyo
PT Pertamina (Persero) MOR V, Surabaya, Indonesia

ABSTRACT: This study will determine the root cause analysis of implementing FAME as a source of renewable energy. Although the governmental policy has been established and a higher increase in demand has been fulfilled, no sanctions have been implemented. Four root causes were analyzed to have an effect on the implementation of renewable energy, method, machine, environment (in terms of policy), and material. Supply and distribution patterns implemented by Pertamina have heaped upon the optimization of the distribution of energy across the region, although the exquisite infrastructure was dedicated to the operations. Besides that, the importance of a policy to have a well-balanced supply and demand of FAME will be a great factor since a higher gap of PPI for FAME as well as gasoil products was observed throughout the past year. Hopefully, a well-balanced composition will maintain a challenge for maintaining provisions in facing challenges of the energy trilemma, as to limiting shortage of energy as well as having extensive value for its consumers.

1 INTRODUCTION

1.1 *Background*

Commodity trading depends not only upon the existence of the product itself but also upon how to meet the number of production and its distribution. (Karya et al., 2020) To ensure that the national energy security will be fulfilled, the government is obliged to organize activity provision, exploitation, and utilization of energy as well as strategic reserves for the conservation of energy resources to provide enough amount. As Sa'adah (2017) pointed out, the growth of fuel production in Indonesia increased only by 1%, but the level of fuel consumption grew to an estimated 7.93% per year. The energy sector still has an important as well as a very strategic factor for the national economic cycle. Over time, the energy demand continues to grow, while non-renewable energy resources are increasingly depleting. Thus, energy management has become one of the "performance indicators", packaged in a term called "energy trilemma" to promote regulations of safeness, affordability, and good use of its systems. (Song et al. 2017). Energy challenges are defined as "energy trilemma" to maintain energy equity, environmental sustainability, and energy security. The main challenge is to maintain the security demand of energy, mitigating climate change that can affect stock resilience, so that energy poverty does not occur. The price of petroleum is influenced by political conditions, weather-related factors, and market prices, which play a key role in the wheel of the world economy, even though the trend changes from time to time (Karya et al., 2020).

The focus of the research subject of biofuels will mainly be on the use of fatty acid methyl ester (known as FAME), which were to be mandatorily used by all types of consumers, as stated in the Minister of ESDM Regulations (PerMen) No. 12/2015: Stages of Gasoil Mix Obligation to Use FAME. In general, the total production of FAME had met the composition stated within PerMen

 DOI 10.1201/9781003196013-57

ESDM No. 12/2015, which is illustrated in Figure 1 (consumption of FAME vs. demand and actual distribution of FAME). The gap has increasingly improved between 2016 and 2019, amounting only up to −8,93%. The security of energy availability has become one of the main concerns of all countries due to the limited energy sources, increasing population, fluctuating energy prices, and limited energy supply (Zhu et al., 2020).

2 LITERATURE REVIEW

The root cause analysis method is one of the most important elements of problem solving because it will help identify and organize the causative factors which may lead to an event, separating the root causes, so it will not occur again. Several tools can be used in the process, and this research will focus on the use of the fishbone diagram. According to Barraza-Suarez (2017) and Jacobs (2018), the fishbone diagram is a method/tool known to improve quality.

There are several stages of achievements for sustainable energy according to Ramadhanie (2017). It includes having energy security, that is, having sufficient energy availability, at an affordable price, with the protection of infrastructure, social issues, environment, efficiency, demand, governance/public policy, and quantification, and emphasis on social, environmental, political and social issues, which must be adapted to climate change and the importance of biodiversity. Energy equity is the access to energy, often discussed in the rubric of the wealth/availability of energy resources, especially the topic of the availability of fuel (Pliousis et al., 2017). It is usually related to the lack of distribution of energy sources, poor infrastructure, and conditions of high energy costs relative to the income level of the local communities. Song et al. (2017) also explained that energy equity is the accessibility and affordability of energy supplies by all the people. Thus, the importance of the availability and distribution of energy sources to meet community needs is one of the important focuses that must be implemented to maintain the availability of energy supplies for the country. To achieve energy efficiency, a balanced demand and supply of energy sources will be needed for the development of provision in a sustainable energy supply (Ramadhanie, 2017). There are long-term benefits of the investment in renewable energy. (Busu, 2019) As described by Pliousis (2019), technology innovation continues to increase energy efficiency, so that less CO_2 energy is wasted to lead to a higher level of people's gross income (GDP).

Key factors for sustainable economic development are securing energy supplies at "acceptable" prices, while being environmentally friendly. According to Plisousis et al. (2019), there are seven dimensions of energy security, namely, energy availability, infrastructure, energy prices, social effects, environment, governance, and energy efficiency.

Successful implementation of FAME for gasoil products from 2009 to 2012 only reached to maximum compliance of 27.66% (Sadewo, 2012). Research conducted by both Sadewo (2012) and ESDM (2015) revealed several dominant factors that had influenced the optimization of FAME mixing as renewable energy. The facts that conveyed the role of FAME blending as a gasoil product include the following:

1. The availability of FAME in Indonesia, mostly influenced by prices of crude palm oil (CPO), FAME domestic prices, and previous FAME product offerings.
2. The main drivers of the FAME demand in Indonesia are largely the implementation of policies, the FAME market price, the price of BBM (gasoil), the availability of infrastructure, and the distribution of FAME across Indonesia.

3 RESULTS AND DISCUSSIONS

Biofuels are one of the potential solutions to address the challenges of sustainable energy management, especially in the fields of industry, power generation, and transportation (Silalahi, 2020). Indonesia is the second largest producer of palm oil in the world. Strategically, it has decreased imported fuel oil up to USD 4.8 billion or equivalent to Rp. 67.2 Trillion in 2019 (CNBC, 2020). It had suppressed gasoil imports since the mandatory B20 program was implemented. Based on the data in 2019, the decline in imports reached up to 94%. The consumption of FAME had decreased in 2020, giving a larger reserve of energy. However, stockout of inventory of FAME still occurred in the middle of 2020. As shown in Figure 2, there had been consistency in the implementation of the demand and supply of FAME in the market. As shown in Figure 3, the availability of FAME as biofuel had also increased significantly due to a large decrease in consumption in the market. The success rate of implementing FAME for the MOR V only reached up to 73.60%. Several factors explained why the achievement cannot be higher throughout 2019–2020: the relaxation policies of the government for consumers of power plants and users of Alutsista, machine performance, delay of FAME supply to several supply point terminals in the MOR V, and a sudden increase in gasoil sales during certain periods (Task Force period).

Based on the data from 2019 and 2020 obtained from Pertamina on the implementation of B0 (pure gasoil without FAME), it was quite large in early 2019, amounting to 7.5% of the total gasoil distribution, and by QIV 2019, the total was 6.63%. The deliverance was for special services of power plants and the National Army Weapon System (Alutsista). A major problem that conducted non-compliance implications of renewable energy obligations was due to the transition period when Pertamina decreased the number of supply points for FAME. To uniformly implement the policy, there was a cut-off period agreement for consignments of fuel oil due to the ending period of contracts. The supply points in the East Java region decreased from 6 points to only 3 points. This was done to optimize transportation costs that must be borne by the Fund Management Agency (BPDPKS). This policy may reduce the distribution cost of up to Rp. 473 million/month.

Based on the focus group discussion held by the researcher, the main discussion of the root cause analysis on the level of success in distributing FAME as a renewable energy source was more dominated by constraints in Pertamina's terminal infrastructure that did not meet the specific handling of FAME. Moreover, the lack of FAME stock inventory in Kupang is still the dominant root cause of B0 consumption in 2020, and the implementation growth of B20/B30 had improved. Kupang experienced stockout consumptions of FAME in July 2020 because of the lack of supply chain management by suppliers. To support this condition, there should also be specific facilities dedicated to the product since there were more varieties of products being handled. However, policies had not led to the inconsistency of sanctions addressed by the government (Migas). No real administration sanctions were being imposed; the reports clarified the conditions.

4 CONCLUSION

Based on the research results that have been carried out, as mentioned in the previous chapter, several factors dominated the level of achievement of FAME's consumption as a renewable energy source, including the following:

a. Monitoring the policies implemented between suppliers, the government, and also BU BBM (Pertamina) needs to be consistently done.
b. The price discrepancy between bio gasoil and non-bio fuel tends to be huge for biofuel products. However, single pricing policies have already been implemented in the form of price compensation for domestic purchases. Such funding is being managed by the BPDPKS. As a result, it was very vulnerable. Strict findings and supervision were due to the quality of FAME mixing and accepting the product to all types of consumers. The price gap between renewable and non-renewable fuel products is so high that the government agreed to make a single pricing policy. However, it is important to consider the total material balance due to such a policy. Other countries should also consider a similar policy of renewable mixing to increase the role of renewable energy widely.
c. The FAME specification material greatly determines the consumer's willingness to consume B20/B30 products. The water content of FAME's specification has been reduced to 350 ppm,

minimizing the absorption of excess oxygen throughout its handling process. The hygroscopic nature of FAME (susceptible to oxidation) affects the quality of B20/B30 composed of monoglycerides. Due to such conditions, it is important for fuel suppliers to periodically check the quality of B20/B30.

d. Environmentally, the use of fuel oil that is more acceptable by the environment and that provides enough energy is still much expected.

For a continuous source of energy to exist, it is important to manage energy challenges as stipulated in the energy trilemma. Supply chain management must also meet the security criteria by paying attention to the safety stock level of fuel inventory, both for producers and consumers, so that FAME and gasoil prices are stable. Policies which are to be extended are as follows:

1. Optimizing a safety stock policy for FAME, avoiding stockout of inventory.
2. In line with supply chain management, it is important to have the flexibility of supply points, as long as it is within one regional area.
3. There is a clear monitoring system for the suitability of the distribution of B20/B30, which has been running so far, but with balanced rewards and consequences under Permen No. 41/2015.

To meet the next composition stage for biofuel in Indonesia, it was stipulated by the government to implement B40, which consists of a consistent composition of B30, and increasing it to B100 (pure biofuel), which will be produced by Pertamina's refineries (up to 10% from the existing target of 12 million kiloliters of biofuel); it will still support higher production of biofuel energy to maintain the energy source, preventing it from becoming extinct.

REFERENCES

Anshori, MY. Herlambang, T. Karya, DF. Rahmalia, D. Inawati PA. (2020) H-Infnitiy for world crude oil price estimation. *Journal of Physics*: Conference Series 1563 *012016*. DOI:10.1088/1742-6596/1563/1/012016.

Azmi, R. Amir, H. (2014) Energy Security: Concepts, Policies and Challenges for Indonesia. Jakarta, Ministry of Finance. (https://www.kemenkeu.go.id/ /files/energy).

Busu, M. (2019) Applications of TQM Processes to Increase the Management Performance of Enterprises in the Romanian Renewable Energy Sector. *Journal Article of Processes*. Bucharest, Bucharest University of Economic Studies.

BPDPKS. (2020) *Palm Oil Fund Support in Implementation of the Mandatory Biodiesel by the Oil Palm Plantation* Fund Management Agency (BPDPKS): Monitoring & Evaluation. Jakarta, EBTKE.

BPS. (2020, June 17th). *Mining*. https://www.bps.go.id/statictable/ produksi-beberapa-hasil-kilang-minyak-dan-gas-menurut-jenis-hasil-kilang-barel-2000-2019.html.

CNBC Indonesia. (2020, 05 Agustus). Post-COVID-19 Post-Pandemic Biodiesel Program, Will it continue or Stop? (https://www.youtube.com/watch?v=96XUKeakEVc).

Fungenzi, T. (2015) Biomass as an opportunity to solve Indonesia's energy challenge. *Module – Principles of Sustainability*. United Kingdom, Cranfield University. https://www.researchgate.net/publication/280075489.

Hitt, M; Ireland R; Hoskisson, R. (2020) *Strategic Management Competitiveness & Globalization 13th edition*. Singapore, Cengage Learning Asia Pte Ltd.

Karya, *DF.* Anshori, MY. Rizqina, R. Katias, P. Muhith, A. Herlambang, T. (2020) Estimation of Crude Oil Price using Unscented Kalman Filter. *Journal of Physics: Conference Series*: ICCGANT 2019. DOI:10.1088/1742-6596/1538/1/012049.

Pliousis, A. (2019) A Multicriteria Assessment Approach to the Energy Trilemma. *The Energy Journal, Vol. 40, SI1 : International Association for Energy Economics.*

Sa'adah, Ana F., Fauzi, A., Juandab, B. (2017) Prediction of Fuel Supply and Consumption in Indonesia with System Dynamics Model. *Journal of Indonesian Economics and Development*. Vol. 17 (No. 2 January 2017) 118–137

p-ISSN 1411-5212; e-ISSN 2406-9280. DOI: 0.21002/jepi.v17i2.661.

Sadewo, H. (2012) *Analysis of Mandatory Policies for Biodiesel Utilization in Indonesia*. Jakarta, Faculty of Economics University of Indonesia.

Silalahi, F. Simatupang, TM. Siallagan, MP. (2020) Biodiesel Produced from Palm Oil in Indonesia: Current Status and Opportunities. *AIMS Energy*. DOI 10.3934/energy.2020.1.81.

Song, L. Fu, Y. Zhou, P. Lai, K. K. (2017) Measuring National Energy Performance via Energy Trilemma Index: A Stochastic Multicriteria Acceptability *Analysis. Energy Economics*. doi:10.1016/j.eneco.2017.07.004.
Ramadhanie, A. (2017) The Evolution of the Concept of Energy Security. *Global: International Political Journal* Vol. 19 (No. 2). Page 98–120.
Zhu, D. Mortazavi, SM. Maleki, A. Aslani, A. Hossein, Y. (2020) Analysis of the robustness of energy supply in Japan: Role of Renewable Energy. *Energy Reports*. Page 378–391. https://doi.org/10.1016/j.egyr.2020.01.011.

Contemporary Research on Business and Management – Noviaristanti (Ed.)

Factors affecting online impulse buying on social commerce in Indonesia: The moderation role of perceived financial risk

V.P. Cecianti & S.R. Hijrah Hati
Faculty of Economics and Business, Universitas Indonesia, Indonesia

ABSTRACT: In recent years, social commerce, which combines social and commercial activities, has been considered as a great innovation, one that has been particularly useful during the COVID-19 pandemic, which has affected various sectors. Many social commerce platforms have updated their features to facilitate commercial transaction. Indonesian consumers tend to exhibit impulse buying behavior that is affected by situational factors, such as interpersonal influence, information fit-to-task, visual appeal, portability, and time pressure. This study sought to analyze factors and motivations that influenced online impulse buying. The results indicate that the situational factors differently affected hedonic motivation and utilitarian motivation. Hedonic motivation directly and positively influenced impulse buying. This study also argues that financial risk influenced online impulse buying and mitigated perceived value's role as a motivator.

1 INTRODUCTION

Recently, social media has not only been used for social interaction and communication but has also been used in the business sector, so that everyone can meet their needs and support their business. Han and Kim (2016) affirm that social media currently has increasing business activities, which is known as social commerce. In recent years, social commerce, which combines social and commercial activities, has been considered as a great innovation, and it has been useful during the COVID-19 pandemic, which has affected various sectors around the world (Siagian, 2020). However, this activity has a weakness in terms of insecurity when transacting on social media due to the lack of capacity for payments in joint accounts, such as in e-commerce. Therefore, the risk of financial loss certainly makes consumers be more careful to do online shopping on social commerce.

During this pandemic, consumer behavior has shifted significantly to online buying and impulse buying. Moreover, these motivations can be the underlying consumer behavior, and they also have encouraged people to engage in certain behaviors (Kim et al., 2007). This study argues that perceived financial risk plays a role in moderating the influence of motivations on online impulse buying. Also, in practice, online retailers need to know the situational factors, such as interpersonal influence, information fit-to-ask, visual appeal, portability, and time pressure, which significantly influence people to make online impulse buying. This study seeks to find out the most important situational factors that influence the online impulsive buying process.

2 LITERATURE REVIEW

According to Solomon and Rabolt (2009), impulse buying is an individual condition in making a purchase that is influenced by feelings of urgency and difficulty to refuse or postpone. This study adopts urge to buy to measure individuals' impulsivity because it is problematic to observe the actual behavior in a controlled environment (Luo, 2005). The impulsive buying process is motivated by several behaviors, namely, hedonic motivation and utilitarian motivation. A hedonic motivation is

DOI 10.1201/9781003196013-58

the willingness to initiate pleasurable or favorable behavior in order to reduce negative experiences (Kaczmarek, 2017), while a utilitarian motivation is a consumer behavior based on objective and rational values (Hanzaee et al., 2011). Both hedonic and utilitarian motivations have encouraged people to engage in certain behaviors, such as shopping (Kim et al., 2007). The consumer feels or pays attention to the stimulus, assimilates it, and then reacts to do impulse buying. According to Piron (1991), the stimulus (situation factor) can be in the form of an actual product, a shopping environment, or a person accompanying shopping.

In this study, impulse buying on social commerce is affected by the situational factors, namely, interpersonal influences, information fit-to-task, visual appeal, portability, and time pressure. According to Lee and Kacen (2008), interpersonal influence is important for individuals' buying behavior, especially in collectivist countries. Information fit-to-task is linked to the quality of information from a product on a particular site, such as accuracy, reliability, completeness, timeliness, and relevance of the information (Peter and McLean, 2009). Visual attractiveness relates to the user interface of a website which initially attracts the users' attention. Varela et al. (2013) state that the visual attractiveness of a website refers to the level of consumer confidence in the pleasing appearance of the site that stimulates the desire to browse. Portability is important, especially for people with high mobility; the longer the time spent browsing, the more is the contact with various stimuli and the more likely it is to induce the user's hedonic and utilitarian motivation. Consumers who usually have time pressure tend to think less carefully when making purchases (Howard and Sheth, 1969). Social commerce does not provide financial transactions directly integrated with the system, such as e-commerce; payment methods on social commerce for consumers are confusing and difficult. Biswas and Biswas (2004) revealed that the high financial risk to online purchases is caused by conditions in which buyers cannot directly contact the seller, so they perceive that the seller lacks credibility.

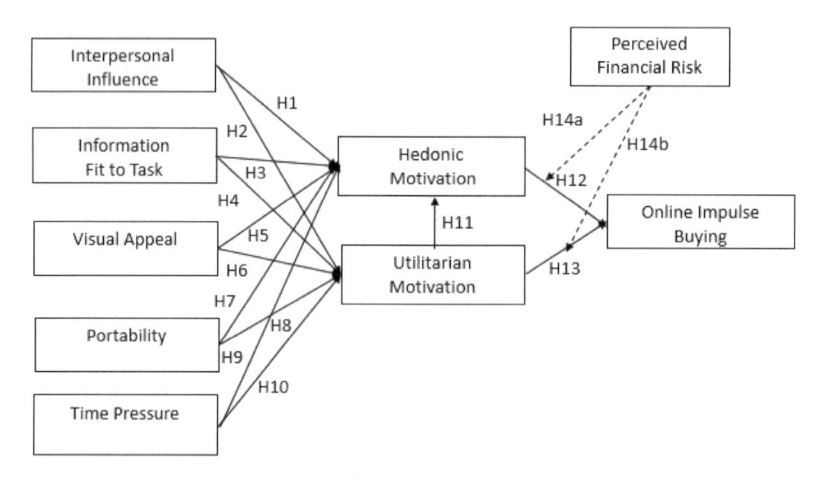

Figure 1. Conceptual framework.

3 RESEARCH METHODS

This study employs a quantitative approach using a survey method through online questionnaires with the Likert scale 1–5 (1 = strongly disagree; 5 = strongly agree). This study was conducted in Indonesia by using a purposive sampling technique that determined the respondents' characteristics, namely, Indonesian consumers who are over 17 years old and have been exposed to social commerce on the most popular social media, such as Instagram, Facebook, WhatsApp, Twitter, and TikTok, in the last 6 months. Based on the data collection, 331 respondents filled out the survey with 326 valid responses obtained for the final data analysis.

4 DATA ANALYSIS AND RESULTS

This study used Lisrel 8.54, and the data were then processed using the structural equation model (SEM) to analyze and evaluate the measurements and structural models of this study. Based on confirmatory factor analysis, all indicators and variables were valid and reliable (SLF $\geq$ 0.50; t-value $\geq$1.645; CR $\geq$ 0.70; VE $\geq$ 0.50). This study had a good fit model with good absolute fit, incremental fit, and parsimony fit indices. Based on 15 hypotheses, eight hypotheses were supported, including H1 (t = 2.47), H4 (t=2.77), H5 (t = 2.62), H8 (t = 5.07), H9 (t = 3.08), H11 (t = 2.67), H12(t = 7.05), and H14b (t = −2.72). The result of the structural model is presented in Table 1.

Table 1. Path diagram results for the structural model.

H	Path	T-value	Evaluation	H	Path	T-value	Evaluation
1	IF–HD	2.47	Supported	9	TP–HD	3.08	Supported
2	IF–UT	−0.27	Not Supported	10	TP–UT	1.15	Not Supported
3	IFT–HD	1.03	Not Supported	11	UT–HD	2.67	Supported
4	IFT–UT	2.77	Supported	12	HD–IB	7.05	Supported
5	VA–HD	2.62	Supported	13	UT–IB	0.25	Not Supported
6	VA–UT	1.27	Not Supported	14	FR (HD–IB)	6.36	Not Supported
7	PRT–HD	0.07	Not Supported	15	FR (UT–IB)	−2.72	Supported
8	PRT–UT	5.07	Supported				

Hedonic and utilitarian motivations underlie the situational factors that influence impulse buying in this study. In line with Xiang et al. (2016), information fit-to-task and portability were variables that affected utilitarian motivation. In other words, when more information is available, consumers will likely assume that the shopping platforms are useful and also portability will increase the utilitarian value in making portability purchases. Also, interpersonal influence in social commerce had a significant effect on consumer hedonic motivation. This result was in line with the research conducted by Arnold and Reynolds (2003), who reported on where and who people seek advice from when they are going to shop. Time pressure in social commerce had a significant effect on consumer hedonic motivation but an insignificant effect on consumer utilitarian motivation. Chang and Chen (2015) consider that time pressure has a negative impact, especially on utilitarian motivation. Time pressure will limit individuals to consider all the information needed when making decisions and lead them to make irrational decisions. Furthermore, a visual appeal had a significant effect on consumer hedonic motivation. This result is in line with Tractinsky et al. (2000), who illustrated that a website interface will affect the first impression and help the user to evaluate.

In comparison to utilitarian motivation, hedonic motivation had a strong and positive effect on the consumers' impulse buying. This is supported by Zheng et al. (2019), who identified hedonic motivation as the main determinant of consumer purchasing behavior (Kukar-Kinnery & Close, 2010). Utilitarian motivation had an insignificant effect on impulse buying but had an indirect influence on online impulse buying by affecting hedonic motivation. This result is in accordance with the research conducted by Babin et al. (2004), who stated that the utilitarian value is expected to complement hedonic values and not to be a contrasting value. The moderation test shows that perceived financial risk did weaken the relationship between consumers' utilitarian value. This condition reduced the desire to make an unplanned online buying. This is consistent with the research conducted by Biswas and Biswas (2004), who revealed that the high financial risk to online purchases is caused by conditions in which buyers cannot directly contact the seller, so they perceive the seller to lack credibility. Consumers will rethink and be more careful in online shopping. However, the perceived financial risk did not weaken the relationship between consumers' hedonic value because consumers had experiences in purchasing products from social commerce (Chang & Tseng, 2010) and need to experience varied, novel, and complex sensations (Zuckerman, 1979).

5 CONCLUSION AND SUGGESTIONS

This study shows how the situation factors, such as interpersonal influences, information fit-to-task, visual appeal, portability, and time pressure induced consumers' impulse buying behavior in social commerce in Indonesia. In addition, it revealed interesting findings in the social commerce that hedonic motivation was the most underlying situational factor that influenced impulse buying compared to utilitarian motivation, which had an indirect influence on online impulse buying by affecting hedonic motivation. Thus, social commerce retailers need to design more attractive interfaces in order to encourage consumers to browse and buy or make a good marketing strategy based on hedonic value, and social commerce needs to provide good quality information and the mobility of their market target. The moderation test shows that perceived financial risk did weaken the relationship of consumers' utilitarian value. Therefore, online retailers may provide mechanisms for secure payment and after-sale evaluations to reduce customers' perceived risk.

REFERENCES

Arnold, M.J. and Reynolds, K.E. 2003. "Hedonic shopping motivations", *Journal of Retailing*,Vol. 79, pp. 77–95.

Babin, B. J., Chebat, J. C., & Michon, R. 2004. Perceived appropriateness and its effecton quality, affect and behavior. *Journal of Retailing and Consumer Services*, 11(5),287–298

Biswas, D., Biswas, A. and Das, N. 2006, "The differential effects of celebrity and expertendorsements on consumer risk perceptions", *Journal of Advertising*, Vol. 35 No. 2,pp. 17–31

Chang, C. C., & Chen, C. W. 2015. Examining hedonic and utilitarian bidding motivations in online auctions: Impacts of time pressure and competition. *Internal Journal of Electronic Commerce,* 19(2), 39e65.

Close, A. G., & Kukar-Kinney, M. 2009. Beyond buying: Motivations behindconsumers' online shopping cart use. *Journal of Business Research*, 63(9–10),986–992.

Han, M. C., & Kim, Y. (2016). Can Social Networking Sites Be E-commerce Platforms? *Pan-Pacific Journal of Business Research, 7*(1), 24.

Hanzaee, K. H., & Khonsari, Y. 2011. A Review of The Role of Hedonic and Utilitarian Values on Customer's Satisfaction and Behavioral Intentions.*Interdisciplinary Journal of Research in Business*, Vol. 1, Issue. 5, May (pp.34–45).

Howard, J. A. and Sheth, J. N. 1969. "*The theory of buyer behavior*," John Wiley and Sons, Inc.

Kaczmarek,Lukasz D. 2017. *Hedonic Motivation.* Poland :Springer International Publishing

Lee, J. & Kacen, J.J. 2008. Cultural influences on consumer satisfaction with impulse and planned purchase decisions, *Journal of Business Research* 61(3), pp.265–272.

Luo, X. 2005. How does shopping with others influence impulsive purchasing? *Journal of Consumer Psychology*, 15(4), 288–294.

Petter, S. & Mclean, E.R., 2009. A Meta–Analytic Assessment of Delone and Mclean IS Success Model: An Examination of IS Success At The Individual Level. *Elsevier,* 46(3), pp. 159–166

Varela, M., Maki, T., Kapov, L., & Hobfel. T. 2013. *Toward an Understanding of Visual Appeal in Website Design. COST QUALINET* Action IC1003

Park, E.J., Kim, E.Y., Funches, V.M. and Foxx, W. 2012, "Apparel product attributes, webbrowsing, and e-impulse buying on shopping websites", *Journal of Business Research*,Vol. 65, pp. 1583–1589 Piron, F. (1991). Defining impulse purchasing.ACR North American Advances, 18,509–514.

Solomon, M.R. & Rabolt, N. 2009. *Consumer Behaviour in Fashion,* 2nd. Edition. USA: Prentice Hall

Tractinsky, N., Katz, A.S., & Ikar, D. 2000. What is beautiful is usable. *Interacting with Computers*, 13, 127–145

Xiang. L., Zheng. X., Lee. M.K.O.,Zhao.D.2016. Exploring Consumers' Impulse Buying Behavior on Social Commerce Platform: The Role of Parasocial Interaction. *International Journal of Information Management* 36, 333–347

Zheng. X., Men. J., Yang. F., Gong. X. (2019) Understanding Impulse Buying in Mobile Commerce: An investigation into hedonic and utilitarian browsing. *International Journal of Information Management* 48, 351–160.

Zuckerman, M. .1979.Sensation Seeking: Beyond the Optimal Level of Arousal. *Lawrence Erlbaum Associates,* Hillsdale, NJ, USA.

Contemporary Research on Business and Management – Noviaristanti (Ed.)

The effect of purchasing strategy on supply management performance through supplier integration and supplier relationship management

N. Wasthy, Z.J.H. Tarigan & H. Siagian
Faculty of Business and Economics, Petra Christian University, Surabaya, Indonesia

ABSTRACT: The purchasing strategy is based on building mutual relationships with suppliers. This study examines the effect of purchasing strategy on supply management performance through supplier integration and supplier relationship management. The respondents are employees involved with the suppliers of the company daily. The findings demonstrate that purchasing strategy affects supplier relationship management and improves supplier integration and supply management performance. Consequently, supplier relationship management affects two aspects, namely supplier integration and supply management performance. The latter is also significantly affected by the former. Lastly, the finding shows that supplier relationship management and supplier integration mediate the influence of purchasing strategy on supply management performance. In conclusion, purchasing strategy improves the supply management performance both directly and indirectly.

1 INTRODUCTION

Companies need to capitalize on the benefits of working with suppliers by practicing a supply management approach. Particularly during the current pandemic era, delivery sustainability has become a critical issue due to the supply chain disruption. Companies are required to use more resources to ensure the supply of material runs effectively. Hence, cooperation with the supplier becomes more vital to provide quality products at competitive prices. Tarigan et al. (2019b) stated that organizations should focus on their core business and allow suppliers, as partners, to provide quality materials at a low cost rather than to produce their source of supply. Organizations need to manage the supply network to optimize overall company performance and align the overall demand channel to ensure efficient delivery.

Companies' relationship with suppliers is the strongest cooperation in the value chain or supply chain (Tarigan et al. 2019a). Companies' purchasing strategy impacts the sustainability of processes on an ongoing basis. Hence, companies' purchasing strategy must maintain effective communication with the material supplier. According to Al-Shboul et al. (2018), the integration built by cooperation with external parties will provide companies adequate flexibility. Tarigan et al. (2019a) state that purchasing strategy can positively impact process integration and improve both parties' performance. Companies determine the purchasing strategy by taking into account the strength of the relationship with their suppliers. Furthermore, companies' ability to coordinate with suppliers or material providers can help companies to increase their competitiveness and profit in the long-term through operational cost reduction, quality improvement, and accurate delivery (Sarang et al. 2018).

Supplier integration can provide a better and more sustainable relationship between companies and their suppliers as trust between the two parties leads to an efficient and effective material flow (Zhang & Huo 2013). Kumar & Rahman (2106) state that integration with suppliers enables companies to select the best suppliers, develop new products, and improve firm performance. Moreover, suppliers' involvement in the company's product design process can improve supplier

DOI 10.1201/9781003196013-59

performance and eventually improves company performance (Sánchez-Rodríguez 2009). Semuel et al. (2018) suggest excellent cooperation through involving suppliers in its internal processes and sharing knowledge can enhance company performance.

This study investigates one of the largest mining companies in the nickel mining industry, located in Sorowako, South Sulawesi, Indonesia. This company cooperates with many suppliers to support its operation. The price of nickel significantly affects the mining company's financial condition and operation results. Unfortunately, nickel prices fluctuate and are strongly influenced by global demand and supply as well as nickel substituted product availability and prices. About 65% of company production cost is determined by the supply of goods and services cost (fuel, supporting materials, and service costs). Thus, supply management function and role are crucial to improving business margins through efficiency and cost optimization to achieve a long-term sustainable growth strategy.

This study explores some of the antecedents of supply management performance and understands how it affects supply management performance. Understanding these antecedents is expected to improve existing management performance to achieve corporate strategy success. This study has two primary objectives: first, to measure the purchasing strategy's impact on supplier integration, supplier relationship management, and supply management performance; second, to measure supplier integration's impact on supplier relationship management and supply management performance.

2 RELATIONSHIP BETWEEN RESEARCH CONCEPTS

The primary purchasing concerns with the supply market environment and activity are securing the materials, components, and equipment by selecting capable suppliers. Research conducted by Tarigan et al. (2020) establishes indicators for purchasing strategies in manufacturing companies, namely estimating potential suppliers, developing suppliers in companies, making contracts with company suppliers, making long-term plans, and evaluating suppliers' ongoing basis. Companies can build relationships with their suppliers who are fully committed to providing value for companies to produce products at relatively lower costs (Yoon & Moon 2019). Hence, based on the theory above, the first hypothesis as follows:

H_1: Purchasing strategy affects supplier relationship management.

Broad supplier integration is recommended to ensure accurate and quick delivery of material, parts, or modules from suppliers. Since new critical technologies or knowledge of components or materials for specific product types are often controlled by suppliers, integration offers a more efficient way of accessing this knowledge than traditional market relationships (Shou et al. 2017). Also, integration positively affects the enterprise and various functional performances, including production costs, product quality, on-time delivery, delivery times, inventory turnover times, inventory turnover rates, new product introductions, adjustment performance, logistics service performance, customer satisfaction, sales growth, sales returns, and ROI (Ellegaard & Koch 2014). Hence, based on the description above, the second hypothesis is as follows:

H_2: Purchasing strategy affects supplier integration.

Evaluation of suppliers through field visits and the use of a supplier reward and reward system improves supplier performance. Buyer involvement in the supplier's new product design process results in better performance for both suppliers and buyers. The development of suppliers in improving supply management performance, in turn, will result in improvements in the purchasing performance of the company (Sánchez-Rodríguez 2009). Based on the theory, the third hypothesis is as follows:

H_3: Purchasing strategy affects supply management performance.

Supplier relationship management is characterized by structural interdependence and asymmetrical distribution of resources, which creates dependency among members. The management reflects bargaining power and the importance of the relationship between partners. Trust and dependence exist simultaneously in supplier relationship management. The integration of influence through understanding the independent and shared effects of dependence as well as trust on supplier relationship management can help practitioners build and manage social ties efficiently and effectively. Several empirical studies (see, for example, Zhang & Huo 2013) have confirmed the impact of dependence and trust in supplier integration. The fourth hypothesis is formulated as follows:

H_4: *Supplier relationship management affects supplier integration.*

Suppliers are involved in identifying buyer expectations. Companies' relationship with suppliers is valuable and useful as suppliers become tools that help companies achieve their goals (Tarigan et al. 2019b). Meeting buyer expectations, which are quality, quantity, delivery, service, and price, can help companies improve overall quality to reduce costs and competition (Al-Shboul et al. 2017). Robust cooperation between companies and their suppliers, such as involving suppliers in internal processes and knowledge sharing, can improve construction companies' performance (Samuel et al. 2018). This argument postulates the fifth hypothesis below:

H_5: *Supplier relationship management affects supply management performance.*

Supplier integration and coordination are often seen as drivers for better performance as alliance mechanisms can generate revenue for both parties (González-Benito et al. 2016). Supplier integration, which positions the supplier as a manufacturer's long-term strategic collaborator, can improve operational performance. Effective supplier management can reduce variance in incoming materials and parts and ensure that suppliers meet specifications and quality standards. This management reduces process variability and can positively affect product delivery times, reliability, quality, and potential defects. These benefits can only be obtained if suppliers are involved early in the process (He et al. 2014). This description determines the last hypothesis as follows:

H_6: *Supplier integration affects supply management performance.*

3 METHODS

This study's population is a supplier company that cooperates with a Nickel mining company in South Sulawesi, Indonesia. This study's unit analysis is suppliers cooperating with one mining company in Sorowako, South Sulawesi, Indonesia. The respondents are 37 full-time employees of the mining company whose main jobs are directly related to suppliers (employees engaged in the inbound logistic department).

The research framework and the relationship between concepts are demonstrated in Figure 1.

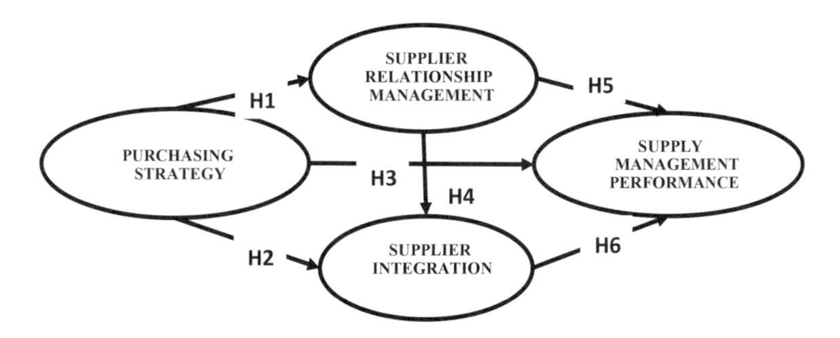

Figure 1. Research framework and concepts relationship.

4 FINDINGS AND DISCUSSION

Of the total employees, 24% have been working for 1 to 3 years, 14% for 3 to 5 years, and 62% for more than five years. Meanwhile, based on the jobs of the respondents, 22 employees (59%) are from the supply management department, 6 (16%) from the operational procurement department, and 9 (25%) are from procurement support, projects, export-imports, goods spots, logistics, shipping, and material management.

Purchasing strategy influences the supplier relationship management relationship (H_1). This result proves that purchasing strategy influences the establishment of supplier relationship management. This finding indicates that the company's ability to build communication and negotiation with suppliers effectively increases company suppliers' alignment and helps the company meet its needs.

The findings also prove the effect of purchasing strategy on supplier integration (H_2). The result suggests that to build effective communication and negotiation with suppliers, the company should involve suppliers in the product development process. Planning is an integrated process as it controls the raw materials needed for production. This study's findings are in line with several researchers that suggest the need of suppliers to involve in the product development process to maintain effective communication and negotiation with suppliers. The results also indicate the effect of purchasing strategy on improving supply management performance (H_3). This finding proves that the company's ability to establish effective communication and negotiation with suppliers is essential in reducing inventory and increasing internal efficiency. This finding supported previous researchers who found that purchasing strategy improves the supply management performance.

Furthermore, the results also show that supplier relationship management practices enhance supplier integration (H_4). This finding indicates that the company can build alignment with suppliers and meet their needs. This process requires suppliers to involve in product development, especially when company planning is integrated with the supplier. This study supports past research which found that supplier relationship management practices affect supplier integration. The result shows that the supplier relationship management improves supply management performance (H_5). This value indicates that the company's ability to build alignment with suppliers is valuable and useful since it enables suppliers to meet their needs to reduce inventory and improve internal efficiency. This study supports the research that supply management performance improves when supplier relationship management runs well.

Lastly, this study found that supplier integration affects supply management performance. This finding indicates that the company's ability to develop products and mechanisms to reduce inventory and increase internal efficiency can reduce transaction costs, generating relational resources and capabilities to maintain a competitive advantage. This study supports the research that examined the effect of supplier integration on management performance. As this research model involves the mediating variable, supplier integration, and supplier relationship management, the result of the analysis regarding the mediating role is interpreted as follows. The mediating role's result can be analyzed by firstly looking at the result of those hypotheses. The mediating role is present when each relationship between the two variables is significant or supported. As those hypotheses are significant, the mediating role of supplier integration and supplier relationship management exists. Hence, purchasing strategy improve supply management performance both directly and directly. This finding reveals that the purchasing strategy establishment is essential in improving supply management performance by implementing supplier integration and supplier relationship management.

5 CONCLUSION

This research primarily examines the effect of purchasing strategy on supply management performance through supplier integration and supplier relationship management. This research concludes that (1) the purchasing strategy has a significant effect on supplier relationship management, (2) the purchasing strategy has a significant effect on supplier integration and supply management performance, (3) supplier relationship management has a significant effect on supplier integration and supply management performance, and (4) supplier integration has a significant effect on supply management performance. Another important finding is that supplier integration and supplier relationship management mediate purchasing strategy on supply management performance. Purchasing strategy improves the supply management performance both directly and indirectly. Hence, purchasing strategy is crucial in improving supply management performance through supplier integration and supplier relationship management. This research contributes to the ongoing research in supply chain management theory by providing essential insight for the practitioner in improving the supply management performance.

REFERENCES

Al-Shboul, M.A.R., Barber, K.D., Garza-Reyes, J.A., Kumar, V., & Abdi, M.R. (2017). The effect of supply chain management practices on supply chain and manufacturing firms' performance. *Journal of Manufacturing Technology Management*, 28(5), 577–609. doi:10.1108/jmtm-11-2016-0154.

Ellegaard, C. and Koch, C. (2014), A model of functional integration and conflict: The case of purchasingproduction in a construction company, *International Journal of Operations & Production Management*, 34(3), 325–346. https://doi.org/10.1108/IJOPM-03-2012-0108.

Fynes, B. and Voss, C. (2002), The moderating effect of buyersupplier relationships on quality practices and performance, *International Journal of Operations & Production Management*, 22(6), 589–613. https://doi.org/10.1108/01443570210427640.

He, Y., Keung Lai, K., Sun, H., & Chen, Y. (2014). The impact of supplier integration on customer integration and new product performance: The mediating role of manufacturing flexibility under trust theory, *International Journal of Production Economics*, 147, 260–270. doi:10.1016/j.ijpe.2013.04.044.

Kumar, D., and Rahman, Z. (2016). Buyer supplier relationship and supply chain sustainability: empirical study of Indian automobile industry, *Journal of Cleaner Production*, 131, 836–848, https://doi.org/10.1016/j.jclepro.2016.04.007.

SánchezRodríguez, C. (2009), Effect of strategic purchasing on supplier development and performance: a structural model, *Journal of Business & Industrial Marketing*, 24(3/4), 161–172. https://doi.org/10.1108/08858620910939714.

Sarang, J., Pankaj, S., Rajendra, C., and Joshi. P.P., (2018). Strategies for buyer supplier relationship improvement: scale development and validation, Procedia Manufacturing 20, 470–476.

Semuel, H., Siagian, H., Arnius, R. (2018). The effects of strategic purchasing on organization performance through negotiation strategy and buyer-supplier relationship, *International Journal of Business and Society*, 19 (2), 323–334.

Tarigan, Z.J.H., Siagian, H., Basana, R.S., and Jie, F., (2019a). Effect of key user empowerment, purchasing strategy, process integration, production system to operational performance, *E3S Web of Conferences*, 130, 01042, https://doi.org/10.1051/e3sconf/201913001042.

Tarigan, Z.J.H., Siagian, H. and Bua, R.R. (2019b). The impact of information system implementation to the integrated system for increasing the supply chain performance of manufacturing companies, *IOP Conference Seri: Material Science and Engineering*, 473, 012050, doi:10.1088/1757-899X/473/1/012050.

Tarigan, Z.J.H., Siagian, H., and Jie, F. (2020). The role of top management commitment to enhancing the competitive advantage through ERP integration and purchasing strategy. *International Journal of Enterprise Information Systems*, 16(1), 53–68.

Yoon, J., and Moon, J. (2019). The moderating effect of buyer purchasing strategy on the relationship between supplier transaction-specific investment and supplier firm performance, *Journal of Business Research*, 99, 516–523.

Zhang, M. and Huo, B. (2013). The impact of dependence and trust on supply chain integration, *International Journal of Physical Distribution & Logistics Management*, 43(7), 544–563. https://doi.org/10.1108/IJPDLM-10-2011-0171.

Contemporary Research on Business and Management – Noviaristanti (Ed.)

Capital investment, internationalization, and firm performance: An empirical study of listed manufacturing firms in the Indonesian stock exchange in 2014–2019

Yovita & D. Marciano
University of Surabaya, Surabaya, Indonesia

ABSTRACT: This study aims to answer whether capital investment of a firm affects the firm's degree of internationalization and to test the effect of internationalization on the firm performance. This research focuses solely on 72 Indonesian manufacturing firms that have been actively engaging in export activities from 2014 to 2019, making up a total of 360 observation points. Capital investment is measured using CAPEXTA, while internationalization is measured using FSTS and the internationalization level change is measured using FSG. In addition, to measure the effect of internationalization on the firm performance, this study uses ROA and Tobin's Q. The control variables include industry return, firm size, leverage, tangible asset ratio, market-to-book ratio, gross profit margin, and sales growth. The hypotheses are confirmed using simple linear regression and ordinary least-squares. The results show that CAPEXTA and internationalization (FSTS) have a significantly negative relationship, and there is an evidence of non-linear relationship. The impact of CAPEXTA on FSG is also shown to be significantly negative and non-linear. While FSTS and ROA have a positive U-shaped relationship, and FSTS shows that it has a directly positive relationship to Tobin's Q.

1 INTRODUCTION

Internationalization is a complex process led by the globalization effect all over the countries. Firms will surely adapt to the globalization through their operational process. One of the processes could be expanding to the international markets, doing foreign direct flow through investing to the host country, and many more. Riahi-Belkaoui (1994) had done a research regarding the effect of internationalization to firms in the United States of America. The result shows that internationalization did have a significant effect, which could be seen from the aspects of sales, profit, and their assets in foreign countries. It is said that the effect of internationalization was ranging from 10% to over 90% of the firm's total operations. Seeing how the internationalization has much impact in the USA, does it also have the same effect in the developing countries, especially Indonesia? Indonesia has been showing great growth in terms of GDP per year compared to other countries, and it is expected to be one of the leading big economy countries in 2024 (IMF and World Bank).

Internationalization simply becomes unavoidable, as some researchers consider internationalization as a firm's strategy. Strategy can be divided into two: (1) growth strategy (Capar & Kotabe; Kylaheiko *et al.* in Vithessonthi, 2017) and (2) diversification strategy (Gulamhussen *et al.* in Vithessonthi, 2017). These two strategies positively support the firm performance to generate more income and operate efficiently. In addition, some firms choose to expand by planting their firms abroad as one of their strategies to mitigate the investment risk of having big capital investment (Vithessonthi, 2017). This raises the question of whether internationalization does work as a strategy and improve the firm performance.

Meanwhile, the relationship of capital investment and internationalization also piqued the interest of researchers. One of the driving factors for the degree of internationalization by the firm is

 DOI 10.1201/9781003196013-60

their capital investment. Firms with higher capital investment would find expanding abroad benefiting to them. Aside from their strategy to mitigate the investment risk, Vithessonthi (2017) said that firms with large capital investment which intend to capitalize the resources and capabilities in order to gain competitive advantage should expand abroad and enter foreign markets. This summarizes how internationalization depends on the firm's long-term assets and capital investments. Further investigation regarding the relationship of capital investment and internationalization will be discussed in this study. The proposed measurement to accurately depict internationalization that will be used is foreign sales to total sales (FSTS). FSTS has been widely used in previous studies regarding internationalization. Nevertheless, this study also includes foreign sales growth (FSG) to identify if foreign sales is a capital investment's function. FSG is used to know if there is even a slight change in the internationalization level.

Further discussion about the relationship between internationalization and firm performance shows diverse results. Sun *et al.* (2018) and Hsu *et al.* (2013) found a significantly positive relationship of internationalization and firm performance, while other studies from Singla and George (2013) and Salim *et al.* (2018) show a negative relationship. This study will also add another variable to measure the firm performance, aside from ROA, which is widely used in other previous studies, that is, Tobin's Q. Tobin's Q is used to represent firm performance in the long run and could also be defined as the firm value in other studies. Based on the aforementioned reasons, there are two hypotheses that will be tested: (H1) capital investment positively impacts internationalization and (H2) internationalization positively impacts firm performance.

2 DATA

The data consist of publicly listed manufacturing firms in Indonesian Stock Exchange from 2014 to 2019. The firms need to be actively engaging in export activities. For this study, the data of 72 public manufacturing firms are collected from their audited financial statement each year. In total, there are 360 observation points to be analyzed in this study. The dependent variables used are FSTS and FSG for internationalization, and both ROA and Tobin's Q as proxy for firm performance. We use FSTS and FSG to test the capital investment to internationalization model. FSTS is measured as firm's foreign sales to firm's total sales, while FSG is measured as the first difference in the natural logarithm of firm's foreign sales. In the internationalization (FSTS) to firm performance model, we use ROA and Tobin's Q. ROA is measured as the ratio of EBIT to total assets. Tobin's Q is computed as the ratio of the sum of the market capitalization and firm's total debt to firm's total asset.

The independent variable used in the capital investment and internationalization model is CAPEXTA, as a proxy to capital investment. CAPEXTA is measured as the ratio of capital expenditure to total asset of a firm. The independent variable used in internationalization and firm performance model is FSTS, as a proxy to internationalization. FSTS is measured as the ratio of foreign sales to total sales, as stated in the previous section. The macroeconomics variable that we use to control the internationalization is the GDP growth rate (GDPGROWTH). Aside from the macroeconomic control variable, this study also applies several firm-specific control variables. These control variables include (1) firm Size (FSIZE), measured as the natural logarithm of total assets, (2) leverage (LEV), measured as the ratio of total debt to total asset, (3) tangible asset ratio (PPETA), defined as the ratio of property, plant, and equipment of a firm to total assets, (4) market-to-book value (MBV), measured as the ratio of market capitalization to book value of total equity, (5) gross profit margin (GPM), calculated by dividing gross profit to total sales, (6) and sales growth (SALESGROWTH), which is computed as the first difference of the natural logarithm of total sales.

We first analyze the impact of capital investment to internationalization into two models; the first model is to detect a linear effect, while the second model is to detect if there is a nonlinear effect. The first part is to use FSTS as a proxy of internationalization. In addition to that, ε is an error term, and i and t stand for firm and time, respectively. The independent variable and control variables are set to one-period lag.

$$
\begin{aligned}
FSTS_{i,t} &= \alpha + \beta_1 CAPEXTA_{i,t-1} + \beta_2 GDPGROWTH_{i,t-1} + \beta_3 FSIZE_{i,t-1} + \beta_4 LEV_{i,t-1} \\
&\quad + \beta_5 PPETA_{i,-1} + \beta_6 MBV_{i,t-1} + \beta_7 GPM_{i,t-1} + \beta_8 SALESGROWTH_{i,t-1} \\
&\quad + \beta_9 ROA_{i,-1} + \varepsilon_{i,t} \qquad (1) \\
FSTS_{i,t} &= \alpha + \beta_1 CAPEXTA_{i,t-1} + \beta_2 CAPEXTA^2_{i,t-1} + \beta_3 GDPGROWTH + \beta_4 FSIZE_{i,t-1} \\
&\quad + \beta_5 LEV_{i,-1} + \beta_6 PPETA_{i,t-1} + \beta_7 MBV_{i,t-1} + \beta_8 GPM_{i,t-1} \\
&\quad + \beta_9 SALESGROWTH_{i,-1} + \beta_{10} ROA_{i,-1} + \varepsilon_{i,t} \qquad (2)
\end{aligned}
$$

The second part is to use the foreign sales growth (FSG) rate as a proxy to internationalization. The idea of using FSG is to address the slight change that goes undetected in FSTS, which could be the result of the capital investment effect.

$$
\begin{aligned}
FSG_{i,t} &= \alpha + \beta_1 CAPEXTA_{i,t-1} + \beta_2 GDPGROWTH_{i,t-1} + \beta_3 FSIZE_{i,t-1} + \beta_4 LEV_{i,t-1} \\
&\quad + \beta_5 PPETA_{i,-1} + \beta_6 MBV_{i,t-1} + \beta_7 GPM_{i,t-1} + \beta_8 SALESGROWTH_{i,t-1} \\
&\quad + \beta_9 ROA_{i,t-1} + \varepsilon_{i,t} \qquad (3) \\
FSG_{i,t} &= \alpha + \beta_1 CAPEXTA_{i,t-1} + \beta_2 CAPEXTA_{2i,t-1} + \beta_3 GDPGROWTH_{i,t-1} + \beta_4 FSIZE_{i,t-1} \\
&\quad + \beta_5 LEV_{i,-1} + \beta_6 PPETA_{i,t-1} + \beta_7 MBV_{i,t-1} + \beta_8 GPM_{i,t-1} \\
&\quad + \beta_9 SALESGROWTH_{i,-1} + \beta_{10} ROA_{i,t-1} + \varepsilon_{i,t} \qquad (4)
\end{aligned}
$$

The impact of internationalization on firm performance is analyzed through two parts, as we replace the proxy of firm performance as ROA and Tobin's Q (TBQ) in each model.

$$
\begin{aligned}
ROA_{i,t} &= \alpha + \beta_1 FSTS_{i,t-1} + \beta_2 GDPGROWTH_{i,t-1} + \beta_3 FSIZE_{i,t-1} + \beta_4 LEV_{i,t-1} + \beta_5 PPETA_{i,t-1} \\
&\quad + \beta_6 MBV_{i,-1} + \beta_7 GPM_{i,t-1} + \beta_8 SALESGROWTH_{i,t-1} + \beta_9 ROA_{i,t-1} + \varepsilon_{i,t} \qquad (5) \\
ROA_{i,t} &= \alpha + \beta_1 FSTS + \beta_2 FSTS_{i2,t-1} + \beta_3 GDPGROWTH_{i,t-1} + \beta_4 FSIZE_{i,t-1} + \beta_5 LEV_{i,t-1} \\
&\quad + \beta_6 PPETA_{i,-1} + \beta_7 MBV_{i,t-1} + \beta_8 GPM_{i,t-1} + \beta_9 SALESGROWTH_{i,t-1} \\
&\quad + \beta_{10} ROA_{i,t-1} + \varepsilon_{i,t} \qquad (6)
\end{aligned}
$$

Then, we analyze the impact of internationalization on Tobin's Q (TBQ) as a proxy to firm performance. All the other variables have the same definition, as has already been stated.

$$
\begin{aligned}
TBQ_{i,t} &= \alpha + \beta_1 FSTS_{i,t-1} + \beta_2 GDPGROWTH + \beta_3 FSIZE_{i,t-1} + \beta_4 LEV_{i,t-1} + \beta_5 PPETA_{i,t-1} \\
&\quad + \beta_6 MBV_{i,-1} + \beta_7 GPM_{i,t-1} + \beta_8 SALESGROWTH_{i,t-1} + \beta_9 ROA_{i,t-1} + \varepsilon_{i,t} \qquad (7) \\
TBQ_{i,t} &= \alpha + \beta_1 FSTS + \beta_2 FSTS_{i2,t-1} + \beta_3 GDPGROWTH_{i,t-1} + \beta_4 FSIZE_{i,t-1} + \beta_5 LEV_{i,t-1} \\
&\quad + \beta_6 PPETA_{i,-1} + \beta_7 MBV_{i,t-1} + \beta_8 GPM_{i,t-1} + \beta_9 SALESGROWTH_{i,t-1} \\
&\quad + \beta_{10} ROA_{i,t-1} + \varepsilon_{i,t} \qquad (8)
\end{aligned}
$$

3 RESULTS AND DISCUSSION: CAPITAL INVESTMENT TO INTERNATIONALIZATION (FSTS)

The data processed provide some evidence that there is indeed a relationship of capital investment to internationalization, but it differs from the hypotheses. This suggests that every increase on capital investment would lessen the degree of internationalization. However, the data indicate that there is a non-linear relationship of capital investment to internationalization. The result is in line with the previous research by Salim et al. (2017). The U-shaped relationship confirms that mostly Indonesian public firms still target their domestic market, as they already have competitive advantages in their home country. Moreover, the negative relationship could be a result of the one-period time-lagged variables in the regression. The assumption made is that the capital investment budget would not be immediately used and brings out a positive return in the same year (Vithessonthi, 2016). Internationalization is defined solely by companies' capability of generating foreign sales. Vithessonthi (2017) also argued that the result of negative or non-existent relationship of capital investment on internationalization could be due to how firms are risk-averse. If the firm policy is driven by their risk-averse profile to the internationalization process, then they would hesitate and could not expand the business more into the international market. The U-shaped relationship also implies that the negative effect will be decreased with more increase in the more capital investment; it is conditional related to the size of capital investment (Vithessonthi, 2017).

3.1 *Capital investment to FSG*

It is apparent that there is an impact of capital investment on FSG as a proxy to internationalization. The effect is negative, and there is also an evidence of the non-relationship evidence. Vithessonthi (2016) argued that the coefficient of CAPEXTA should be positive if firms with larger capital investment are to focus on their international business through export. Instead, the following regression results show otherwise. This adds more to the argument that Indonesian public firms are still indeed focusing on their domestic market, as the domestic market alone has so much potential, with Indonesia being one of the largest consumer countries; 74% of the total sales goes to the home country alone. The result could be driven with the downtrend of the Indonesia's export. The result also confirms that FSG is not a function of capital investment. Vithessonthi (2016) in his research also found the evidence of a significantly negative relationship of CAPEXTA to FSG with the moderating variables of firm size.

3.2 *Internationalization (FSTS) to firm performance (ROA)*

The data regressed show no linear relationship detected in model 5. Meanwhile, when the FSTS is computed to square FSTS in model 6, we found an evidence of a non-linear relationship, that is, a positive U-shaped relationship. The U-shaped relationship explains how it is related to the early internationalization. In the early stage of internationalization, firms tend to have low firm performance as to how they are still adapting and get to know about the international market, as they collect foreign resources and new knowledge to be applied within. The firm still faces uncertainty toward the new foreign market. However, once they become familiar with the international markets and know how the target market works, they will start thriving and become positive (Contractor et al., 2007; Capar and Kotabe, 2003; Ruigrok and Wagner, 2003 in Altaf and Shah, 2016). Previous studies show that the benefit of doing internationalization will be experienced once the firm successfully adjusts to the foreign markets. Coombs et al. (2009) in Altaf and Shah (2016) identified one of the benefits mentioned as risk diversification. As stated before, one of the driving factors for firms to do internationalization is risk mitigation. According to the data, there is a negative linear relationship in model 6, alongside the positive U-shaped relationship. It only emphasizes the evidence that the internationalization and firm performance relationship is conditional, related to the stage of internationalization. Because the variables are one-period time-lagged, we can conclude that the time needed for internationalization to benefit the firm could be more than one year.

3.3 *Internationalization (FSTS) to firm performance (Tobin's Q)*

Regressed data indicate that there is a positive direct relationship of internationalization to Tobin's Q as firm performance. A non-linear relationship is not found. The result is in line with those reported by Sun et al. (2018), Riahi-Belakaoui (1999), and Olsen and Elango (2016), who stated that internationalization positively affects the firm performance/firm value. Olsen and Elango (2016) argued that the geography of the sample gathered played an important role. In Indonesia, the firms are located at smaller home markets (considering the U.S. market and international markets), which drives them to do internationalization. This leads to the increase in Tobin's Q as firm performance (or firm value). Investors consider the degree of internationalization as one of the important factors; so they are willing to invest in these firms, and the market capitalization of the firm will grow.

4 CONCLUSION AND IMPLICATION

It can be concluded that capital investment negatively impacts the internationalization level, while there is also an evidence of a U-shaped relationship. On the other hand, the study confirms that internationalization affects firm performance (ROA) in a non-linear model, that is, a U-shaped relationship. However, there is a positive relationship on the impact of internationalization to Tobin's Q as firm performance.

REFERENCES

Altaf, N. and Farooq Ahmad Shah (2015), "Internationalization and firm performance of Indian firms: Does product diversity matter?", *Pacific Science Review B: Humanities and Social Sciences*, Vol. 1, p. 76–84.

Ater, Daniel Kon (2017), "The Joint Effect of Firm Growth, Macroeconomic Factors and Capital Structure on The Value of Nonfinancial Firms Listed On The Nairobi Securities Exchange", *International Journal of Economics, Commerce, and Management,* Vol. 5, Issue 9, p. 618–628.

Billett, M.T. *et al.* (2011), "The influence of governance on investment: Evidence from a hazard model", *Journal of Financial Economics*, Volume 102, Issue 3, Pages 643–670.

Chung, Chris Changwha, *et al.* (2013), "Internationalization and Performance of Firms in China:

Moderating Effects of Governance Structure and the Degree of Centralized Control", *Journal of International Management,* Vol. 19, Issue 2, Pages 118–137.

Duchin, R. *et al.* (2010), "Costly external finance, corporate investment, and the subprime mortgage credit crisis", *Journal of Financial Economics*, Volume 97, Issue 3, Pages 418435.

Khamees, B.A., Al-Fayoumi, N. and Al-Thuneibat, A.A. (2010), "Capital budgeting practices in the Jordanian industrial corporations", *International Journal of Commerce and Management*, Vol. 20 No. 1, pp. 49–63.

Khasawneh, A.Y. and Dasouqi, Q.A. (2017), "Sales nationality and debt financing impact on firm's performance and risk: Evidence from Jordanian companies", *EuroMed Journal of Business*, Vol. 12 No. 1, pp. 103–126.

Luo, Xueming and Zheng, Qinqin (2018), "How firm internationalization is recognized by outsiders: The response of financial analysts", *Journal of Business Research*, Volume 90, Pages 87–106.

Likitwongkajon, N. and Vithessonthi, C. (2020), "Do foreign investments increase firm value and firm performance? Evidence from Japan", *Research in International Business and Finance,* Vol. 51.

Olsen, B. and Elango, B. (2005), "Do Multinational Operations Influence Firm Value? Evidence form the Triad Regions", *International Journal of Business and Economics,* Vol. 4, No. 1, p. 11–29.

Rely, Gilbert and Regina Jansen Arsjah (2018), "An Effecting Ownership Structure in Firm Value Towards Offshore Debt Financing in Manufacturing Firms", *Research Journal of Finance and Accounting,* Vol. 9, No. 12, p. 46–56.

Riahi-Belkaoui, Ahmed (1999), "The degree of internationalization and the value of the firm: theory and evidence", *Journal of International Accounting, Auditing, and Taxation*, Vol. 8, Issue 1, Pages 189–196.

Salim *et al.* (2017), "Capital Investment, Internationalisation, and Firm Performance: An Empirical Study of Listed Manufacturing Firms in The Indonesian Stock Exhange 20112015", *Proceeding The International Conference: The Roles of Social Science and Humanities in Socio-Economic Development and International Integration.*

Tan, Yong and Christos Floros (2012), "Bank profitability and GDP growth in China: A Note", *Journal of Chinese Economics and Business Studies,* Vol. 10, Issue 3, pp. 267–273.

Vithessonthi, C. and Olimpia C. Racela (2016), "Short- and long-run effects of internationalization and R&D intensity on firm performance", *Journal of Multinational Financial Management*, Vol. 34, p. 28-45. Vithessonthi, Chaiporn (2016), "Capital investment, internationalization, and firm performance: Evidence from Southeast Asian countries", *Research in International Business and Finance,* Vol. 38, pp. 393–403.

Vithessonthi, Chaiporn (2017), "Capital investment and internationalization", *Journal of Economics and Business*, Volume 90, Pages 31–48.

Contemporary Research on Business and Management – Noviaristanti (Ed.)
© 2022 copyright the Author(s), ISBN 978-1-032-05097-3.
Open Access: www.taylorfrancis.com, CC BY-NC-ND 4.0 license

Author index